Toxicology of 1→3-Beta-Glucans
Glucans as a Marker for Fungal Exposure

Toxicology of 1→3-Beta-Glucans
Glucans as a Marker for Fungal Exposure

EDITED BY

Shih-Houng Young and Vincent Castranova

informa
healthcare

New York   London

First published in 2005 by CRC Press. CRC Press is an imprint of the Taylor & Francis Group, LLC.

This edition published in 2011 by Informa Healthcare, Telephone House, 69–77 Paul Street, London EC2A 4LQ, UK.

Simultaneously published in the USA by Informa Healthcare, 52 Vanderbilt Avenue, 7th Floor, New York, NY 10017, USA.

Informa Healthcare is a trading division of Informa UK Ltd. Registered Office: 37–41 Mortimer Street, London W1T 3JH, UK. Registered in England and Wales number 1072954.

©2011 Informa Healthcare, except as otherwise indicated

No claim to original U.S. Government works

Reprinted material is quoted with permission. Although every effort has been made to ensure that all owners of copyright material have been acknowledged in this publication, we would be glad to acknowledge in subsequent reprints or editions any omissions brought to our attention.

All rights reserved. No part of this publication may be reproduced, stored in a retrieval system, or transmitted, in any form or by any means, electronic, mechanical, photocopying, recording, or otherwise, unless with the prior written permission of the publisher or in accordance with the provisions of the Copyright, Designs and Patents Act 1988 or under the terms of any licence permitting limited copying issued by the Copyright Licensing Agency, 90 Tottenham Court Road, London W1P 0LP, UK, or the Copyright Clearance Center, Inc., 222 Rosewood Drive, Danvers, MA 01923, USA (http://www.copyright.com/ or telephone 978-750-8400).

Product or corporate names may be trademarks or registered trademarks, and are used only for identification and explanation without intent to infringe.

This book contains information from reputable sources and although reasonable efforts have been made to publish accurate information, the publisher makes no warranties (either express or implied) as to the accuracy or fitness for a particular purpose of the information or advice contained herein. The publisher wishes to make it clear that any views or opinions expressed in this book by individual authors or contributors are their personal views and opinions and do not necessarily reflect the views/opinions of the publisher. Any information or guidance contained in this book is intended for use solely by medical professionals strictly as a supplement to the medical professional's own judgement, knowledge of the patient's medical history, relevant manufacturer's instructions and the appropriate best practice guidelines. Because of the rapid advances in medical science, any information or advice on dosages, procedures, or diagnoses should be independently verified. This book does not indicate whether a particular treatment is appropriate or suitable for a particular individual. Ultimately it is the sole responsibility of the medical professional to make his or her own professional judgements, so as appropriately to advise and treat patients. Save for death or personal injury caused by the publisher's negligence and to the fullest extent otherwise permitted by law, neither the publisher nor any person engaged or employed by the publisher shall be responsible or liable for any loss, injury or damage caused to any person or property arising in any way from the use of this book.

A CIP record for this book is available from the British Library.

Library of Congress Cataloging-in-Publication Data available on application

ISBN-13: 978-0-4157-0037-5

Orders may be sent to: Informa Healthcare, Sheepen Place, Colchester, Essex CO3 3LP, UK
Telephone: +44 (0)20 7017 5540
Email: CSDhealthcarebooks@informa.com
Website: http://informahealthcarebooks.com/

For corporate sales please contact: CorporateBooksIHC@informa.com
For foreign rights please contact: RightsIHC@informa.com
For reprint permissions please contact: PermissionsIHC@informa.com

Preface

The investigation of indoor air quality has shown that microbial contamination is often a problem in buildings. Water damage or dampness in buildings has been associated with respiratory diseases. Recent evidence indicates that fungi are often an essential part of the picture. However, in comparison to other areas of medicine, very few toxicology studies of exposure to fungi have been conducted. (1→3)-β-Glucans are cell wall components of fungi. They have been considered surrogate compounds to monitor environmental exposure to fungi. Although (1→3)-β-glucans have been studied for some time, most of the previous literature has focused on the beneficial effects of (1→3)-β-glucans rather than on their toxicological effects. Consequently, not much toxicology data are available to investigators of indoor air issues. Therefore, we feel that it is important to summarize current knowledge concerning the toxicological effects of (1→3)-β-glucans in the hope that it may inspire more research in this area. This book, *Toxicology of (1→3)-β-Glucans*, intends to provide a comprehensive review by experts of current information regarding the effects of the fungal cell wall component (1→3)-β-glucan on human health.

The first chapter, "Introduction to the Chemistry and Immunobiology of β-Glucans," by Williams, Lowman, and Ensley, reviews the basic chemistry, physicochemical analysis and immunobiology of (1→3)-β-D-glucans. Recent advances in analytical methods for (1→3)-β-D-glucans and the recent discovery of (1→3)-β-D-glucan-specific membrane receptors are covered. Chapter 2, "Health Effects of (1→3)-β-D-Glucans: The Epidemiological Evidence," by Douwes, presents an overview of the epidemiological literature concerning respiratory health effects, such as atopy, airway inflammation, airway symptoms, asthma, and altered lung function, which result from exposure to airborne (1→3)-β-glucans. Both occupational and environmental health studies are reviewed. In addition to population studies, studies in human volunteers experimentally exposed to (1→3)-β-glucan are also described. The strengths and weaknesses of these studies are discussed. Chapter 3, "(1→3)-β-D-Glucan in the Environment: A Risk Evaluation," by Rylander, focuses on the effects of (1→3)-β-D-glucan after inhalation, and extrapolates this information to evaluate the risks involved in occupational and general environments.

Chapter 4, "Animal Model of (1→3)-β-Glucan-Induced Pulmonary Inflammation in Rats," by Young and Castranova, reviews data concerning (1→3)-β-glucan-induced pulmonary inflammation in a rat animal model. Evidence of dose response and the time dependence of this (1→3)-β-glucan-induced pulmonary inflammation are presented. The effects of soluble and particulate forms of β-glucans on animal exposure are compared. The role of conformation on the inflammatory potential of (1→3)-β-glucans is discussed. The implications of these findings and suggestions for future analytic method development are presented. Chapter 5, "β-Glucan Receptor(s) and Their Signal Transduction," by Adachi, reviews the characteristics of fungal β-glucan receptors and recognition mechanisms. Chapter 6, "Fate of β-Glucans *In Vivo*," by Miura, investigates the distribution and degradation mechanisms of β-glucans in the body in both cases of deep mycoses and after using β-glucans as biological response modifiers (BRMs).

Chapter 7, "Adjuvant Effects of β-Glucans in a Mouse Model for Allergy," by Ormstad and Hetland, discusses the adjuvant activity of (1→3)-β-glucan in a mouse model. Results indicate that, whereas (1→3)-β-glucan exhibits little effect on the local inflammatory response, it has strong adjuvant activity on the systemic allergic immune response. Chapter 8, "Endogenous Septic Shock by Combination of β-Glucan and NSAIDs," by Ohno, reports that pretreatment of mice with β-glucan significantly increases mortality in response to subsequent indomethacin administration. Chapter 9, "Particulate and Soluble β-Glucans from *Candida albicans* Modulate Cytokine Release from Human Leukocytes," by Ishibashi, Nakagawa, Ohno, and Murai, summarizes the effect of particulate and soluble forms of (1→3)-β-glucans on the modulation of cytokine release from human leukocytes.

Chapter 10, "Detection and Measurement of (1→3)-β-D-Glucan with Limulus Amebocyte Lysate-Based Reagents," by Finkelman and Tamura, focuses on LAL-based methodologies applied to the detection and quantitation of (1→3)-β-D-glucans. The last chapter, "Clinical Utilization of the Measurement of (1→3)-β-D Glucan in Blood," by Obayashi discusses the clinical utilization of plasma (1→3)-β-D-glucan measurements.

The editors wish to express our gratitude for the excellent contributions and expertise of the authors. We also thank Erika Dery and Stephen Zollo at CRC Press/Taylor and Francis for their expert advice and cooperation. Finally we thank Dr. Sam Kacew for his invitation and encouragement for the concept of this book.

Shih-Houng Young

Vincent Castranova

No official support or endorsement by the National Institute for Occupational Safety and Health/Centers for Disease Control and Prevention is intended or should be inferred.

Editors

Shih-Houng Young, Ph.D., received a B.S. and an M.S. in chemistry from National Tsing Hua University, Hsinchu, Taiwan. After eight years of working in the field of occupational safety and health, he decided to further his education in occupational health. He went to University of Alabama at Birmingham and received a Ph.D. in environmental health sciences and industrial hygiene in 1998. He is a member of the American Industrial Hygiene Association and the American Conference of Governmental Industrial Hygienists. His thesis involved the elucidation of conformational-biological activity relationships of $(1\rightarrow3)$-β-glucans via the fluorescence resonance energy transfer method. He was awarded a National Research Council Associateship at the National Institute for Occupational Safety and Health (NIOSH) to continue his study in $(1\rightarrow3)$-β-glucans.

Dr. Young's research in $(1\rightarrow3)$-β-glucans includes both the chemical characterization of conformation of $(1\rightarrow3)$-β-glucans and the pulmonary toxicology of $(1\rightarrow3)$-β-glucans in rats. His research at UAB involved the use of both external dye (aniline blue) and internal fluorochrome labeling of $(1\rightarrow3)$-β-glucans. These methods enabled him to monitor conformational changes in $(1\rightarrow3)$-β-glucans. Shortly after he joined NIOSH in 1999, he began a series of toxicology experiments to investigate the pulmonary effect of $(1\rightarrow3)$-β-glucans. He has characterized a rat model of zymosan-induced pulmonary inflammation. Using this rat model, he has studied the effect of different physical forms of zymosan and the effect of different conformations on pulmonary responsiveness in rat. In addition, this model was used to evaluate the interaction of endotoxin and zymosan. Furthermore, he also investigated the role of NF-κB in zymosan-induced TNF-α production. His results have been published in the *Journal of Toxicology and Environmental Health*, the *Journal of Biological Chemistry*, *Toxicology and Applied Pharmacology*, *Carbohydrate Research*, and elsewhere.

From 2002 to 2004, Dr. Young held a senior research associate position at the MB Research Laboratory, Spinnerstown, PA. In 2004, Dr. Young was granted a research associate position at NIOSH.

Vincent Castranova, Ph.D., is the Chief of the Pathology and Physiology Research Branch in the Health Effects Laboratory Division of the National Institute for Occupational Safety and Health, Morgantown, WV. He holds the grade of a CDC Distinguished Consultant. He is also an adjunct professor in the Department of Physiology and the Department of Basic Pharmaceutical Sciences at West Virginia University, Morgantown, WV and the Department of Environmental and Occupational Health at the University of Pittsburgh, Pittsburgh, PA. He is a member of the American Physiological Society, the Society of Toxicology, Beta Beta Beta, and the Allegheny-Eric Chapter of the Society of Toxicology, where he once served as president. He is on the editorial board of *Annals of Agricultural and Environmental Medicine*, the *Journal of Toxicology and Environmental Health*, and *Toxicology and Applied Pharmacology*. In addition, he was guest editor for the *Journal of Environmental Pathology, Toxicology and Oncology*.

Dr. Castranova received a B.S. in biology from Mount Saint Mary's College, Emmitsburgh, MD in 1970, graduating magna cum laude. He received a Ph.D. in physiology and biophysics in 1974 from West Virginia University, Morgantown, WV before becoming an NIH fellow and research faculty member in the Department of Physiology at Yale University, New Haven, CT. In 1977, Dr. Castranova received a research staff position at the National Institute for Occupational Safety and Health and an adjunct faculty position at West Virginia University, Morgantown, WV. He has served at these institutions since that time.

Dr. Castranova's research interests have concentrated on two major areas. First is the isolation and characterization of the physiological properties of lung cells; in particular, alveolar macrophages, polymorphonuclear leukocytes, and alveolar type II epithelial cells. Studies with alveolar macrophages and polymorphonuclear leukocytes involve characterization of stimulus-secretion coupling mechanisms for the respiratory burst and activation of phagocytosis, signals involved in activation of transcription factors controlling the production of inflammatory cytokines, and the measurement of the production of reactive oxygen species and nitric oxide using techniques such as chemiluminescence. Studies with type II cells involve measurement of surfactant production, the production of nitric oxide and inflammatory cytokines, xenobiotic metabolism, and water transport to prevent edema. The second area is determining mechanisms involved in disease initiation and progression after exposure to silica, coal mine dust, diesel exhaust particulate, fibers, cotton dust, organic dusts, and microbial products such as endotoxin and β-glucans. These studies have resulted in the publication of four books as co-editor and over 350 manuscripts in peer-reviewed journals and chapters in textbooks.

Contributors

Yoshiyuki Adachi
Tokyo University of Pharmacy and Life Science
Tokyo, Japan

Vincent Castranova
National Institute for Occupational Safety and Health
Morgantown, West Virginia

Jeroen Douwes
Massey University
Wellington, New Zealand

Harry E. Ensley
Tulane University
New Orleans, Louisiana

Malcolm A. Finkelman
Associates of Cape Cod, Inc.,
Falmouth, Massachusetts

Geir Hetland
Norwegian Institute of Public Health
Oslo, Norway

Ken-ichi Ishibashi
Tokyo University of Pharmacy and Life Science
Tokyo, Japan

Douglas W. Lowman
Eastman Chemical Company
Kingsport, Tennessee

Noriko N. Miura
Tokyo University of Pharmacy and Life Science
Tokyo, Japan

Toshimi Murai
National Institute of Health Sciences
Osaka, Japan

Yukari Nakagawa
National Institute of Health Sciences
Osaka, Japan

Taminori Obayashi
Tokyo Metropolitan Komagome General Hospital
Tokyo, Japan

Naohito Ohno
Tokyo University of Pharmacy and Life Science
Tokyo, Japan

Heidi Ormstad
Norwegian Institute of Public Health
Oslo, Norway

Ragnar Rylander
University of Gothenburg
Gothenburg, Sweden

Hiroshi Tamura
Seikagaku Corporation
Tokyo, Japan

David L. Williams
James H. Quillen College of Medicine
East Tennessee State University
Johnson City, Tennessee

Shih-Houng Young
National Institute for Occupational Safety and Health
Morgantown, West Virginia

Table of Contents

Chapter 1
Introduction to the Chemistry and Immunobiology of β-Glucans 1
David L. Williams, Douglas W. Lowman, and Harry E. Ensley

Chapter 2
Health Effects of (1→3)-β-D-Glucans: The Epidemiological Evidence 35
Jeroen Douwes

Chapter 3
(1→3)-β-D-Glucan in the Environment: A Risk Assessment 53
Ragnar Rylander

Chapter 4
Animal Model of (1→3)-β-Glucan-Induced Pulmonary Inflammation in Rats 65
Shih-Houng Young and Vincent Castranova

Chapter 5
β-Glucan Receptor(s) and Their Signal Transduction 95
Yoshiyuki Adachi

Chapter 6
Fate of β-Glucans *In Vivo*: Organ Distribution and Degradation
Mechanisms of Fungal β-Glucans in the Body 109
Noriko N. Miura

Chapter 7
Adjuvant Effects of β-Glucans in a Mouse Model for Allergy 127
Heidi Ormstad and Geir Hetland

Chapter 8
Endogenous Septic Shock by Combination of β-Glucan and NSAIDs 143
Naohito Ohno

Chapter 9
Particulate and Soluble β-Glucans from *Candida albicans* Modulate
Cytokine Release from Human Leukocytes.. 161
*Ken-ichi Ishibashi, Yukari Nakagawa, Naohito Ohno, and
Toshimi Murai*

Chapter 10
Detection and Measurement of (1→3)-β-D-Glucan with Limulus
Amebocyte Lysate-Based Reagents 179
Malcolm A. Finkelman and Hiroshi Tamura

Chapter 11
Clinical Utilization of the Measurement of (1→3)-β-D-Glucan in Blood 199
Taminori Obayashi

Index.. 209

1 Introduction to the Chemistry and Immunobiology of β-Glucans

David L. Williams, Douglas W. Lowman, and Harry E. Ensley

CONTENTS

1.1 Introduction 2
1.2 What Are Fungal β-Glucans? 2
1.3 Physicochemical Characterization of Fungal Glucans 2
    1.3.1 Structural Characterization of (1→3)-β-D-Glucans by
          ^{13}C and ^{1}H NMR 4
    1.3.2 Analysis and Quantification of Glucans in Complex Biomatrices ... 11
    1.3.3 Molecular Weight Analysis of Glucans 11
1.4 Immunobiology of Glucans 14
1.5 Recognition and Binding of Glucans by Membrane Receptors 15
1.6 The Influence of Glucan Polymer Molecular Weight, Structure, and
    Solution Conformation on Binding to (1→3)-β-D-Glucan Receptors 16
1.7 Glucan Receptors Differentially Recognize Glucan Polymers
    Based on Solution Conformation and Molecular Weight 17
1.8 Identification of Dectin-1 and Scavenger Receptors as Glucan-Specific
    Membrane Receptors 20
    1.8.1 The Role of Dectin-1 as a Primary Glucan Receptor 20
    1.8.2 Scavenger Receptors as Glucan Binding Sites 21
1.9 Activation of Proinflammatory and Immunoregulatory Intracellular
    Signaling Pathways by Glucans 22
    1.9.1 Toll-Like Receptor 2 Recognition and Signaling in Response to
          Glucan Exposure 22
    1.9.2 Activation of Nuclear Factor Kappa B (NFkB), Nuclear Factor
          Interleukin 6 (NF-IL6), Nuclear Factor 1 (NF-1), Activator
          Protein 1 (AP-1) and Specificity Protein 1 (SP-1) Signaling
          Pathways by Fungal Glucans.. 23

1.10 Effect of Glucans on Cytokine and Growth Factor Expression 24
1.11 Antiinflammatory Activity of Glucans 25
1.12 Conclusions 27
References 27

1.1 INTRODUCTION

Riggi and Di Luzio (1961) reported that the macrophage stimulating activity of a yeast cell wall fraction, zymosan, was exclusively due to the presence of glucan in the zymosan. Since that time a myriad of publications have appeared in the scientific literature that have focused on the immunobiology of glucans, particularly as it relates to anti-cancer and anti-infective efficacy (Chihara and Mihich, 1985; Williams, 1997; Williams et al., 1996) and, to a lesser extent, as it relates to the potential involvement of glucans in the pathophysiology of mold inhalation in indoor environments (Beijer et al., 2002; Rylander and Lin, 2000). However, many of these studies have been phenomenological and have not addressed the cellular and molecular mechanisms by which glucans modulate cellular activity. Over the past five to six years, there have been several important discoveries which have dramatically increased our knowledge of the cellular and molecular biology of glucans (Brown and Gordon, 2001; Brown et al., 2002; Brown et al. 2003; Kougias et al., 2001). Of equal importance, a number of investigators have developed new technologies or applied existing technologies to better understand the physicochemical characteristics of these intriguing polymer systems (Ensley et al., 1994; Kim et al., 2000; Lowman and Williams, 2001; McIntire and Brant, 1997; Mueller et al., 1995). The purpose of this chapter is to provide an introduction to the field of glucan biology, with special emphasis on the most recent advances in the chemistry, physicochemical characterization, and biology of fungus-derived $(1\rightarrow3)$-β-D-glucans.

1.2 WHAT ARE FUNGAL β-GLUCANS?

The cell walls of saprophytic and pathogenic fungi are composed of proteins, lipids, and carbohydrates, which are structurally distinct from cell wall components produced by higher species, including mammals (Elorza et al., 1989; Kapteyn et al., 1995; San Juan et al., 1995; Yu et al., 1993). Glucans, major carbohydrate constituents of fungal cell walls (Domer, 1989; Elorza et al., 1989; Kapteyn et al., 1995; Nelson et al., 1991; San Juan et al., 1995; Yu et al., 1993), are polymers generally composed of a $(1\rightarrow3)$-β-D-linked linear backbone containing anhydroglucose repeat units (AGRUs) with a glycosidic linkage between the 1 and 3 positions (Figure 1.1, Structure 1) (Chauhan et al., 2002; Kapteyn et al., 2000; Klis et al., 2001; Shepherd, 1987). Some, but not all, glucan polysaccharides exhibit side chain AGRUs which branch exclusively from the 6 position of the backbone AGRU (Figure 1.1, Structures 2 and 3) (Ensley et al., 1994; Kim et al., 2000; Kogan et al., 1988; Lowman and Williams, 2001).

Glucan can assume a number of solution conformations depending upon the solvent system (Henderson, 1996; Saito et al., 1991; Young et al., 1998a; Young and

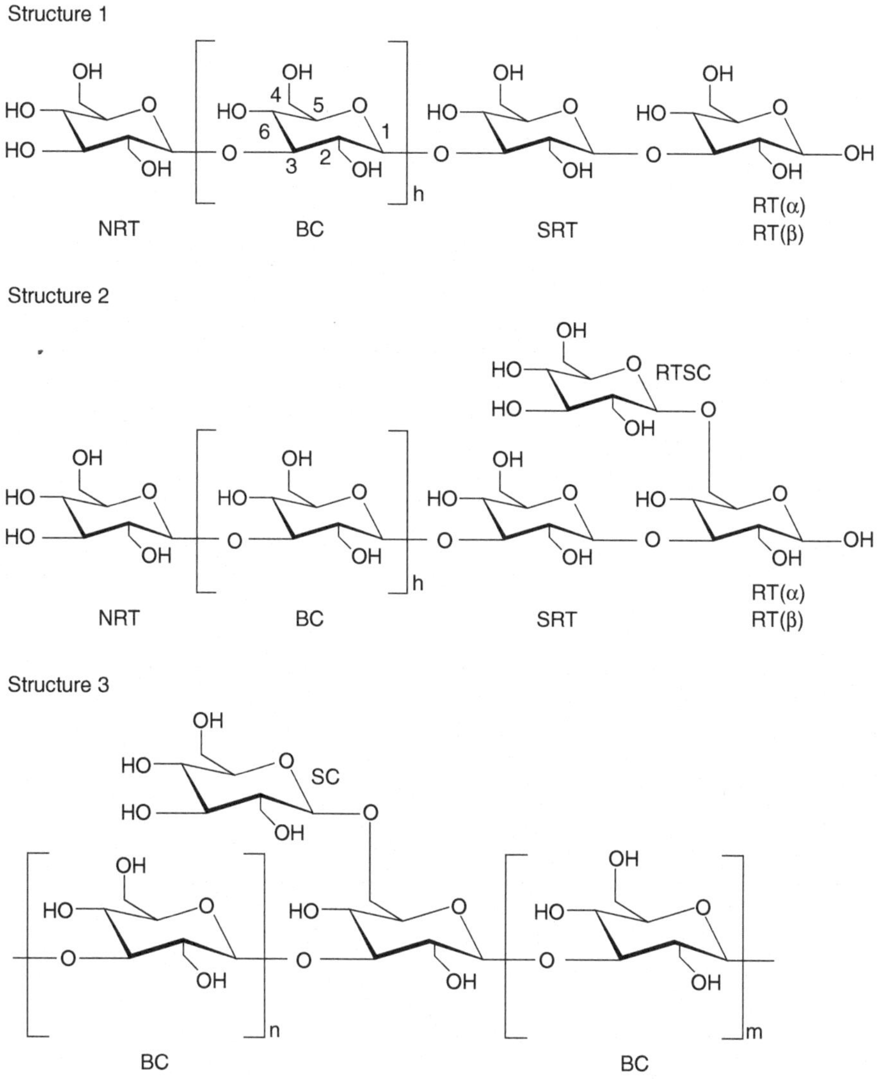

Introduction to the Chemistry and Immunobiology of β-Glucans

Structure 1

Structure 2

Structure 3

FIGURE 1.1 Glucans are polymers composed of a backbone chain (BC) containing linear (1→3)-β-D-linked anhydroglucose repeat units (AGRUs) (Structure 1). Side chain AGRUs branch exclusively from the 6 position of the backbone AGRU (Structures 2 and 3). The glucan polymer has a reducing and a nonreducing terminus (RT and NRT). The RT can exist as α or β anomers. SRT = second AGRU from the RT.

Jacobs, 1998b; Young et al., 2000; Young et al., 2003). For water soluble glucans, the two predominant conformations are single helix and/or triple helix (Figure 1.2). The single helical conformation is characterized as a semi-flexible coil, while the triple helix exists as a complex of three intertwined single helices that are stabilized by extensive hydrogen bonding involving the C2 hydroxyl group, located at the center of the helix (Figure 1.2). In the fungal cell wall, glucans comprise a three-dimensional network of (1→3)- and (1→6)-β-D-linked AGRUs that are connected to other carbohydrates, proteins, and lipids (Stone and Clarke, 1992).

1.3 PHYSICOCHEMICAL CHARACTERIZATION OF FUNGAL GLUCANS

Glucans are produced by a wide range of fungi and, to a lesser extent, bacteria (Stone and Clarke, 1992). Different microbes may produce glucans with varying physicochemical properties (Stone and Clarke, 1992). In addition, most of the glucans that have been studied were isolated from the host organism using a variety of extraction procedures (Adachi et al., 1989; Mueller et al., 1997). It is important to understand that the extraction method influences the nature of the polymer (Mueller et al., 1997). The first step in working with these complex carbohydrates is to characterize the polymers based on accepted physicochemical characteristics, including primary structure, degree of branching (DB), molecular weight (MW), polydispersity (I), degree of polymerization (DP), and purity. The following sections describe recent advances in the analysis, characterization, and quantification of glucans.

1.3.1 STRUCTURAL CHARACTERIZATION OF (1→3)-β-D-GLUCANS BY ^{13}C AND ^{1}H NMR

Carbon-13 (^{13}C) and proton (^{1}H) nuclear magnetic resonance spectroscopy (NMR) have proven to be particularly useful for the characterization of β-(1→3)-glucans (Ensley et al., 1994; Kim et al., 2000; Lowman et al., 1998; Lowman and Williams, 2001; Lowman et al., 2003; Pretus et al., 1991; Williams et al., 1991a; Williams et al., 1991b). Williams (Williams et al., 1991a; Williams et al., 1991b), Pretus (Pretus et al., 1991), Ensley (Ensley et al., 1994), Lowman (Lowman et al., 1998; Lowman and Williams, 2001; Lowman et al., 2003), Kim (Kim et al., 2000) and colleagues have examined a variety of water soluble and insoluble glucans using ^{13}C and ^{1}H NMR. In general, glucans are water insoluble microparticulates upon initial isolation from the cell wall or culture medium. By way of example, the glucan isolated from *Saccharomyces cerevisiae* is a water insoluble microparticulate with MW = 105 kD that can be converted to a water-soluble form with minimal reduction in molecular weight (Williams et al., 1994). Most glucans, including those from *S. cerevisiae*, are soluble in dimethylsulfoxide at concentrations of up to 25 mg/mL (Ohno et al., 1999; Sato et al., 1981; Williams et al., 1991a; Williams et al., 1994). Consequently, DMSO-d_6 is an extremely useful NMR solvent for these higher molecular weight glucans (Williams et al., 1991a; Williams et al., 1994). Lower molecular weight glucans (M_W < 20 kD), or glucans that have been derivatized, may be water soluble (Williams et al., 1991a). We have found that deuterium oxide (D_2O) can be a useful solvent for NMR studies of water-soluble glucans; however, water-soluble glucans can also be analyzed in DMSO-d_6 with even greater resolution of minor peaks (Kim et al., 2000; Pretus et al., 1991). Glucan preparations frequently contain varying quantities of molecular water and yet still appear as dry, free-flowing

Introduction to the Chemistry and Immunobiology of β-Glucans

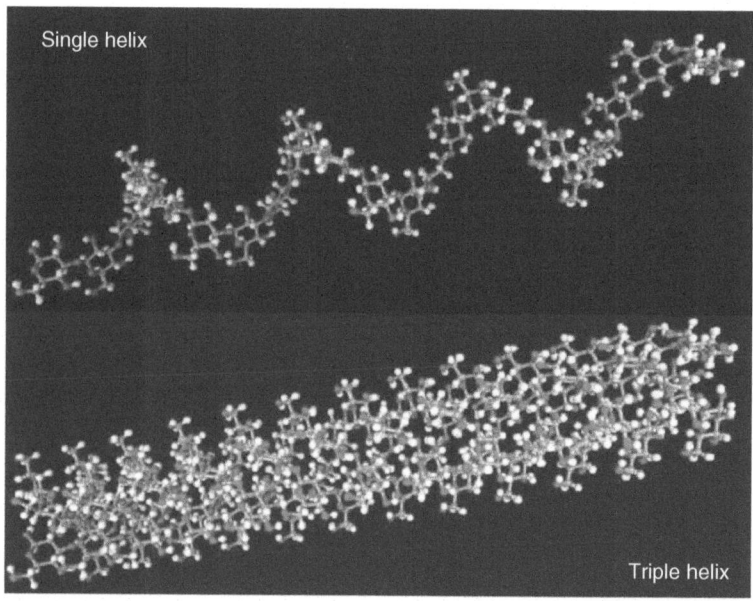

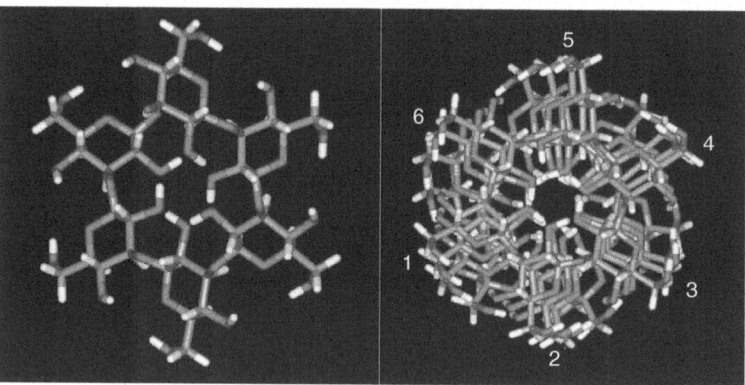

FIGURE 1.2 (See color insert following page 20.) Glucan polymers can assume a number of solution conformations depending upon the polymer and the solvent system. The two predominant conformations are single helix and triple helix (top panel). The single helical conformation is characterized as a semi-flexible coil, while the triple helix exists as a complex of three single helices. Triple helices are stabilized by extensive hydrogen bonding at the C2 hydroxyl, which is located at the center of the helix (bottom panel). The bottom panel shows an end view of a single glucan helix (left) and an end view of a glucan triple helix (right). The C2 hydroxyl is located at the center of the helix. The numbers at the outer edge of the model indicate the position of individual anhydroglucose subunits in the helical structure (lower right). In this model, six glucose subunits are required for one helical turn. To demonstrate this, the single helical end view model (shown on left) is composed of six glucose units, which comprise one helical turn.

powders. To facilitate NMR analysis the water should be removed. This is accomplished by suspending and dissolving the glucan in D_2O, stirring for 15 minutes, freezing, and lyophilizing. This process is repeated three times. D_2O exchanges with almost all of the water adsorbed by the glucan, but does not affect the hydroxyl protons of the water-insoluble glucan. Proton NMR analysis of glucan in DMSO-d_6 shows a marked temperature dependence. At room temperature, the ^1H NMR of β-glucan is a series of ill-defined broad peaks ranging from 5.4 ppm to 3.2 ppm with the residual HOD peak coming at 3.4 ppm (Figure 1.3). As the temperature is increased the protons bonded to oxygen rapidly move upfield, while the protons bonded to carbon move only slightly. The signal for the C_2-OH moves from 5.3 ppm at 20°C to 4.87 ppm at 80°C, while the water peak moves from 3.4 ppm to 3.04 ppm, upfield of all the glucan signals. As a result of these thermal shifts in position, optimum separation of signals occurs at about 80°C. At this temperature, the exchange rate for the hydroxyl protons is sufficiently slow that three-bond H-C-O-H coupling can be observed. As the temperature of the sample is raised even higher, the exchange of the oxygen-bonded protons becomes fast compared to the NMR timescale, resulting in the disappearance of the three-bond H-C-O-H coupling at 120°C. Heating at 120°C for 3 hours in DMSO-d_6 caused some thermal decomposition, as evidenced by darkening of the solution; however, reducing the temperature produced spectra identical to those recorded previously (Ensley et al., 1994).

Heteronuclear correlation spectroscopy (HETCOR) and correlated spectroscopy (COSY) (Figures 1.4 and 1.5) spectra of β-glucan at 80°C permit the unambiguous assignment of all signals. An obvious starting point for the assignment is the C_1 carbon that occurs at 102.7 ppm and correlates with the C_1-H proton signal at 4.52 ppm (doublet, J = 8 Hz). Also, HETCOR shows four peaks in the ^1H NMR spectrum (at 4.87, 4.4, 4.29, and 3.04 ppm) which do not correlate with any carbon signals (Figure 1.4). These are protons bonded to oxygen rather than carbon. COSY shows that the 3.04 peak is not coupled to any other peaks and can be assigned as the water peak. The triplet at 4.29 ppm (J = 6 Hz) must be the C_6-OH signal, since it is the only hydroxyl signal in COSY coupled to two other signals, the two magnetically nonequivalent C_6 geminal hydrogens (Figure 1.5).

C_6-H_a occurs as a doublet of doublets (J_1 = 6 Hz, J_2 = 11 Hz) at 3.7 ppm, while C_6-H_b occurs at 3.46 ppm and overlaps the C_3-H signal. The hydroxyl resonance at 4.87 ppm (doublet, J = 3.5 Hz) is coupled to the same signal that is coupled to the C_1-H signal and thus must be the C_2-OH signal. This allows the assignment of the C_2-H signal at 3.28 ppm (triplet, J = 8 Hz). The remaining hydroxyl signal doublet at 4.4 ppm can be assigned as C_4-OH (J = 1.5 Hz). This permits the assignment of the C_4-H signal at 3.25 ppm, overlapping with the C_5-H signal. The coincidence of C_4-H and C_5-H at 3.25 ppm, and of C_3-H and C_6-H_b at 3.46 ppm, can clearly be seen in the HETCOR spectrum (Figure 1.4).

With the assignment of the ^1H NMR, the six signals that appear in the ^{13}C NMR can be assigned as 102.7 (C_1), 85.98 (C_3), 76.14 (C_5), 72.59 (C_2), 68.23 (C_4), and 60.71 (C_6). The combination of HETCOR and COSY spectra shows that C_1, C_3, and C_5 are the only carbons that lack a hydroxyl group, indicating that the polymer is a 1→3 linked polyglucoside (Figures 1.4 and 1.5). The 8 Hz coupling found for C_1-H shows that the C_1 proton is axial and verifies the β-configuration. The β-(1 →3)

Introduction to the Chemistry and Immunobiology of β-Glucans

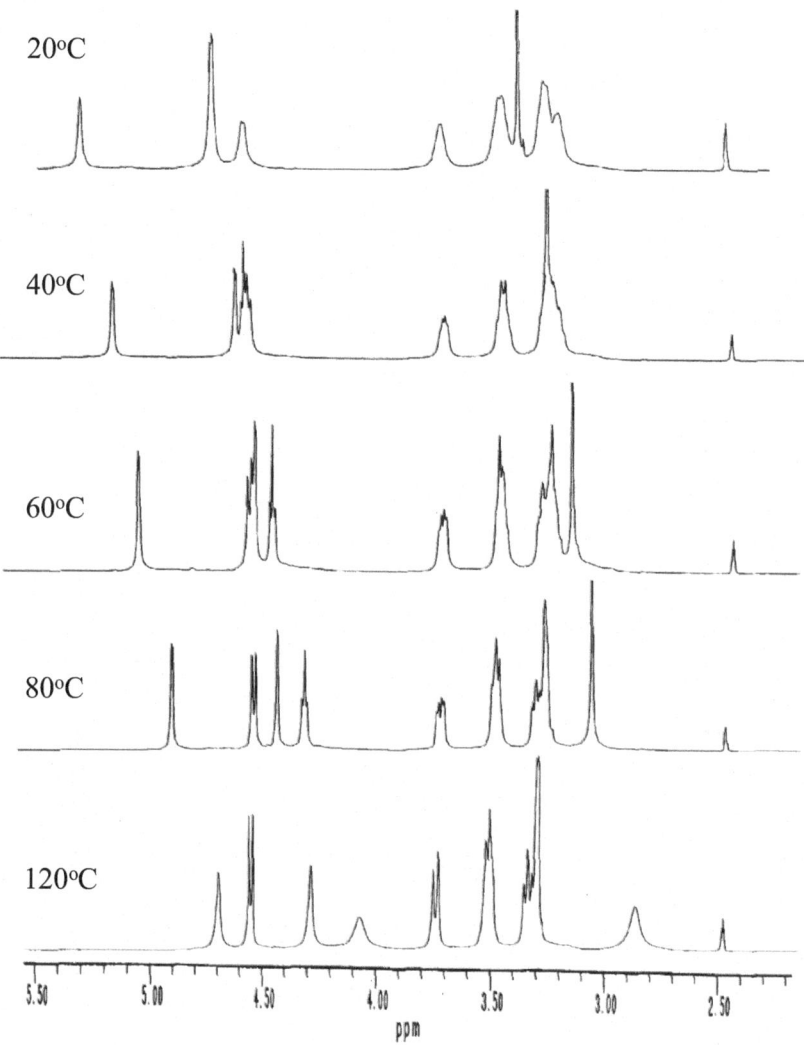

FIGURE 1.3 Proton (^{1}H) NMR analysis of glucan in DMSO- d_6 shows a marked temperature dependence. At room temperature, the ^{1}H-NMR of β-glucan is a series of ill-defined broad peaks. As the temperature is increased the protons bonded to oxygen rapidly move upfield, while the protons bonded to carbon move only slightly. Optimum separation of signals occurs at about 80°C. Thus, NMR analysis of glucans is greatly facilitated by increasing temperature.

linkage was also verified by an heteronuclear multiple bond correlation (HMBC) experiment optimized for a three-bond $^3J_{CH}$ (C_1-O-C_3-H) coupling of 5.5 Hz across the glycosidic oxygen. The 5.5 Hz value minimizes signals coming from the $^3J_{CH}$ coupling within the same ring. The intra-ring C_1-C_3-H dihedral angle is constrained

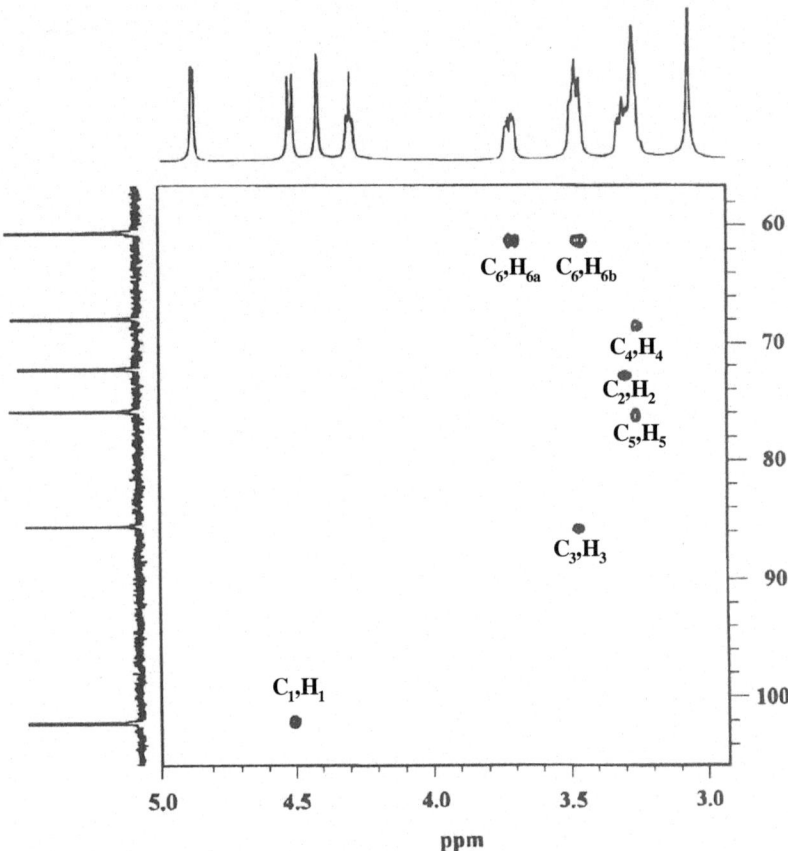

FIGURE 1.4 Heteronuclear correlation spectroscopy (HETCOR) of particulate glucan (20 mg/mL) at DMSO-d_6 at 80°C. The ^{13}C-NMR spectrum is displayed on the vertical axis and the ^{1}H-NMR spectrum on the horizontal axis.

to be about 60°. The Karplus equation provides an estimate of about 1 Hz for the $^3J_{CH}$ coupling constant involved in this constrained dihedral angle (Tvaroska, 1990). This HMBC experiment showed a substantial signal only for the coupling of C_3 and C_1-H, with a much weaker signal for intra-ring C_4 and C_6-H coupling. This approach confirms the (1→3)-β linkage of the glucan polymer backbone.

^{13}C NMR spectra of cell wall glucans isolated from a variety of sources are shown in Figure 1.6. Curdulan and glucan isolated from *S. cerevisiae* are linear glucans with very few, if any, branch points (glucose units attached at the C_6 hydroxyl) (Marchessault and Deslandes, 1979; Saito et al., 1977; Williams et al., 1991a). The ^{13}C NMR spectra of these linear glucans consist of six peaks whose chemical shifts are consistent with the ^{13}C NMR assignments described above. The presence of side chains leads to a more complex ^{13}C NMR spectrum; however, the basic β-(1→3) linear backbone resonances are usually present (Figure 1.6). Highly branched glucans, such as scleroglucans, have a branching frequency of 1:3 (one

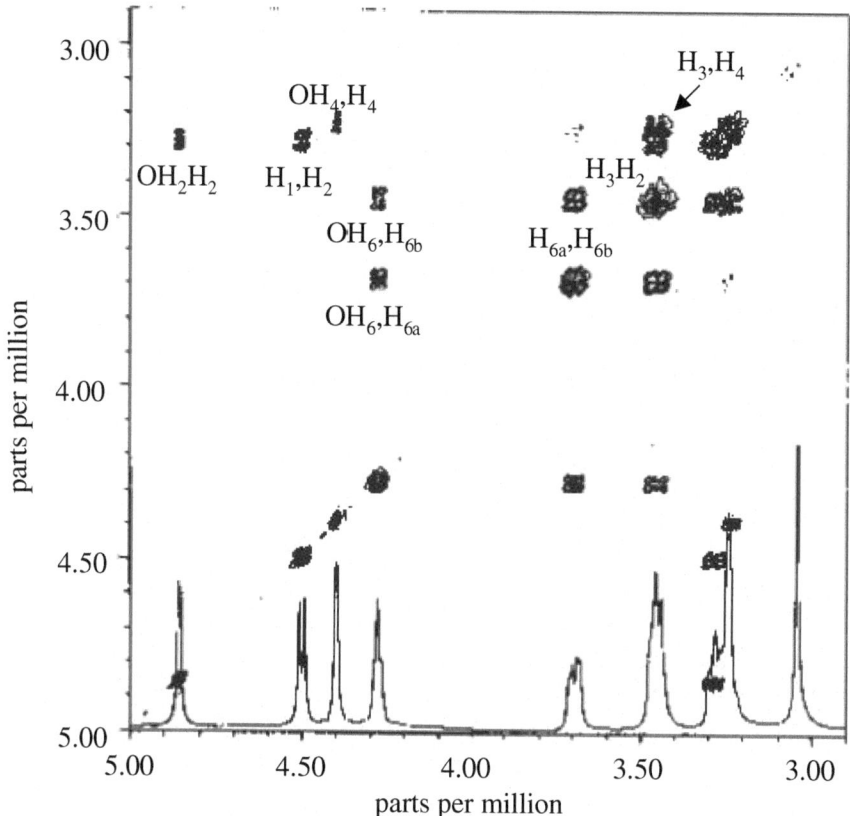

FIGURE 1.5 Phase-sensitive double-quantum-filtered correlated spectroscopy (COSY) of particulate glucan (10 mg/mL, after D_2O exchange) in DMSO- d_6 at 80°C.

side chain branch per three AGRUs in the backbone), while other glucans, like laminarin, have lower branching frequency on the order of 1:10. These branched glucans have very complex ^{13}C NMR spectra (Kim et al., 2000; Pretus et al., 1991). For example, the ^{13}C NMR spectrum of scleroglucan is characterized by three resonances of equal area assigned to carbons in the 3 position of the glucose repeat unit (Figure 1.6). The 3 resonances correspond to carbons in the 3 position of the (1→3)-β-linked AGRUs in the polymer backbone, to the (1→3)-β-linked AGRUs in the backbone to which a (1→6)-β branch is attached, and to the (1→3)-β-linked AGRUs in the side chain branch (Figure 1.6).

In a series of elegant experiments, Kim and colleagues (2000) assigned ^1H NMR chemical shifts for the anomeric protons of several glucose repeat units in different chemical environments. These environments include the AGRU at the reducing terminus (RT), the AGRU second from RT (SRT), the AGRU in the polysaccharide backbone (BC), the AGRU in the nonreducing terminus (NRT), and the branching AGRU attached to the backbone or to RT. Building on the work of Kim and colleagues (2000), Lowman and Williams (2001) used these anomeric proton

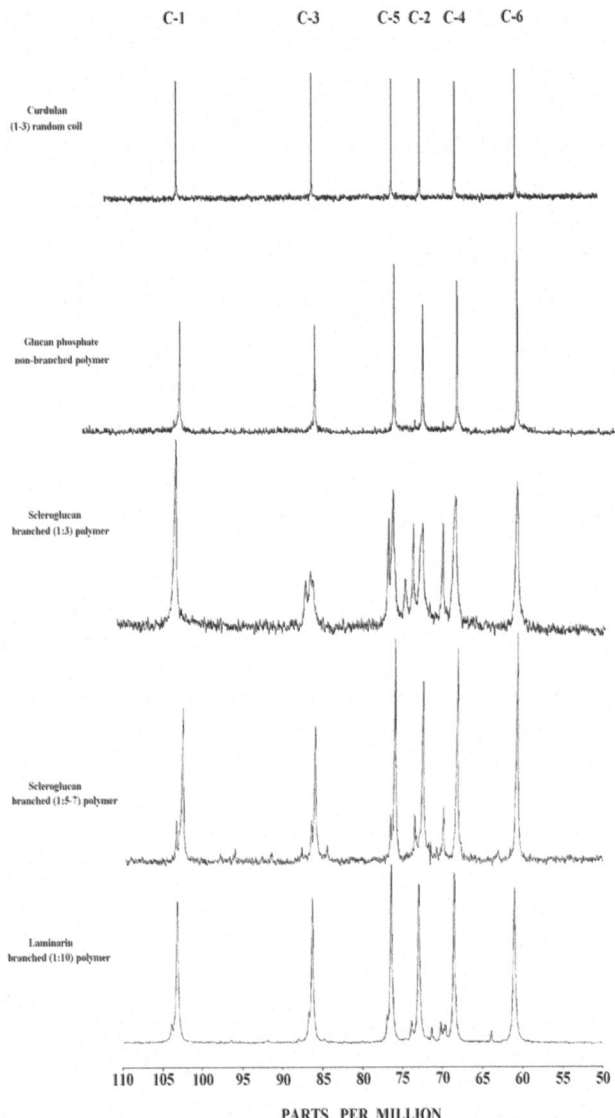

FIGURE 1.6 ^{13}C-NMR analysis of glucans isolated from a variety of sources. Curdulan and glucan isolated from *S. cerevisiae* are linear glucans with very few, if any, branch points (glucose units attached at the C6 hydroxyl). The ^{13}C-NMR spectrum of these glucans consist of six well defined peaks. These spectra are diagnostic for (1→3)-β-D-glucan polymers. The presence of side chains leads to a more complex ^{13}C-NMR spectrum; however, the basic β-(1→3) backbone signals are clearly present in these examples. Highly branched glucans, such as scleroglucans, have very complex NMR spectra which are characterized by a C_3 triplet. The three signals in the triplet correspond to the (1→3)-β-linked glucose subunits in the polymer backbone, to the (1→3)-β-linked glucose subunits in the backbone to which a (1→6)-β branch is attached, and to the (1→3)-β-link in the side chain branch. Laminarin, a low molecular weight (7700 g/mol) glucan has a branching frequency of 1:10.

Introduction to the Chemistry and Immunobiology of β-Glucans

chemical shift assignments to develop a detailed polymer characterization scheme in terms of polymer MW calculated from the DP, the DB, and glucan purity.

1.3.2 Analysis and Quantification of Glucans in Complex Biomatrices

For most biomedical and environmental studies, glucans have been isolated in pure form or derivatized in order to render them water soluble so as to facilitate tissue culture and animal studies (Fogelmark et al., 1992; Rylander, 1997; Williams et al., 1991a). However, in the environment glucans are usually found as macromolecular cell wall complexes containing proteins, lipids, and nonglucan carbohydrates (Kim et al., 2000). Until recently, the methods for the extraction, analysis, identification, and quantification of glucans in complex biomatrices were very time consuming and laborious. In addition, the available methods were based on chemical extraction methodology that did not provide absolute quantification. A simple and rapid assay for the identification and quantification of glucans found in complex mixtures was needed. The NMR studies described above provided the foundation for the development of a rapid and simple assay for the detection and quantification of glucans that are found in association with other macromolecules.

Lowman and Williams (2001) reported the development of NMR-based methodology for assaying the glucan content in solutions or suspensions of complex biomatrices, such as microbial cell wall extracts, plant and cereal grain cell wall extracts, microbial fermentations, and dietary fibers. This methodology, based on the application of internal standard quantitative NMR (Kasler, 1973), allows for the characterization of the constituents in a complex matrix on a solvent-free basis without employing extensive extraction, isolation, and drying protocols (Figure 1.7). Using this approach, the quantification and speciation of the glucan polymers, and other matrix components, such as other polysaccharides, lipids, and residual solvents, can be accomplished within a few hours per sample versus the 1 to 2 days that were previously needed (Mueller et al., 1994; Williams et al., 1991a). Furthermore, this method provides a direct measure of the glucan content on a dry-weight basis, even in the presence of other macromolecules and identifiable solvents in addition to water (Lowman and Williams, 2001).

1.3.3 Molecular Weight Analysis of Glucans

A major barrier to the understanding and ultimate utilization of natural product carbohydrate polymers was the difficulty involved in accurately characterizing carbohydrate polymers with molecular weights ranging from 10^3 to 10^7 g/mol. Over the last decade, there have been significant advances in aqueous and organic high performance gel permeation chromatographic (GPC) analysis of complex carbohydrate polymers (Mueller et al., 1995; Williams et al., 1992; Williams et al., 1994). Specifically, the development and application of multi-detector GPC systems combining multi-angle laser light scattering (MALLS) and differential viscometry (DV) have made it possible to rapidly, simply, and accurately determine a number of important physicochemical characteristics including weight, average molecular weight, polydispersity, root mean

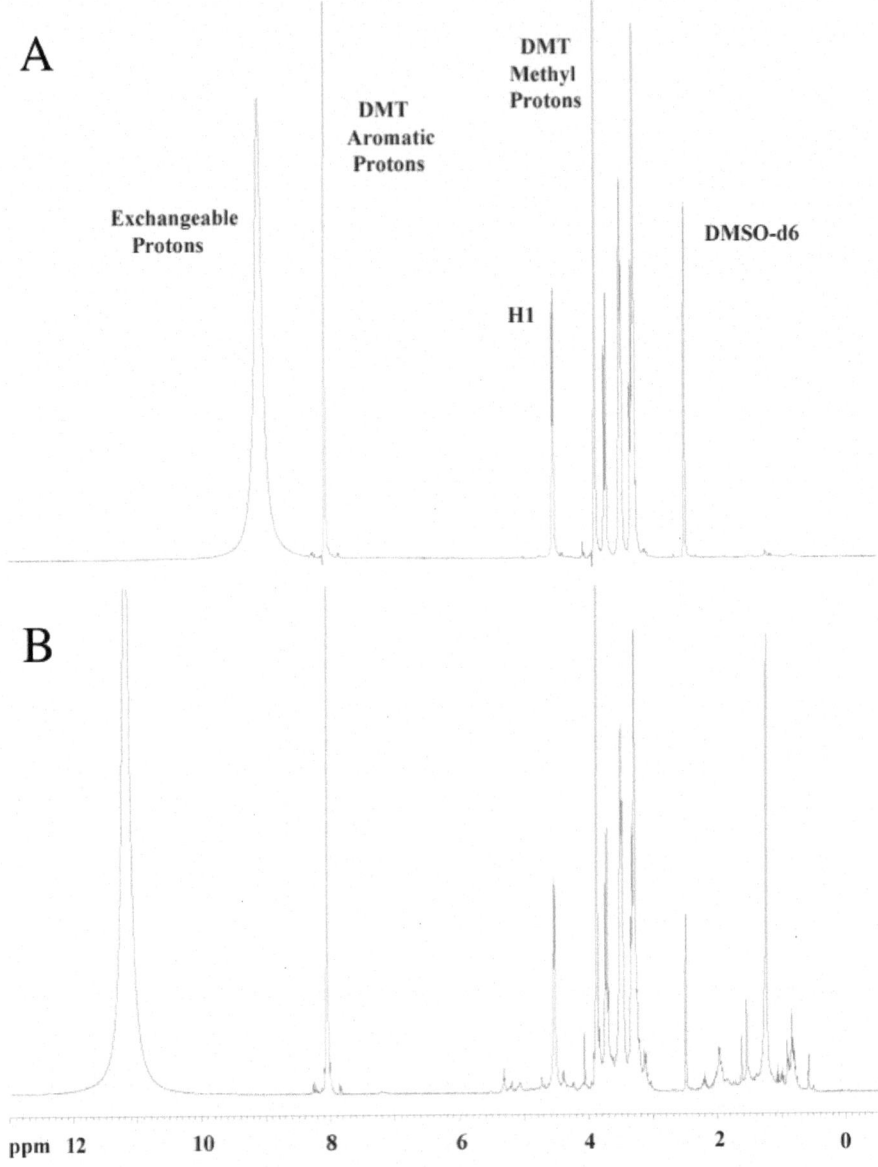

FIGURE 1.7 (A) Proton NMR spectrum of a highly purified (1→3)-β-D-glucan dissolved in DMSO-d_6 containing internal standard, dimetyl terephthalate (DMT), trifluoroacetic acid (TFA), and residual water. (B) Proton NMR spectrum from a crude isolation of (1→3)-β-D-glucan from yeast dissolved in DMSO-d_6 containing DMT, TFA, and residual water.

square radius, and solution conformation (Mueller et al., 1995; Williams et al., 1992; Williams et al., 1994). Williams (Williams et al., 1992; Williams et al., 1994), Mueller (Mueller et al., 1995), Christensen (Christensen et al., 2001) and colleagues have published detailed reports of the application and methods for GPC/MALLS and/or

Introduction to the Chemistry and Immunobiology of β-Glucans

TABLE 1.1
Analysis of Water-Soluble Carbohydrate Polymers as Determined by Gel Permeation Chromatography/Multiple-Angle Laser Light Scattering Photometry/Differential Viscometry

Carbohydrate Polymers	dn/dc[1] (ml/g)	Molecular Mass M_w (g/mol)	Polydispersity I (M_w/M_n)	Radius of the Center of Gravity r_z (nm)	Intrinsic Viscosity [η] dL/g	v^2	α^3
Glucan phosphate	0.122	1.57×10^5	1.67	20.3	0.33	0.302	0.65
Glucan sulfate	0.182	3.70×10^4	1.07	n.d.[4]	0.16	n.d.	n.d.
Schizophyllan (SPG)	0.253	3.06×10^5	1.08	37.3	7.32	0.599	n.d.
Laminarin	0.164	7.70×10^3	1.17	n.d.	0.07	n.d.	0.52
Scleroglucan	0.140	1.02×10^6	3.21	35.4	1.08	1.820	0.80
Barley glucan	0.175	1.91×10^5	1.49	29.6	2.18	0.193	0.49
Dextran	0.186	7.30×10^4	1.16	18.5	0.23	0.1	n.d.
Mannan	0.171	6.84×10^4	1.05	n.d.	0.25	0.2	n.d.

[1]dn/dc: refractive index increment.

[2]v: slope of the linear relationship between the log of the root mean square radius and log of the molecular mass moment ($R_G = K_{v'} \cdot M^v$) has been termed "v".

[3]α: slope of the linear relationship between log intrinsic viscosity and log molecular mass ($[\eta] = K_\alpha \cdot M^\alpha$) is known as the Mark-Houwink or α-value for a polymer system.

[4]n.d.: not detectable.

GPC/DV analysis of water insoluble and water soluble glucans derived from a variety of sources. An example of the type and spectrum of data that can be derived from GPC/MALLS/DV analyses is shown in Table 1.1.

It is noteworthy that many water-insoluble glucans exist as stable triple helices (Bluhm and Sarko, 1977; Bluhm et al., 1982; Norisuye et al., 1980; Young et al., 1998a; Young et al., 2000). We and others have reported that DMSO is a useful mobile phase for the GPC analysis of water insoluble glucans (Williams et al., 1994). However, there is an important caveat to this type of analysis. When triple helical glucans are dissolved in DMSO, they undergo a helical dissolution (Williams et al., 1994; Young et al., 1998a; Young et al., 2000). Following helical dissolution, the individual glucan polymer strands assume a random coil conformation (Williams et al., 1994; Young et al., 1998a; Young et al., 2000). The end result is that the molecular weight of a triple helical glucan, when analyzed in DMSO, reflects only one third of the natural molecular weight due to the disruption of the helical structure (Williams et al., 1994).

In the preceding discussion, we have addressed some of the major issues in the characterization of glucans and related carbohydrate polymers. These data are important not only for establishing the specific characteristics of a given polymer system, but they are also critical to understanding how the polymer interacts with biological systems. By way of example, NMR analysis provides the investigator with information about the primary structure, purity, and composition of the glucan or macromolecular complex that is being studied. With this information, investigators can determine, with a high degree of confidence, whether they are working with a pure glucan, a glucan preparation that contains trace contaminants, a glucan containing a macromolecular complex, or a nonglucan polymer system. Obviously, the interpretation of biological data is critically dependent upon knowing whether the material being studied is pure or not. Accurate molecular weight analysis is critical to establishing the interactions (affinity K_D and receptor number B_{max}) of the polymer with pattern recognition receptors (Mueller et al., 2000). Similarly, establishing the ν and α values (Table 1.1) provides information regarding the solution conformation of the polymer, i.e., the predominant tertiary form of the polymer in a given solvent system (Mueller et al., 2000). Such data are highly relevant because recent evidence suggests that human macrophage receptors recognize glucans, in part, based on solution conformation (Mueller et al., 2000). Thus, it is no exaggeration to say that understanding the structure of glucans is an absolute prerequisite for deciphering their biological activity. Therefore, we extensively characterize glucan polymer systems, using the techniques described above, before initiating biological studies.

1.4 IMMUNOBIOLOGY OF GLUCANS

There is an extensive literature describing the effect of glucans on various biological systems (Bohn and BeMiller, 1995; Williams et al., 1996). An exhaustive review of that literature is beyond the scope of this work. In addition, the published results are sometimes conflicting and even contradictory, making it difficult to draw overall conclusions. This may be due to differences in the glucan preparations that have been studied, the biological systems that were analyzed or a combination of these factors. Interpretation of these data is further complicated by the fact that many of the published works do not provide details about the nature or purity of the polymer systems being studied. In the discussion that follows, we have focused on published results in which we could confirm that the glucan employed was chemically pure. This is not intended to diminish the importance or relevance of other published reports; rather it is intended to present an accurate evaluation of the effect of glucans of known structural composition on biological systems.

An important caveat to the following discussion is the fact that glucan exposure in environmental settings is almost certainly not as pure glucans, but as macromolecular cell wall complexes that contain varying amounts of glucan. In light of this, it would seem prudent to study the effect of these macromolecular complexes on biological systems. However, it is virtually impossible to establish the effect of a single component, such as glucan, on a biological system when one is working with a complex biomatrix, such as a fungal cell wall. Our approach has been to isolate the glucans and other cell wall components in pure form and to study the biological effects of the individual components. Once this is accomplished, it is possible to combine the pure components and study their combined effects on biological systems. In other words, studying the individual components, such as glucan, provides the reference data for studying multiple-component macromolecules.

1.5 RECOGNITION AND BINDING OF GLUCANS BY MEMBRANE RECEPTORS

The interaction of microbial products and host cells is a major determinant of the innate immune response. This interaction evokes acute and chronic inflammatory responses and, in some cases, cellular infiltration at the site of recognition. The innate immune system has evolved a complex network of receptors, which rapidly identify microorganisms based on invariant molecular structures (lipids, proteins, and/or carbohydrates) that are shared by a variety of microbes (Akira, 2001; Brown and Gordon, 2001; Gough et al., 2000). These invariant molecular structures are called pathogen-associated molecular patterns (PAMPs). They include glucan, lipopolysaccharide (LPS), lipoteichoic acid, peptidoglycan, lipoproteins, lipoarabinomannans, and other products (Janeway and Medzhitov, 1999; Medzhitov and Janeway, 1997). PAMPs are recognized by pattern recognition receptors (PRRs). The first step in the modulation of cellular activity by glucans is the recognition and binding of the glucan polymer by pattern recognition receptors located on cell membranes (Williams et al., 1991b; Williams et al., 1996).

Many species have receptors or binding proteins that recognize $(1\rightarrow 3)$-β-D-glucans (Cosio et al., 1992; Engstad and Robertsen, 1994; Schmidt and Ebel, 1987; Vargas-Albores and Yepiz-Plascencia, 2000; Willment et al., 2001). The nature of the mammalian $(1\rightarrow 3)$-β-D-glucan receptors has been the subject of controversy. For a number of years, it was thought that the Type 3 complement receptor (CR3) was the β-glucan binding site on macrophages, neutrophils, and NK cells in mammals (Ross et al., 1987; Thornton et al., 1996; Vetvicka et al., 1996). Studies by Ross (1987), Thornton (1996), Vetvicka (1996) and colleagues reported the existence of a glucan binding site located outside the I domain of CR3 (CD11b/CD18). In contrast, Michalek and colleagues (1998) have reported the binding of glucan to a non-CR3 site on a neutrophil cell line. In 1996, we reported the existence of a specific glucan binding site on undifferentiated human promonocytes that did not express detectable levels of CR3 (Battle et al., 1998; Mueller et al., 1996). In addition, we reported that glucan ligand binding to the promonocyte receptor stimulated intracellular signaling pathways for immunoregulatory and proinflammatory transcriptional activator proteins (Battle et al., 1998). This suggested that ligation of a non-CR3 receptor had functional consequences that were consistent with modulation of immune function (Brown et al., 2002; Kougias et al., 2001; Lowe et al., 2002). Zimmerman and colleagues (1998) have reported that lactosylceramide binds glucans, and that this glycosphingolipid may be a leukocyte glucan binding moiety, although the role of lactosylceramide as a glucan binding site is unproven. When taken together these data argue for the existence of multiple glucan binding sites.

In 2000, Kougias and colleagues (2001) reported the discovery of β-glucan specific receptors on primary cultures of normal human dermal fibroblasts. This was the first report of glucan specific receptors on cells outside the immune system. Since this report, glucan specific receptors have been identified on normal human vascular endothelial cells (Brown et al., 2002; Lowe et al., 2002), human epithelial cells (Ahren et al., 2001) and human anterior pituitary cells (Breuel et al., 2004). Of potentially greater significance, direct coincubation of glucan with human fibroblasts (Kougias et al., 2001; Wei et al., 2002a; Wei et al., 2002b), vascular endothelial cells (Lowe et al., 2002), or anterior pituitary cells (Breuel et al., 2004) modulated cellular function. The potential ramifications of these data were significant because they required a re-examination of the existing hypotheses regarding the mechanisms by which the host recognizes and responds to these fungal glucans. By way of example, glucans have been reported to exert a plethora of systemic effects (Bohn and BeMiller, 1995; Williams et al., 1996). The presumed mechanism was that glucans interacted with leukocytes and other elements of innate immunity resulting in either a primed or activated state (Adams et al., 2000; Falch et al., 2000; Soltys and Quinn, 1999). The systemic effects were attributed to release of proinflammatory or immunoregulatory mediators, which served as second messengers, i.e., the systemic effects were indirect. While this was a reasonable explanation for the observed effects, the data of Kougias (2001), Ahren (2003), Lowe (2002), Breuel (2004), and colleagues indicate that glucans may also directly interact with and modify the functional state of various cell types throughout the body. Thus, it is reasonable to speculate that some of the effects that have been ascribed to glucans may be mediated through direct interaction with a variety of cells distributed throughout the body, not just immune competent cells. These observations also cast doubt on the central role of CR3 as the primary glucan binding site. CR3 is a β-integrin, which is leukocyte restricted (Ross et al., 1987). Consequently, CR3 is not expressed on fibroblasts, endothelial cells, epithelial cells, or anterior pituitary cells. The fact that CR3 is not expressed on these cells indicates that CR3 participation is not an absolute requirement for modulation of cellular function by glucans.

1.6 THE INFLUENCE OF GLUCAN POLYMER MOLECULAR WEIGHT, STRUCTURE, AND SOLUTION CONFORMATION ON BINDING TO (1→3)-β-D-GLUCAN RECEPTORS

Several reports have suggested that specific physicochemical parameters, such as primary structure, solution conformation, molecular weight, and polymer charge, are important determinants of the affinity with which (1-3)-β-D-glucans bind to receptor(s) and modulate cellular functions (Falch et al., 2000; Kulicke et al., 1997; Mueller et al., 2000; Saito et al., 1991). However, the precise relationship between (1-3)-β-D-glucan physicochemical parameters and receptor ligand interaction remains to be fully defined. This is due, in part, to a lack of well-characterized (1H3)-β-D-glucan polymers with varying molecular weights and conformational structures. Several recent publications have begun to shed light on the structure-activity relationships of (1-3)-β-D-glucans, particularly as it relates to interaction with membrane receptors.

Janusz and colleagues (1989) reported that a (1→3)-β-linked yeast heptaglucoside (DP = 7 glucose repeat units) inhibited monocyte phagocytosis of zymosan particles. They concluded that this glucan was the unit ligand for human monocyte β-glucan receptors. While this was an intriguing observation, these results were equivocal, because they did not characterize the heptaglucoside. Therefore, they could not confirm that the carbohydrate employed was a (1→3)-β-*D*-linked glucan. They did not determine whether the polymer was linear or branched, nor did they employ oligosaccharides with DP other than seven to confirm that the heptaglucoside was the minimum binding unit (Janusz et al., 1989). In addition, Janusz and colleagues (1989) employed a phagocytosis inhibition assay which did not directly assess ligand receptor interactions, nor did they evaluate the immunobiological properties of the heptaglucoside. Consequently, a number of important questions remained to be answered. In 2001, Lowe and colleagues (2001) confirmed and extended the observations of Janusz and colleagues (1989) by demonstrating that a heptasaccharide (1-3)-β-*D*-glucan was the minimum unit ligand for glucan receptors on human U937 promonocytic cells. These investigators reported that a heptasaccharide glucan interacted with glucan specific membrane receptors at a K_D of 31 μ*M* (95% CI 20–48 μ*M*) and 100% inhibition. Penta- and hexasaccharide glucans were not recognized by the glucan receptors. While the heptasaccharide glucan did bind to the receptors, it did not stimulate human monocyte intracellular signaling, nor did it increase survival in a murine model of polymicrobial sepsis (Lowe et al., 2001). Laminarin, a larger and more complex glucan polymer (MW = 7700 g/mol, DP = 41), only partially inhibited binding (61 ± 4%) at a K_D of 2.6 μ*M* (99% CI 1.7–4.2 μ*M*) with characteristics of a single binding site (Figure 1.8). Interestingly, laminarin also did not stimulate monocyte intracellular signaling (Lowe et al., 2001). Thus, the studies of Lowe and colleagues (2001) demonstrated that a heptasaccharide is the smallest unit ligand recognized by human macrophage glucan receptors. The heptasaccharide was a low affinity receptor antagonist, but it was not a glucan receptor agonist with respect to activation of NFκB-dependent signaling pathways or protection against experimental sepsis. Their data also indicated the presence of at least two glucan binding sites on human monocytes and that the binding sites on human monocyte/macrophages can discriminate between glucan polymers. It is not clear why the heptasaccharide glucan does not modulate cellular function; however, Michalek and colleagues (1998) have reported that binding and functional activation of macrophages requires cross-linking of membrane receptors that are distinct from CR3. Based on these data, Lowe and colleagues (2001) speculated that glucan polymers larger than 7700 g/mol were required in order to crosslink spatially separated receptors and alter cellular function.

1.7 GLUCAN RECEPTORS DIFFERENTIALLY RECOGNIZE GLUCAN POLYMERS BASED ON SOLUTION CONFORMATION AND MOLECULAR WEIGHT

Mueller and colleagues (2000) characterized a variety of (1-3)-β-*D*-glucan and nonglucan polymers in order to accurately establish molecular mass and to gain insights into the solution conformation of these polymers (Table 1.1). This polysaccharide library was then employed to compare and contrast the effect of glucan polymer structure, molecular weight, and solution conformation on receptor binding affinity (Table 1.1 and Table 1.2). They reported that the receptors on the human

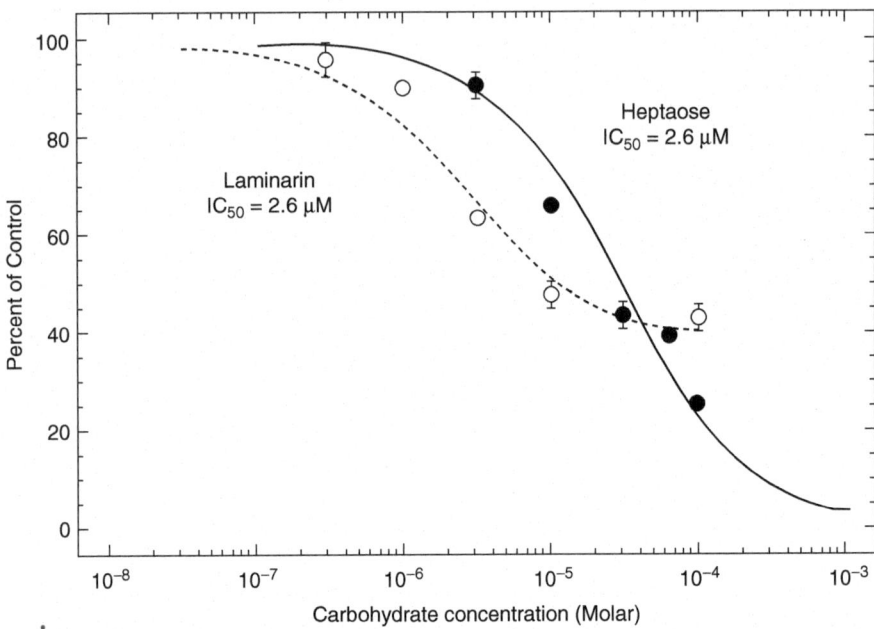

FIGURE 1.8 Differential recognition of glucans by human promonocyte receptors. The heptasaccharide glucan completely inhibited U937 membrane binding to immobilized glucan in a concentration dependent manner. Heptasaccharide competition fits a model for inhibition at a single binding site. Complete inhibition of binding was observed at competitor concentrations of ~10^{-3} M. The K_D of the heptaose was 5 µM. Laminarin partially inhibited U937 membrane binding to immobilized glucan. Binding of laminarin was concentration dependent and fit a model for inhibition at one of two binding sites. The K_D of laminarin was 2.6 µM. Only 61 ± 4% of binding could be inhibited over a dose range of 10^{-8} to 10^{-3} M. Approximately 39% of the U937 membrane binding could not be inhibited by laminarin even at high concentrations.

promonocytic cell line (U937) were specific for (1→3)-β-D-glucan binding, since mannan, dextran, and barley glucan did not bind (Mueller et al., 2000). Scleroglucan, a high molecular weight glucan, exhibited the highest binding affinity with a IC_{50} of 23 nM. The rank order competitive binding affinities for the glucan polymers were scleroglucan >>> schizophyllan > laminarin > glucan phosphate > glucan sulfate. Scleroglucan also exhibited a triple helical solution structure (v = 1.82, α = 0.8). These investigators also noted that there were two different binding/uptake sites on U937 cells. Glucan phosphate and schizophyllan interacted nonselectively with the two sites (Figure 1.9). Scleroglucan and glucan sulfate interacted preferentially with one site, while laminarin interacted preferentially with the other site. The authors concluded that human monocytic cells have at least two glucan receptor(s) that specifically interact with (1-3)-β-D-glucans and that the triple helical solution conformation and molecular weight of the glucan polymer are important determinants in receptor ligand interactions.

TABLE 1.2
Inhibitory Concentration (IC_{50}) of Different (1→6→3)-β-D-Glucans on the Binding of Glucan Phosphate to U937

Glucan	IC_{50} (μM)	Displacement (%)	P Value
Glucan phosphate	35	100	C[1]
Glucan sulfate	43	37	< 0.01
Scleroglucan	0.023	40	< 0.005[2,3]
Schizophyllan (SPG)	11	100	< 0.01[2,4]
Laminarin	21	57	< 0.01[2]

[1]Glucan phosphate binding was employed as the control (displacement) to which all other glucans were compared.
[2]$p < 0.05$ vs. glucan phosphate and glucan sulfate.
[3]$p < 0.05$ vs. all other glucans.
[4]$p < 0.05$ vs. laminarin.

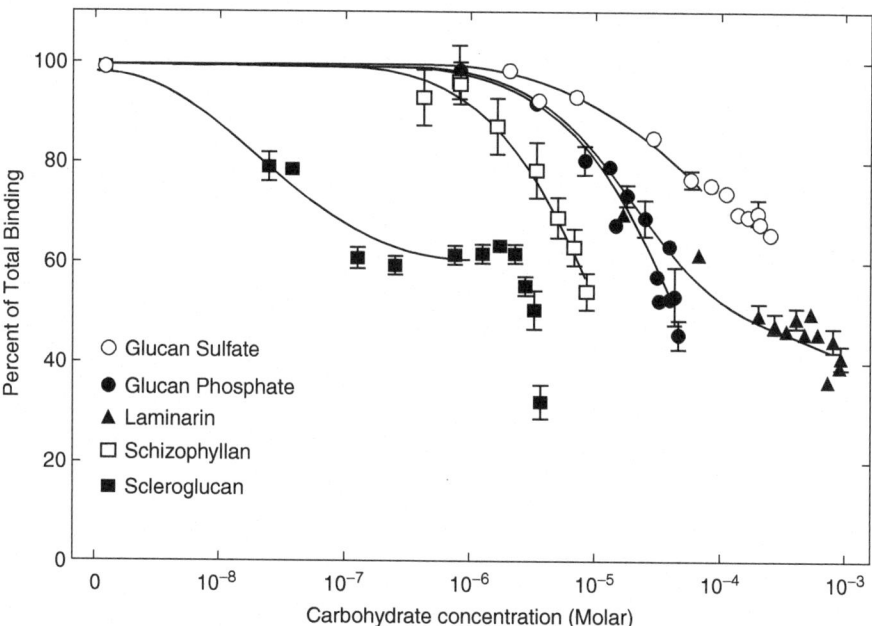

FIGURE 1.9 Competitive displacement of tritiated glucan phosphate by unlabeled glucan sulfate, glucan phosphate, schizophyllan, laminarin, and scleroglucan. The rank order competitive binding affinities for the glucan polymers were scleroglucan >>> schizophyllan > laminarin > glucan phosphate > glucan sulfate. The data are expressed as mean ± SEM and represent at least four replicates with four to eight data points/concentration/replicate.

1.8 IDENTIFICATION OF DECTIN-1 AND SCAVENGER RECEPTORS AS GLUCAN-SPECIFIC MEMBRANE RECEPTORS

As noted above, evidence suggests that there are multiple glucan binding sites distributed on several cell types. The nature of the glucan receptors and which receptors were responsible for modulating innate host defenses has been the subject of controversy. Some of these controversies have largely been laid to rest by the identification and characterization of Dectin-1 and scavenger receptors as glucan binding sites.

1.8.1 THE ROLE OF DECTIN-1 AS A PRIMARY GLUCAN RECEPTOR

Brown and Gordon (2001) reported the identification of Dectin-1 as a glucan receptor. These investigators screened a murine macrophage cDNA expression library for the ability to bind zymosan, a glucan containing yeast cell wall extract. They identified Dectin-1 as the glucan binding moiety (Brown and Gordon, 2001). Yokota (2001), Hermanz-Falcon (2001), and coworkers had previously isolated Dectin-1 from a dendritic cell cDNA subtraction library and showed that Dectin-1 recognized a ligand on T cells. Dectin-1 possesses a single C-type lectin-like carbohydrate recognition domain connected to a transmembrane region by a stalk (Brown and Gordon, 2001; Brown et al., 2002; Brown et al., 2003; Taylor et al., 2002; Willment et al., 2001). The cytoplasmic tail possesses a immunotyrosine-based activation motif (ITAM) (Willment et al., 2001). Brown, Willment, Taylor, and colleagues have demonstrated that Dectin-1 fulfills all of the characteristics of the previously unidentified β–glucan receptor, and it appeared to act as a classic pattern recognition receptor (PRR) (Brown and Gordon, 2001; Brown et al., 2002; Brown et al., 2003; Taylor et al., 2002; Willment et al., 2001). Dectin-1 recognizes $(1\rightarrow3)$-β and $(1\rightarrow6)$-β-linked glucans, and it also recognizes intact *S. cerevisiae* and *Candida albicans* in a glucan-dependent fashion (Brown and Gordon, 2001a; Brown et al., 2002; Brown et al., 2003; Taylor et al., 2002; Willment et al., 2001). Dectin-1 is broadly expressed with the highest expression levels found on monocytes and neutrophils in the blood, bone marrow, and spleen (Taylor et al., 2002). Dendritic cells and a subpopulation of T cells also express Dectin-1 but at rather low levels (Hermanz-Falcon et al., 2001; Yokota et al., 2001). It was also noted that Dectin-1 was highly expressed on alveolar and inflammatory macrophages, which may be indicative of a role for this receptor in immune surveillance (Brown and Gordon, 2001). As noted above, CR3 was originally thought to be the primary glucan receptor. However, studies by Brown and colleagues employing CR3 knockout mice have clearly demonstrated that Dectin-1 is the primary receptor for binding and internalization of glucan in macrophages (Brown et al., 2002). Studies from the same laboratory have identified the human homologue of Dectin-1, and reported it to be a type II transmembrane receptor with a single carbohydrate recognition domains (CRD) and an ITAM motif (Willment et al., 2001). They have also shown that the human homologue is functionally similar to the mouse receptor; however, the human receptor mRNA is alternatively spliced, resulting in two major and six minor isoforms (Willment et al., 2001). The two major isoforms differ in the presence of a stalk region

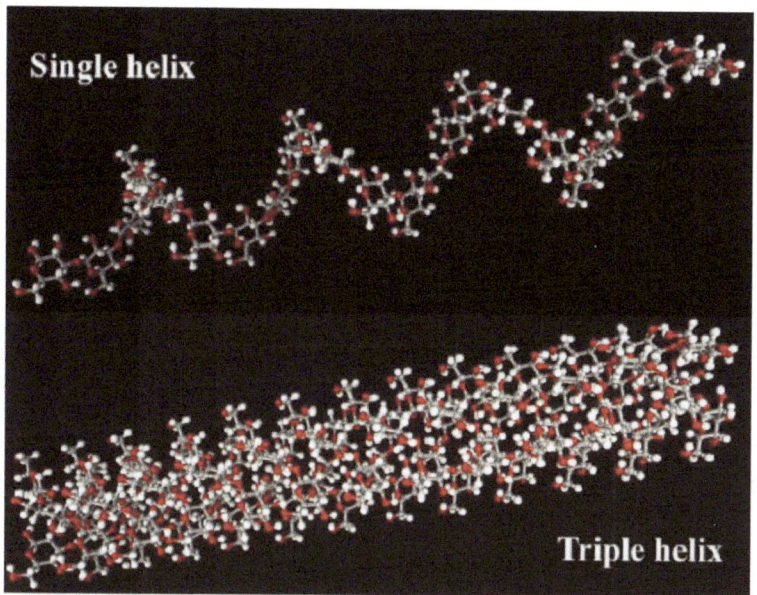
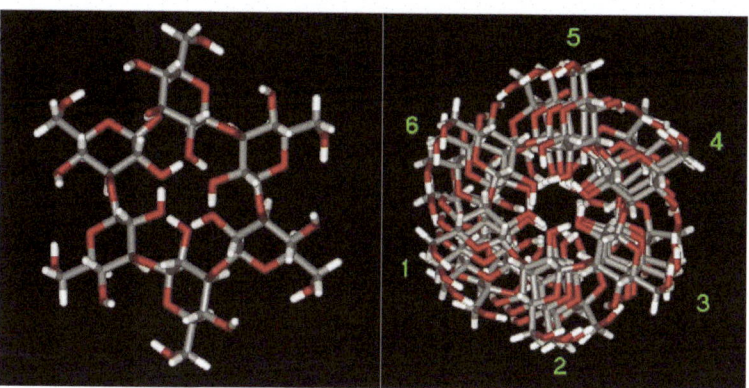

FIGURE 1.2 Glucan polymers can assume a number of solution conformations depending upon the polymer and the solvent system. The two predominant conformations are single helix and triple helix (top panel). The single helical conformation is characterized as a semi-flexible coil, while the triple helix exists as a complex of three single helices. Triple helices are stabilized by extensive hydrogen bonding at the C2 hydroxyl, which is located at the center of the helix (bottom panel). The bottom panel shows an end view of a single glucan helix (left) and an end view of a glucan triple helix (right). The C2 hydroxyl is located at the center of the helix. The numbers at the outer edge of the model indicate the position of individual anhydroglucose subunits in the helical structure (lower right). In this model, six glucose subunits are required for one helical turn. To demonstrate this, the single helical end view model (shown on left) is composed of six glucose units, which comprise one helical turn.

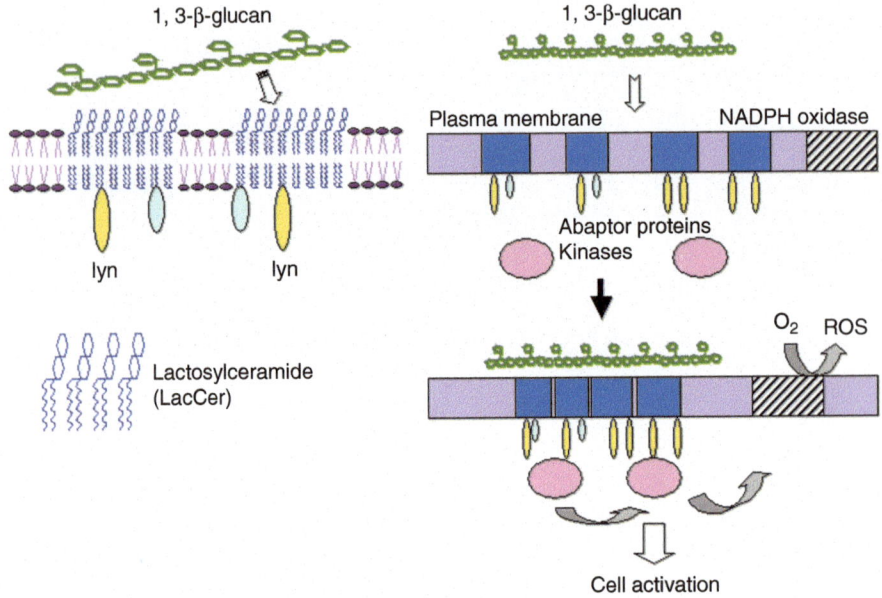

FIGURE 5.1 LacCer can transmit signals for activation of neutrophils by 1,3-β-glucans.

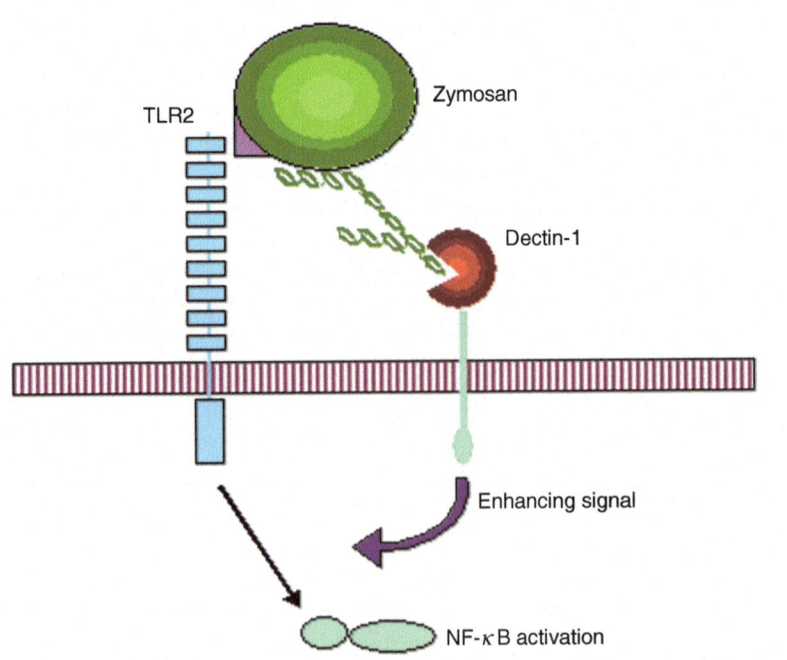

FIGURE 5.2 Collaborative effect of Dectin-1 on TLR2-mediated NF-B activation by zymosan.

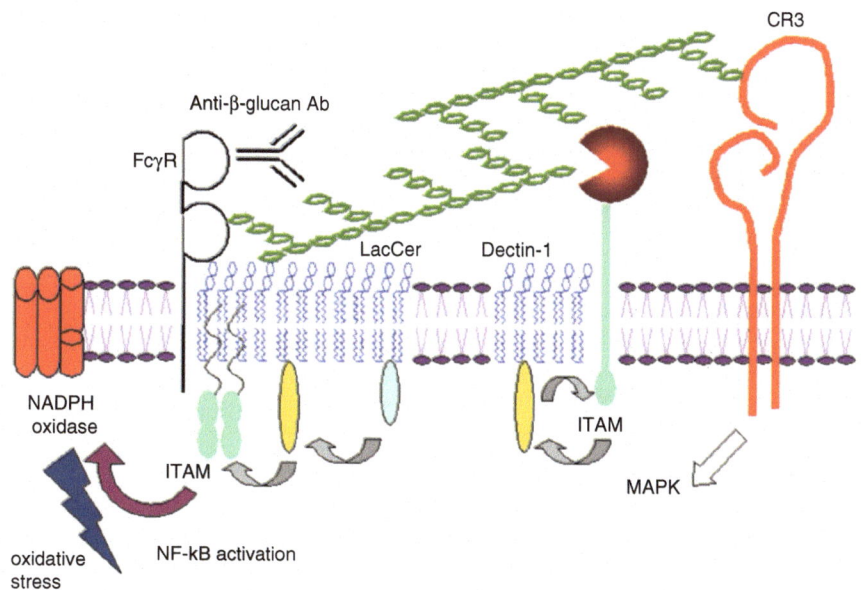

FIGURE 5.3 Possible mechanisms in leukocyte activation signaling through multiple receptors.

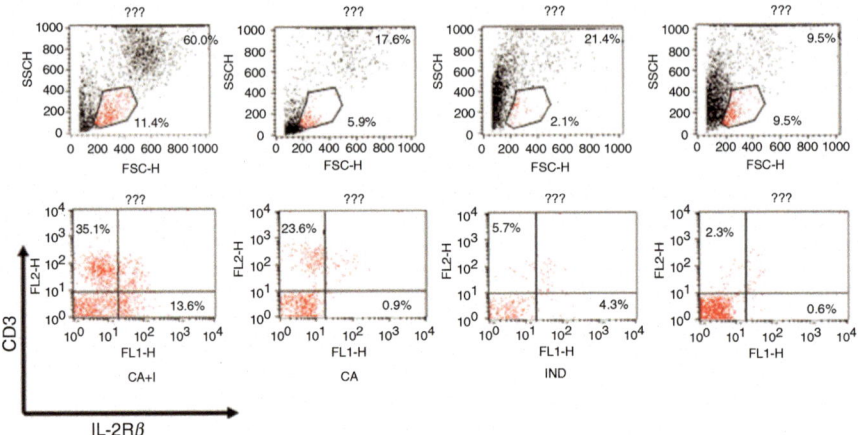

FIGURE 8.7 Lymphocyte phenotype in lung (1). Cells were collected from BAL of mice which were administered CA/IND. The surface phenotypes of these cells were analyzed using a two-color immunofluorescence test. FITC anti-mouse CD122 (IL-2 Receptor -chain) mAbs and R-PE anti-mouse CD3e (CD3 ε–chain) mAbs were used.

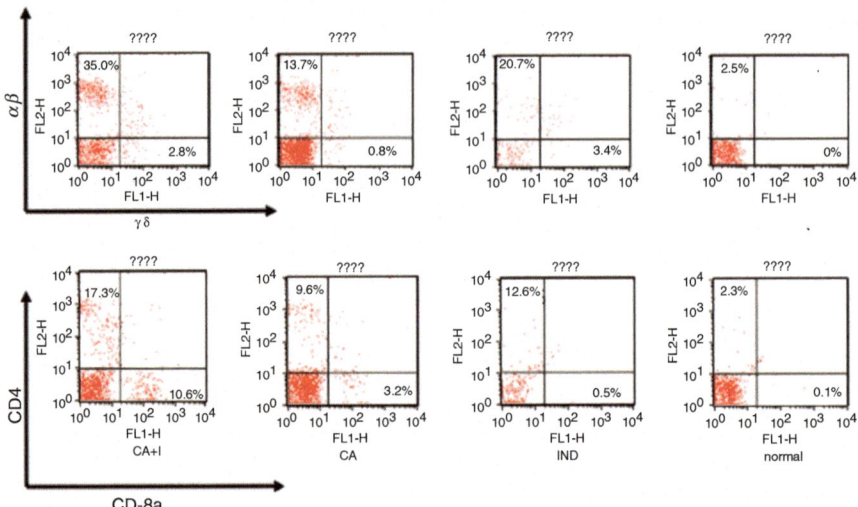

FIGURE 8.8 Lymphocyte phenotype in lung (2). Cells were collected from BAL of mice which were administered CA/IND. The surface phenotypes of these cells were analyzed using a two-color immunofluorescence test. FITC anti-mouse δ T cell receptor mAbs and R-PE anti-mouse TCR chain (α) mAbs (upper side), and FITC anti-mouse CD8a mAbs and R-PE anti-mouse CD4 mAbs (lower side) were used.

Introduction to the Chemistry and Immunobiology of β-Glucans

separating the CRD from the transmembrane region (Willment et al., 2001). The discovery of Dectin-1 as a glucan-specific receptor will almost certainly result in major advances in our understanding of how the host immune system recognizes and responds to glucans.

1.8.2 Scavenger Receptors as Glucan Binding Sites

Dushkin (1996), Vereschagin (1998), and colleagues have reported that a water soluble, polyanionic, carboxymethylated glucan (CM-glucan) binds to mouse peritoneal macrophages via the class A scavenger receptors. Vereschagin and colleagues (1998) have extended this observation by reporting that glucan interaction with macrophage scavenger receptors protects against endotoxic shock. This suggested that macrophage scavenger receptors may be involved in glucan recognition and signaling. The data of Pearson and colleagues (1995) support the results of Dushkin (1996), Vereschagin (1998), and colleagues by demonstrating that the Drosophila CI (class C) scavenger receptor can bind laminarin and other microbial cell wall components. Rice and colleagues (2002) have extended these observations to demonstrate that human promonocytic cells specifically bind glucans via scavenger receptors. These investigators employed surface plasmon resonance to characterize the interaction (Kougias et al., 2001; Lowe et al., 2002; Rice et al., 2002). Specifically, they immobilized acetylated-LDL (AcLDL) on a biosensor surface and used surface plasmon resonance to examine the binding of several glucans, differing in fine structure and charge density, to scavenger receptors on the human monocyte cell line (Rice et al., 2002). They compared the binding of glucan phosphate, schizophyllan, laminarin, two carboxymethyl glucan preparations, and carboxymethyl cellulose (CM-cellulose) to scavenger receptors (Table 1.3). The CM-glucans showed comparable affinities (14 and 22 nM) and inhibited 58–67% of U937 membrane binding to AcLDL. CM-cellulose showed a comparable degree of inhibition (56%), but at lower affinity (360 M). Glucan phosphate, a weak polyelectrolyte, completely inhibited binding of

TABLE 1.3
Dissociation Constants of Polysaccharides Employed as Competitors of Human Promonocyte Scavenger Receptor Binding to AcLDL Surface

Carbohydrate	K_D	95% Confidence Interval	B_{max} (%)[1]
Glucan phosphate	8.4 μM	4.0–18 μM	100/0
CM glucan (Swiss)	14 nM	8.0–26 nM	58/42 ± 3
CM glucan (Slovak)	22 nM	4.9–99 nM	67/33 ± 8
CM cellulose	360 nM	160–870 nM	56/44 ± 3
Schizophyllan (SPG)	>10 μM	—	—
Laminarin	6.2 μM	4.0–9.9 μM	59/41 ± 4

[1]Data are expressed as the percentage displaced divided by the percentage remaining.

U937 membranes to AcLDL (K_D = 8.4 μM). Laminarin inhibited 59% of binding (K_D = 6.2 μM). Schizophyllan, a high molecular weight, neutral glucan, showed minimal interaction with the scavenger receptor. They concluded that there were at least two AcLDL binding sites on human monocytic cells that interacted with glucans (Rice et al., 2002). They further concluded that the multiple binding sites may reflect interactions of AcLDL with scavenger receptor subtypes, most likely involving class A scavenger receptors (Rice et al., 2002). Glucan phosphate interacted with all sites, while the CM-glucans and laminarin interact with a single site or subset of sites. Not surprisingly, they also reported that the charge of the glucan polymer had a dramatic effect on the affinity with which it is recognized by macrophage scavenger receptors (Rice et al., 2002). However, it was also clear that human monocyte scavenger receptors recognize the basic glucan structure irrespective of polymer charge (Rice et al., 2002). The fact that scavenger receptors are widely distributed may provide additional insights into the ability of fibroblasts, endothelial cells, and other cell types to recognize and respond to glucans.

1.9 ACTIVATION OF PROINFLAMMATORY AND IMMUNOREGULATORY INTRACELLULAR SIGNALING PATHWAYS BY GLUCANS

As noted above, there is now substantive evidence for the recognition, binding, and internalization of glucans by multiple cell surface receptors (Brown and Gordon, 2001; Rice et al., 2002). However, receptor–ligand interaction is merely the first step in the activation cascade. The second step in the modulation of cellular function by glucans involves signal transduction into the cell. The third step involves activation of intracellular signaling pathways that are involved in regulating gene expression and cellular function. Numerous reports demonstrate that glucans are pleiotrophic agents, i.e., compounds that activate an array of intracellular signaling processes (Adams et al., 1997; Adams et al., 2000; Battle et al., 1998; Wei et al., 1997; Wei et al., 2002a; Wei et al., 2002b). The following section reviews the recent data on glucan induced signal transduction and the activation of intracellular signaling pathways following ligation of glucan by membrane receptors.

1.9.1 TOLL-LIKE RECEPTOR 2 RECOGNITION AND SIGNALING IN RESPONSE TO GLUCAN EXPOSURE

The recognition of microorganisms by immune competent cells is known to involve a family of pattern recognition receptors designated Toll-like receptors (TLRs) (Anderson, 2000; Janeway and Medzhitov, 1999). TLRs are an ancient and evolutionarily conserved family of signaling molecules, which regulate innate immunity and antimicrobial responses. TLRs are distinguished from other pattern recognition receptors by their ability to recognize different classes of pathogens (Anderson, 2000; Janeway and Medzhitov, 1999). TLR2 recognizes gram-positive bacteria, peptidoglycans, lipoproteins, fungal mannans, and glucans (Anderson, 2000; Opitz et al., 2001). Stimulation of TLR2 leads to activation of a series of signaling proteins, including MyD88, IRAK, and TRAF6. This cascade activates the NFκB signaling pathway by stimulating IκBα degradation and nuclear translocation of NFκB

Introduction to the Chemistry and Immunobiology of β-Glucans

(Anderson, 2000; Arbibe et al., 2000). NFκB activates a wide array of proinflammatory genes leading to synthesis of cytokines, chemokines, arachidonic acid metabolites, and nitric oxide (NO) (Barnes, 1997; Gantner et al., 2003). Several investigators had speculated that glucans were TLR2 agonists; however, there was no empirical evidence to support that speculation (Janeway and Medzhitov, 1999). In May of 2003, two research groups, working independently, simultaneously published evidence for Dectin–TLR2 cooperation in glucan recognition and signaling (Brown et al., 2003; Gantner et al., 2003). These simultaneous publications have provided compelling evidence that glucans are recognized by Dectin-1 and the glucan stimulatory signal is transduced into the cell through TLR2-dependent mechanisms. The data presented in these two papers indicate that Dectin-1 and TLRs trigger independent and cooperative inflammatory responses (Brown et al., 2003; Gantner et al., 2003). A specific example is reactive oxygen species, which can be triggered by Dectin-1 independent of TLR activation (Gantner et al., 2003).

The reports of Brown (2003), Gantner (2003), and coworkers demonstrating the cooperation of Dectin-1 and TLR2 employed glucan or zymosan treatment of cultured cells. An important question was what, if any, effect glucan would have on *in vivo* modulation of TLR expression. Our laboratory has observed that systemic administration of water-soluble glucan to normal mice will stimulate TLR2 and TLR4 expression in the liver and lung in a time-dependent fashion (Williams et al., 2003). Taken together, these data strongly implicate TLR2 signaling in the response to glucan. However, it is important to note that the results of Brown and colleagues (2003) suggest that glucan may also stimulate TLR-independent signaling pathways. This would be consistent with the pleiotrophic nature of glucans.

1.9.2 Activation of Nuclear Factor Kappa B (NFκB), Nuclear Factor Interleukin 6 (NF-IL6), Nuclear Factor 1 (NF-1), Activator Protein 1 (AP-1) and Specificity Protein 1 (SP-1) Signaling Pathways by Fungal Glucans

Adams and colleagues (1997) reported that a branched (1-3)-β-*D*-glucan activated NFκB- and NF-IL6-like transcription factors in cultured cells. These investigators also speculated that, with regard to transcription factor activation, glucans may utilize "signal transduction pathways different from those used by LPS" (Adams et al., 1997). Battle and colleagues (1998) confirmed and extended these observations by demonstrating that low and high affinity glucan ligands increased NFκB binding activity in a human macrophage cell line. Williams and colleagues (1999b) extended these observations to a mouse model in which systemic administration of water-soluble glucans stimulated tissue NFκB and NF-IL6 expression.

Wei and colleagues (2002a; 2002b) have reported that direct coincubation of glucan with primary cultures of normal human dermal fibroblasts results in a time-dependent increase in NF-1, AP-1, and SP-1 nuclear translocation. In this study, glucan was shown to stimulate fibroblast alpha 1(I) and α1 (III) procollagen.mRNA, when compared to the untreated control cells. Type I and type III collagen synthesis were increased by 720% (24 h) and 390% (48 h) following glucan treatment of fibroblasts (Wei et al., 2002b). Glucan directly stimulated fibroblast collagen biosynthesis through an NF-1 dependent mechanism. This was demonstrated by treating the cells with pentifylline, an inhibitor of NF-1 (Wei et al., 2002b). Down-regulation of NF-1 by pentifylline inhibited glucan induced procollagen mRNA expression (Wei et al., 2002b). In subsequent studies, Wei and coworkers (2002a) examined the effect of glucan on activation of the transcription factors AP-1 and SP-1 in normal human dermal fibroblasts. Glucan stimulated fibroblast AP-1 and SP-1 activation in a time-dependent manner, although the temporal kinetics varied between the two transcription factors. AP-1 binding activity was increased at early time intervals (1, 2, 4, 8, and 12 h), while SP-1 nuclear binding activity was increased at later time intervals (12, 24, 36, and 48 h). It is likely that other transcription factors and/or signaling pathways are activated by glucan or glucan-containing complexes. Additional studies are required to elucidate the spectrum of glucan activity, particularly as it relates to signal transduction.

1.10 EFFECT OF GLUCANS ON CYTOKINE AND GROWTH FACTOR EXPRESSION

One of the most controversial aspects of glucan activity is the effect of these carbohydrate polymers on induction of cytokine, chemokine, and growth factor release. There are a number of reports which suggest that glucans will stimulate expression of various cytokines (Engstad et al., 2002; Falch et al., 2000; Okazaki et al., 1995; Wei et al., 2002a). However, there are equally credible reports indicating that glucans will stimulate intracellular signaling pathways, but will not stimulate cytokine gene or protein expression (Adams et al., 1997; Adams et al., 2000). This has led some investigators, such as Adams and colleagues (1997; 2000), to speculate that glucans may prime cells for subsequent stimuli; e.g., a second hit effect. The precise reasons for confounding and conflicting data on glucans and cytokine expression are not known. However, there is reason to speculate that the nature (for example, molecular weight and branching frequency) and physical state (soluble versus insoluble) of the glucan employed and the biological system in which it is studied may be critical in determining whether cytokine expression occurs. By way of example, Kulicke and colleagues (1997) have studied the correlation between induction of immunological activity and the molar mass and molecular structure of glucans. They reported that glucans with molecular weights of ~5×10^5 g/mol were the most potent stimuli of tumor necrosis factor α (TNFα) in human peripheral blood monocytes. Interestingly, glucans with molecular weights in excess of 1×10^6 g/mol were much less potent stimuli for TNFα production. In addition, Kulicke and colleagues (1997) concluded that helically ordered glucan structures were not essential or even advantageous for immunologic activity. Data from our laboratory tend to support this conclusion with respect to induction of survival in a murine sepsis model (Williams et al., 1999b). Our data show that single helical glucans are just as effective in protection as triple helical glucans with molecular weights in excess of 1×10^6 g/mol (Williams et al., 1999b). However, we have also

Introduction to the Chemistry and Immunobiology of β-Glucans 25

shown that triple helical glucans are recognized by glucan receptors with much higher affinity than the single helical forms, although neither of these glucans stimulated cytokine expression (Mueller et al., 2000).

Kougias (2001), Lowe (2002), Wei (2002a; 2002b), and colleagues have studied the effect of a water-soluble glucan on interleukin 6 (IL-6), interleukin 8 (IL-8), vascular endothelial cell growth factor (VEGF) and wound growth factor gene and protein expression in primary cultures of normal human dermal fibroblasts or vascular endothelial cells. Interleukin-8 gene and protein expression were observed in vascular endothelial cells treated with a water-soluble glucan (Lowe et al., 2002). Glucan-stimulated fibroblast expression of IL-6; neurotrophin 3 (NT-3); platelet-derived growth factor A or B (PDGFA or PDGFB); fibroblast growth factor, acidic or basic (FGFa or FGFb); transforming growth factor alpha or beta (TGFα or TGFβ); and VEGF mRNA. However, glucan had no effect on epidermal growth factor (EGF), nerve growth factor (NGF) or ciliary neurotrophic factor (CNTF) mRNA expression in primary fibroblast cultures. Glucan also had no effect on VEGF expression in vascular endothelial cells (Wei et al., 2002a).

1.11 ANTIINFLAMMATORY ACTIVITY OF GLUCANS

The data presented above show that significant strides have been made in understanding the cellular and molecular mechanisms of glucans. There is a substantial literature that demonstrates that glucans are potent immune stimulators, and that they will upregulate proinflammatory responses. While these data are clearly valid, they also present an interesting paradox in light of recent data. Williams et al. (1999b) and others (Hara et al., 1982; Masihi et al., 1997; Ukai et al., 1983) have reported that various glucans will stimulate proinflammatory and immunoregulatory mechanisms. However, we have also shown that glucans will decrease morbidity and increase survival in a murine model of fulminating polymicrobial sepsis (Williams et al., 1999b). The pathophysiology of this experimental infection is known to involve the host over-expressing inflammatory mediators, which result in a systemic inflammatory response that culminates in severe shock, multiple organ failure, and death (Williams et al., 1999a). Thus, a major question was how a proinflammatory agent, such as glucan, might ameliorate a disease that had a significant inflammatory component. Indeed, agents such as glucans would seem to be contraindicated in patients with diseases that have a significant inflammatory process. We have obtained data that shed new light on this paradox and, more importantly, suggest an entirely new and novel mechanism of action for glucans in certain inflammatory and infectious diseases. In the normal host, glucan binding to the (1→3)-β-D-glucan receptor stimulates a mild inflammatory and nonspecific immunostimulatory event (Williams et al., 2000; Williams et al., 1999b; Williams et al., 2003). In contrast, glucan ligands administered in the presence of an inflammatory/septic insult blunt the early increase in tissue transcription factor activity and cytokine transcription that is associated with the septic challenge, thereby limiting the host inflammatory response to the injury (Williams et al., 1999b). Preventing early activation of NFκB and NF-IL6 positively correlated with improved long-term survival (Figure 1.10) (Williams et al., 1999b). It is important to note that glucan did not suppress NFκB or NF-IL6 levels;

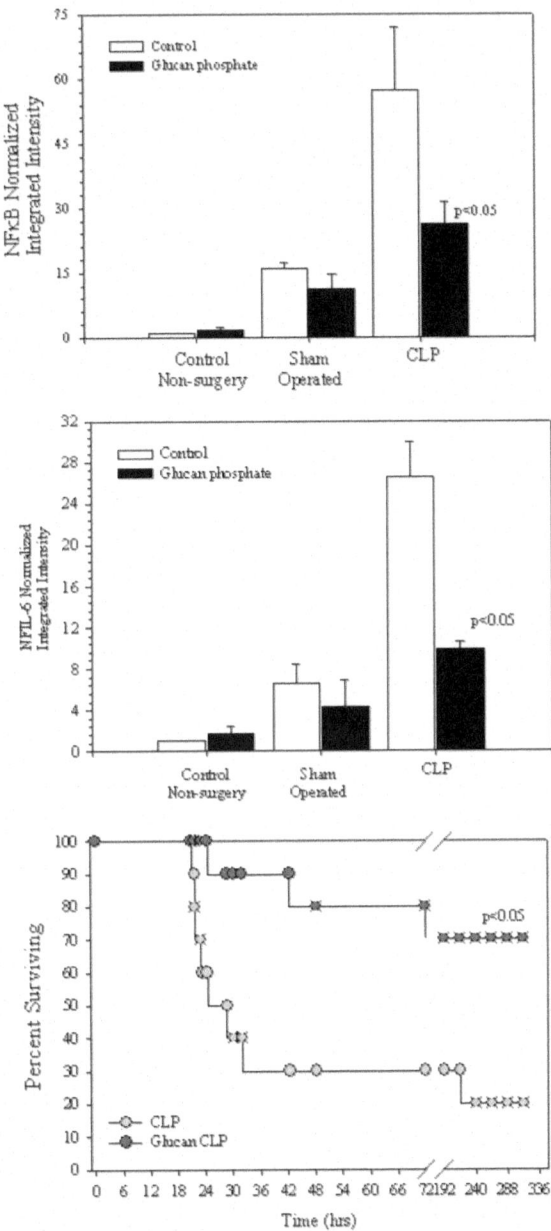

FIGURE 1.10 Decreased hepatic NFκB and NF-IL6 nuclear binding activity correlates with increased survival in glucan phosphate (50 mg/kg, IV) treated septic mice. Sepsis was induced by cecal ligation and puncture (CLP). Tissue samples were harvested 3 h following CLP. The data were derived from gel shift assays where the autoradiograms were quantified by scanning densitometry. N = 4 – 20/group.

rather, it prevented the dramatic increase in transcription factor activation that is observed in septic mice. Thus, while the effect may be viewed as anti-inflammatory, it is not immunosuppressive, since normal or above normal levels of these important factors are maintained. In fact, immune competence is increased, as demonstrated by our clinical observation that glucan administration stimulated conversion from anergy (immunosuppression) in trauma patients (Browder et al., 1990).

The observation that glucans may exert an anti-inflammatory effect seems to be in conflict with the toxicology of these compounds in environmental settings. However, there is some evidence that glucans may have a similar effect in certain environmental settings. Rylander and colleagues have extensively studied the relationship of glucan exposure to indoor-air-related symptoms of allergy and asthma (Rylander, 1996; Rylander et al., 1992; Rylander and Lin, 2000; Thorn et al., 2001). Their results indicate that the response to inhaled glucans in humans is quite different from that observed following inhalation of bacterial endotoxin (Thorn et al., 2001). By way of example, endotoxin causes a predominately neutrophilic inflammatory response, while glucan does not induce neutrophil recruitment. Indeed, Rylander and Lin (2000) speculate that glucans may prime airway cells for simultaneous or subsequent exposure to another agent. They reported that glucans potentiated ovalbumin-induced eosinophilia and IgE responses. At the same time, some inflammatory responses have been shown to be down-regulated (Rylander and Lin, 2000). Specifically, TNFα production is decreased under certain conditions (Rylander and Lin, 2000). The precise reasons for this differential response to glucans are unclear.

1.12 CONCLUSIONS

A number of recent advances have dramatically increased our knowledge of the chemistry and biology of (1→3)-β-D-glucans. The chapters that follow have been written by eminent scientists who will focus on a variety of topics related to the toxicology of glucans. It is the hope of the authors that this introductory chapter will provide a foundation that enhances the readers' understanding of the chapters that follow.

REFERENCES

Adachi, Y., Ohno, N., Ohsawa, M., Sato, K. D., Oikawa, S., and Toshiro, Y. (1989). Physicochemical properties and antitumor activities of chemically modified derivatives of antitumor glucan "Grifolan LE" from *Grifola frondosa*. *Chem. Pharm. Bull.* **37**, 1838–1843.
Adams, D. S., Pero, S. C., Petro, J. B., Nathans, R., Mackin, W. M., and Wakshull, E. (1997). PGG-glucan activates NF-κB-like and NF-IL-6-like transcription factor complexes in a murine monocytic cell line. *J. Leuko. Biol.* **62**, 865–873.
Adams, D. S., Nathans, R., Pero, S. C., Sen, A., and Wakshull, E. (2000). Activation of a Rel-A/CEBP-beta-related transcription factor heteromer by PGG-glucan in a murine monocytic cell line. *J. Cell Biochem.* **77**, 221–233.

Ahren, I. L., Williams, D. L., Rice, P. J., Forsgren, A., and Riesbeck, K. (2001). The importance of a beta-glucan receptor in the nonopsonic entry of nontypeable Haemophilus influenzae into human monocytic and epithelial cells. *J. Infect. Dis.* **184**, 150–158.

Ahren, I. L., Eriksson, E., Egesten, A., and Riesbeck, K. (2003). Non-typeable Haemophilus influenzae activates human eosinophils through beta-glucan receptors. *Am. J. Respir. Cell Mol. Biol.* **29**, 598–605.

Akira, S. (2001). Toll-like receptors and innate immunity, *Adv. Immunol.* **78**, 1–56.

Anderson, K. V. (2000). Toll signaling pathways in the innate immune response. *Curr. Opin. Immunol.* **12**, 13–19.

Arbibe, L., Mira, J.-P., Teusch, N., Kline, L., Guha, M., Mackman, N., Godowski, P. J., Ulevitch, R. J., and Knaus, U. G. (2000). Toll-like receptor 2-mediated NF-κB activation requires a RAC I-dependent pathway. *Nature Immunol.* **1**, 533–540.

Barnes, P. J. (1997). Nuclear factor-κB, *Int. J. Biochem. Cell Biol.* **29**, 867–870.

Battle, J., Ha, T., Li, C., Della Beffa, V., Rice, P., Kalbfleisch, J., Browder, W., and Williams D. (1998). Ligand binding to the (1→3)-beta-D-glucan receptor stimulates NFκB activation, but not apoptosis in U937 cells. *Biochem. Biophys. Res. Commun.* **249**, 499–504.

Beijer, L., Thorn, J., and Rylander, R. (2002). Effects after inhalation of (1-3)-β-D-glucan and relation to mould exposure in the home. *Mediators Inflam.* **11**, 149–153.

Bluhm, T. L. and Sarko, A. (1977), The triple helical structure of lentinan, a linear β-(1-3)-D-glucan. *Can. J. Chem.* **55**, 293–299.

Bluhm, T. L., Deslandes, Y., Marchessault, R. H., Perez, S., and Rinaudo, M. (1982). Solid-state and solution conformation of scleroglucan. *Carbo. Res.* **100**, 117–130.

Bohn, J. A. and BeMiller, J. N. (1995) (1-3)-Beta-D-glucans as biological response modifiers: a review of structure-functional activity relationships. *Carbo. Polym.* **28**, 3–14.

Breuel, K. F., Kougias, P., Rice, P. J., Wei, D., De Ponti, K., Wang, J., Laffan, J. J., Li, C., Kalbfleisch, J., and Williams, D. L. (2004). Anterior pituitary cells express pattern recognition receptors for fungal glucans: implications for neuroendocrine immune involvement in response to fungal infections. *Neuroimmunomodulation* **11**, 1–9.

Browder, W., Williams, D., Pretus, H., Olivero, G., Enrichens, F., Mao, P., and Franchello, A. (1990). Beneficial effect of enhanced macrophage function in the trauma patient. *Annals Surg.* **211**, 605–613.

Brown, G. D. and Gordon, S. (2001). Immune recognition A new receptor for beta-glucans. *Nature* **413**, 36–37.

Brown, G. D., Taylor, P. R., Reid, D. M., Willment, J. A., Williams, D. L., Martinez-Pomares, L., Wong, S. Y. C., and Gordon, S. (2002). Dectin-1 is a major β-glucan receptor on macrophages. *J. Exper. Med.* **196**, 407–412.

Brown, G. D., Herre, J., Williams, D. L., Willment, J. A., Marshall, A. S. J., and Gordon, S. (2003). Dectin-1 mediates the biological effects of β-glucans. *J. Exp. Med.* **197**, 1119–1124.

Chauhan, N., Li, D., Singh, P., Calderone, R., and Kruppa, M. (2002). The cell wall of *Candida* spp. In *Candida and Candidiasis,* 1st ed., R. A. Calderone, ed., ASM Press, Washington, DC, pp. 159–175.

Chihara, G. and Mihich, E. (1985). Glucans as immunomodifier II. In *Advances in Immunopharmacology,* L. Chedid and J. W. Hadden, eds., Pergamon Press, New York, pp. 397–402.

Christensen, B. E., Ulset, A. S., Beer, M. U., Knuckles, B. E., Williams, D. L., Fishman, M. L., Chau, H. K., and Wood, P. J. (2001). Macromolecular characterisation of three barley β-glucan standards by size-exclusion chromatography combined with light scattering and viscometry: an inter-laboratory study. *Carbo. Polymers* **45**, 11–22.

Cosio, E. G., Frey, T., and Ebel, J. (1992). Identification of a high-affinity binding protein for a hepta-β-glucoside phytoalexin elicitor in soybean. *Eur. J. Biochem.* **204**, 1115–1123.

Domer, J. E. (1989). Candida cell wall mannan: a polysaccharide with diverse immunologic properties. *Crit. Rev. Microbiol.* **17**, 33–51.

Dushkin, M. I., Safina, A. F., Vereschagin, E. I., and Schwartz, Y. S. (1996). Carboxymethylated beta-1,3-glucan inhibits the binding and degradation of acetylated low density lipoproteins in macrophages *in vitro* and modulates their plasma clearance *in vivo*. *Cell Biochem. Funct.* **14**, 209–217.

Elorza, M. V., Mormeneo, S., Garcia de la Cruz, F., Gimeno, C., and Sentandreu, R. (1989). Evidence for the formation of covalent bonds between macromolecules in the domain of the wall of *Candida albicans* mycelial cells. *Biochem. Biophys. Res. Commun.* **162**, 1118–1125.

Engstad, R. E. and Robertsen, B. (1994). Specificity of a beta-glucan receptor on macrophages from Atlantic salmon (Salmo salar L.). *Dev. Comp. Immunol.* **18**, 397–408.

Engstad, C. S., Engstad, R. E., Olsen, J.-O., and Osterud, B. (2002). The effect of soluble β-1, 3-glucan and lipopolysaccharide on cytokine production and coagulation activation in whole blood. *Inter. Immunopharmacol.* **2**, 1585–1597.

Ensley, H. E., Tobias, B., Pretus, H. A., McNamee, R. B., Jones, E. L., Browder, I. W., and Williams, D. L. (1994). NMR spectral analysis of a water-insoluble (1-3)-β-D-glucan isolated from *Saccharomyces cerevisiae*. *Carbo. Res.* **258**, 307–311.

Falch, B. H., Espevik, T., Ryan, L., and Stokke, B. T. (2000). The cytokine stimulating activity of (1-3)-β-D-glucans is dependent on the triple helix conformation. *Carbo. Res.* **329**, 587–596.

Fogelmark, B., Goto, H., Yuasa, K., Marchat, B., and Rylander, R. (1992). Acute pulmonary toxicity of inhaled β-1,3-glucan and endotoxin. *Agents Actions* **35**, 50–56.

Gantner, B. N., Simmons, R. M., Canavera, S. J., Akira, S., and Underhill, D. M. (2003). Collaborative Induction of Inflammatory Responses by Dectin-1 and Toll-like Receptor 2. *J. Exp. Med.* **197**, 1107–1117.

Gough, P. J. and Gordon, S. (2000). The role of scavenger receptors in the innate immune system. *Micr. Infect.* **2**, 305–311.

Hara, C., Kiho, T., Tanaka, Y., and Ukai, S. (1982). Anti-inflammatory activity and conformational behavior of a branched (1→3)-beta-D-glucan from an alkaline extract of *Dictyophora indusiata* fisch. *Carbo. Res.* **110**, 77–87.

Henderson, S. J. (1996). Monte Carlo modeling of small-angle scattering data from non-interacting homogeneous and heterogeneous particles in solution. *Biophys. J.* **70**, 1618–1627.

Hermanz-Falcon, P., Arce, I., Roda-Navarro, P., and Fernandez-Ruiz, E. (2001). Cloning of human DECTIN-1, a novel C-type lectin-like receptor gene expressed on dendritic cells. *Immunogen.* **53**, 288–295.

Janeway, C. A., Jr. and Medzhitov, R. (1999). Lipoproteins take their toll on the host. *Curr. Biol.* **9**, R879–R882.

Janusz, M. J., Austen, K. F., and Czop, J. K. (1989). Isolation of a yeast heptaglucoside that inhibits monocyte phagocytosis of zymosan particles. *J. Immunol.* **142**, 959–965.

Kapteyn, J. C., Montijn, R. C., Dijkgraaf, G. J. P., Van den Ende, H., and Klis, F. M. (1995). Covalent association of β-1, 3-glucan with β-1, 1-6 glucosylated mannoproteins in cell walls of *Candida albicans*. *J. Bacteriol.* **177**, 3788–3792.

Kapteyn, J. C., Hoyer, L. L., Hecht, J. E., Muller, W. H., Andel, A., Verkleij, A. J., Makarow, M., Van den Ende, H., and Klis, F. M. (2000). The cell wall architecture of *Candida albicans* wild-type cells and cell wall-defective mutants. *Mol. Microbiol.* **35**, 601–611.

Kasler, F. (1973). *Quantitative Aanalysis by NMR Spectroscopy.* Academic Press, New York.

Kim, Y. T., Kim, E., Cheong, C., Williams, D. L., Kim, C. W., and Lim, S. T. (2000). Structural characterization of beta-D-(1-3, 1-6) glucans using NMR spectroscopy. *Carbo. Res.* **328**, 331–341.

Klis, K. M., de Groot, P., and Hellingwerf, K. (2001). Molecular organization of the cell wall of Candida albicans. *Med. Mycol.* **39**, 1–8.

Kogan, G., Alfoldi, J., and Masler, L. (1988). ^{13}C-NMR spectroscopic investigation of two yeast cell wall β-D-glucans. *Biopolymers* **27**, 1055–1063.

Kougias, P., Wei, D., Rice, P. J., Ensley, H. E., Kalbfleisch, J., Williams, D. L., and Browder, I. W. (2001) Normal human fibroblasts express pattern recognition receptors for fungal (1→3)-β-D-glucans. *Infect. Immun.* **69**, 3933–3938.

Kulicke, W.-M., Lettau, A. I., and Thielking, H. (1997). Correlation between immunological activity, molar mass, and molecular structure of different (1-3)-beta-D-glucans. *Carbo. Res.* **297**, 135–143.

Lowe, E., Rice, P., Ha, T., Li, C., Kelley, J., Ensley, H., Lopez-Perez, J., Kalbfleisch, J., Lowman, D., Margl, P., Browder, W., and Williams, D. (2001). A (1-3)-β-D-linked heptasaccharide is the unit ligand for glucan pattern recognition receptors on human monocytes. *Microbes Infect.* **3**, 789–797.

Lowe, E. P., Wei, D., Rice, P. J., Li, C., Kalbfleisch, J., Browder, I. W., and Williams, D. L. (2002). Human vascular endothelial cells express pattern recognition receptors for fungal glucans which stimulates nuclear factor κB activation and interleukin 8 production. *Amer. Sur.* **68**, 508–517.

Lowman, D., Ensley, H., and Williams, D. (1998). Identification of phosphate substitution sites by NMR spectroscopy in a water-soluble phosphorylated (1-3)-β-D-glucan. *Carbo. Res.* **306**, 559–562.

Lowman, D. W. and Williams, D. L. (2001). A proton nuclear magnetic resonance method for the quantitative analysis on a dry weight basis of (1→3)-β-D-glucans in a complex, solvent-wet matrix. *J. Agric. Food Chem.* **49**, 4188–4191.

Lowman, D. W., Ferguson, D. A., and Williams, D. L. (2003). Structural characterization of (1→3)-beta-D-glucans isolated from blastospore and hyphal forms of *Candida albicans*. *Carbo. Res.* **388**, 1491–1496.

Marchessault, R. H. and Deslandes, Y. (1979). Fine structure of (1→3)-D-glucans: curdlan and paramylon. *Carbo. Res.* **75**, 231–242.

Masihi, K. N., Madaj, K., Hintelmann, H., Gast, G., and Kaneko, Y. (1997). Down-regulation of tumor necrosis factor-alpha, moderate reduction of interleukin-1beta, but not interleukin-6 or interleukin-10, by glucan immunodulators curdlan sulfate and lentinan. *Int. J. Immunopharmacol.* **19**, 463–68.

McIntire, T. M. and Brant, D. A. (1997). Imaging of individual biopolymers and supramolecular assemblies using noncontact atomic force microscopy. *Biopolymers* **42**, 113–146.

Medzhitov, R. and Janeway, Jr., C. A. (1997). Innate immunity: impact on the adaptive immune response. *Curr. Opinion Immunol.* **9**, 4–9.

Michalek, M., Melican, D., Brunke-Reese, D., Langevin, M., Lemerise, K., Galbraith, W., Patchen, M., and Mackin, W. (1998). Activation of rat macrophages by betafectin PGG-glucan requires cross-linking of membrane receptors distinct from complement receptor three (CR3). *J. Leuko. Biol.* **64**, 337–344.

Mueller, A., Mayberry, W., Acuff, R., Thedford, S., Browder, W., and Williams, D. (1994). Lipid content of macroparticulate (1→3)-beta-D-glucan isolated from *Saccharomyces cerevisiae*. *Microbios.* **79**, 253–261.

Mueller, A., Pretus, H., McNamee, R., Jones, E., Browder, I., and Williams, D. (1995). Comparison of the carbohydrate biological response modifiers Krestin, schizophyllan and glucan phosphate by aqueous size exclusion chromatograpthy with in-line argon-ion multi-angle laser light scattering photometry and differential viscometry detectors. *J. Chromatog.* **666**, 283–290.

Mueller, A., Rice, P. J., Ensley, H. E., Coogan, P. S., Kalbfleisch, J. H., Kelley, J. L., Love, E. J., Portera, C. A., Ha, T., Browder, I. W., and Williams, D. L. (1996). Receptor binding and internalization of water-soluble (1→3)-beta-D-glucan biologic response modifier in two monocyte/macrophage cell lines. *J. Immunol.* **156**, 3418–3425.

Mueller, A., Ensley, H., Pretus, H., McNamee, R., Jones, E., McLaughlin, E., Chandley, W., Browder, W., Lowman, D., and Williams, D. (1997). The application of various protic acids in the extraction of (1-3)-β-D-glucan from *Saccharomyces cerevisiae*. *Carbo. Res.* **299**, 203–208.

Mueller, A., Raptis, J., Rice, P. J., Kalbfleisch, J. H., Stout, R. D., Ensley, H. E., Browder, W., and Williams, D. L. (2000). The influence of glucan polymer structure and solution conformation on binding to (1→3)-β-D-glucan receptors in a human monocyte-like cell line. *Glycobiol.* **10**, 339–346.

Nelson, R. D., Shibata, N., Podzorski, R. P., and Herron, M. J. (1991). *Candida mannan*: chemistry, suppression of cell-mediated immunity, and possible mechanisms of action. *Clin. Microbiol. Rev.* **4**, 1–19.

Norisuye, T., Yanaki, T., and Fujita, H. (1980). Triple helix of a Schizophyllum commune polysaccharide in aqueous solution. *J. Polymer Science* **18**, 547–558.

Ohno, N., Uchiyama, M., Tsuzuki, A., Tokunaka, K., Miura, N. N., Adachi, Y., Aizawa, M. W., Tamura, H., Tanaka, S., and Yadomae, T. (1999). Solubilization of yeast cell-wall β-(1→3)-D-glucan by sodium hypochlorite oxidation and dimethyl sulfoxide extraction. *Carbo. Res.* **316**, 161–172.

Okazaki, M., Adachi, Y., Ohno, N., and Yadomae, T. (1995). Structure-activity relationship of (1→3)-β-D-glucans in the induction of cytokine production from macrophages, in vitro. *Biol. Pharm. Bull.* **18**, 1320–1327.

Opitz, B., Schroder, N. W., Spreitzer, I., Michelsen, K. S., Kirschning, C. J., Hallatschek, W., Zahringer, U., Hartung, T., Gobel, U. B., and Schumann, R. R. (2001). Toll-like receptor-2 mediates Treponema glycolipid and lipoteichoic acid-induced NF-kappaB translocation. *J. Biol. Chem.* **276**, 22041–22047.

Pearson, A., Lux, A., and Kreieger, M. (1995). Expression cloning of dSR-CI, a class C macrophage-specific scavenger receptor from *Drosophila melanogaster*. *Proc. Nat. Acad. Sci. U.S.A.* **92**, 4056–4060.

Pretus, H. A., Ensley, H. E., McNamee, R. B., Jones, E. L., Browder, I. W., and Williams, D. L. (1991). Isolation, physicochemical characterization and preclinical efficacy evaluation of soluble scleroglucan. *J. Pharmacol. Exp. Ther.* **257** (1), 500–510.

Rice, P. J., Kelley, J. L., Kogan, G., Ensley, H. E., Kalbfleisch, J. H., Browder, I. W., and Williams, D. L. (2002). Human monocyte scavenger receptors are pattern recognition receptors for (1-3)-β-D-glucans. *Leuko. Biol.* **72**, 140–146.

Riggi, S. J. and Di Luzio, N. R. (1961). Identification of a reticuloendothelial stimulating agent in zymosan. *Amer. J. Physiol.* **200**, 297–300.

Ross, G. D., Cain, J. A., Myones, B. L., Newman, S. L., and Lachmann, P. J. (1987). Specificity of membrane complement receptor type three (CR_3) for β-glucans. *Complement* **4**, 61–74.

Rylander, R. (1996). Airway responsiveness and chest symptoms after inhalation of endotoxin or (1→3)-beta-D-glucan. *Indoor Built Environ.* **5**, 106–111.

Rylander, R. (1997). Investigations of the relationship between disease and airborne (1→3)-β-D-glucan in buildings. *Mediators Inflam.* **6**, 275–77.

Rylander, R. and Lin, R. H. (2000). (1-3)-β-D-glucan — relationship to indoor air-related symptoms, allergy and asthma. *Toxicol.* **152**, 47–52.

Rylander, R., Persson, K., Goto, H., Yuasa, K., and Tanaka, S. (1992). Airborne beta-1,3-glucan may be related to symptoms in sick buildings. *Indoor Environ.* **1**, 263–267.

Saito, H., Ohki, T., and Sasaki, T. (1977). A 13C nuclear magnetic resonance study of gel-forming (1-3)-beta-D-glucans. Evidence of the presence of single-helical conformation in a resilient gel of a curdlan-type polysaccharide 13140 from *alcaligenes faecalis* var. myxogenes IFO 13140. *Biochem.* **16**, 908–914.

Saito, H., Yoshioka, Y., and Uehara, N. (1991). Relationship between conformation and biological response for (1-3)-β-D-glucan in the activation of coagulation Factor G from Limulus amebocyte lysate and host-mediated antitumor activity. Demonstration of single-helix conformation as a stimulant. *Carbo. Res.* **217**, 181–190.

SanJuan, R., Zueco, J., and Stock, R. (1995). Identification of glucan-mannoprotein complexes in the cell wall of *Candida albicans* using monoclonal antibody that reacts with a (1,6)-β-glucan epitope. *Microbiol.* **141**, 1545–1551.

Sato, T., Norisuye, T., and Fujita, H. (1981). Melting behavior of *Schizophyllum commune* polysaccharides in mixtures of water and dimethyl sulfoxide. *Carbo. Res.* **95**, 195–204.

Schmidt, W. E. and Ebel, J. (1987). Specific binding of a fungal glucan phytoalexin elicitor to membrane fractions from soybean *Glycine max. Proc. Nat. Acad. Sci. U.S.A.* **84**, 4117–4121.

Shepherd, M. G. (1987). Cell envelope of *Candida albicans*. *Crit. Rev. Microbiol.* **15**, 7–25.

Soltys, J. and Quinn, M. T. (1999). Modulation of endotoxin- and enterotoxin-induced cytokine release by *in vivo* treatment with beta-(1,6)-branched beta-(1,3)-glucan. *Infect. Immun.* **67**, 244–252.

Stone, B. A. and Clarke, A. E. (1992). *Chemistry and Biology of (1-3)-β-d-Glucan*. La Trobe University Press, Melbourne.

Taylor, P. R., Brown, G. D., Reid, D. M., Willment, J. A., Martinez-Pomares, L., Gordon, S., and Wong, S. W. C. (2002). The β-glucan receptor, Dectin-1, is predominantly expressed on the surface of cells of the monocyte/macrophage and neutrophil lineages. *J. Immunol.* **169**, 3876–3882.

Thorn, J., Beijer, L., and Rylander, R. (2001). Effects after inhalation of (1-3)-β-D-glucan in healthy humans. *Mediators Inflam.* **10**, 173–178.

Thornton, B. P., Vetvicka, V., Pitman, M., Goldman, R. C., and Ross, G. D. (1996). Analysis of the sugar specificity and molecular location of the beta-glucan-binding lectin site of complement receptor type 3 (CD11b/CD18). *J. Immunol.* **156**, 1235–1246.

Tvaroska, I. (1990). Dependence on saccharide conformation of the one-bond and three-bond carbon-proton coupling constants. *Carbo. Res.* **206**, 55–64.

Ukai, S., Kiho, T., Hara, C., Kuruma, I., and Tanaka, Y. (1983). Polysaccharides in fungi. XIV. Anti-inflammatory effect of the polysaccharides from the fruit bodies of several fungi. *J. Pharmacobio-Dynamics* **6**, 983–990.

Vargas-Albores, F. and Yepiz-Plascencia, G. (2000). Beta-glucan binding protein and its role in shrimp immune response. *Aquaculture* **191**, 13–21.

Vereschagin, E. I., Van Lambalgen, A. A., Dushkin, M. I., Schwartz, Y. S., Polyakov, L., Heemskerk, A., Huisman, E., Thijs, L. G., and Van den Bos, G. C. (1998). Soluble glucan protects against endotoxin shock in the rat: the role of the scavenger receptor. *Shock* **9**, 193–198.

Vetvicka, V., Thornton, B. P., and Ross, G. D. (1996). Soluble beta-glucan polysaccharide binding to the lectin site of neutrophil or natural killer cell complement receptor type 3 (CD11b/CD18) generates a primed state of the receptor capable of mediating cytotoxicity of iC3b-opsonized target cells. *J. Clin. Invest.* **98**, 50–61.

Wei, D., Ge, S., Chen, Y., Dai, F., and Su, B. (1997). Expression of endogenous transforming growth factor-β and its type I and type II receptors in rat burn wounds. *Wound Repair Regen.* **5**, 229–234.

Wei, D., Williams, D., and Browder, W. (2002a). Activation of AP-1 and SP1 correlates with wound growth factor gene expression in glucan-treated human fibroblasts. *Int. Immunopharm.* **2**, 1163–1172.

Wei, D., Zhang, L., Williams, D. L., and Browder, I. W. (2002b). Glucan stimulates human dermal fibroblast collagen biosynthesis through a nuclear factor-1 dependent mechanism. *Wound Repair Regen.* **10**, 161–168.

Williams, D. L., McNamee, R. B., Jones, E. L., Pretus, H. A., Ensley, H. E., Browder, I. W., and Di Luzio, N. R. (1991a). A method for the solubilization of a (1-3)-β-D-glucan isolated from *Saccharomyces cerevisiae*. *Carbo. Res.* **219**, 203–213.

Williams, D. L., Pretus, H. A., McNamee, R. B., Jones, E. L., Ensley, H. E., Browder, I. W., and Di Luzio, N. R. (1991b). Development, physicochemical characterization and preclinical efficacy evaluation of a water soluble glucan sulfate derived from *Saccharomyces cerevisiae*. *Immunopharmacol.* **22**, 139–56.

Williams, D. L., Pretus, H. A., and Browder, I. W. (1992). Application of aqueous gel permeation chromatography with in-line multi-angle laser light scattering and differential viscometry detectors for the characterization of natural product carbohydrate pharmaceuticals. *J Liq. Chromatog.* **15**, 2297–2309.

Williams, D. L., Pretus, H. A., Ensley, H. E., and Browder, I. W. (1994). Molecular weight analysis of water-insoluble yeast-deived (1-3)-β-D-glucan by organic phase size exclusion chromatography. *Carbo. Res.* **253**, 293–298.

Williams, D. L., Mueller, A., and Browder, W. (1996). Glucan-based macrophage stimulators: a review of their anti-infective potential. *Clin. Immunother.* **5**, 392–399.

Williams, D. L. (1997). Overview of (1→3)-β–D-glucan immunobiology. *Mediators Inflam.* **6**, 247–250.

Williams, D. L., Ha, T., Li, C., Kalbfleisch, J. H., and Ferguson, D. A., Jr. (1999a). Early activation of hepatic NFkB and NF-IL6 in polymicrobial sepsis correlates with bacteremia, cytokine expression and mortality. *Annals of Surg.* **230**, 95–104.

Williams, D. L., Ha, T., Li, C., Kalbfleisch, J. H., Laffan, J. J., and Ferguson, D. A. (1999b). Inhibiting early activation of tissue nuclear factor-κB and nuclear factor interleukin 6 with (1→3)-β-D-glucan increases long-term survival in polymicrobial sepsis. *Surg.* **126**, 54–65.

Williams, D. L., Ha, T., Li, C., Laffan, J., Kalbfleisch, J., and Browder, W. (2000). Inhibition of LPS induced NFκB activation by a glucan ligand involves down regulation of IKKβ kinase activity and altered phosphorylation and degradation of IκBα. *Shock* **13**, 446–452.

Williams, D. L., Ha, T., Li, C., Kalbfleisch, J. H., Schweitzer, J., Vogt, W., and Browder, I. W. (2003). Modulation of tissue Toll-like receptor 2 and 4 during the early phases of polymicrobial sepsis correlates with mortality. *Crit. Care Med.* **31**, 1808–1818.

Willment, J. A., Gordon, S., and Brown, G. D. (2001). Characterization of the human β-glucan receptor and its alternatively spliced isoforms. *J. Biol. Chem.* **276**, 43818–43823.

Yokota, K., Takashima, A., Bergstresser, P. R., and Ariizumi, K. (2001). Identification of a human homologue of the dendritic cell-associated C-type lectin-1, dectin-1. *Gene* **272**, 51–60.

Young, S. H., Dong, W. J., Williams, D. L., and Jacobs, R. R. (1998a). The solvent dependence of *schizophyllan* change in triple-helix glucan. Unpublished data.

Young, S. H. and Jacobs, R. R. (1998b). Sodium hydroxide-induced conformational change in schizophyllan detected by the florescence dye, aniline blue. *Carbo. Res.* **310**, 91–99.

Young, S. H., Dong, W. J., and Jacobs, R. R. (2000). Observation of a partially opened triple-helix conformation in 1-3-β-glucan by fluorescence resonance energy transfer spectroscopy. *J. Biol. Chem.* **275**, 11874–11879.

Young, S. H., Robinson, V. A., Barger, M., Frazer, D. G., Castranova, V., and Jacobs, R. R. (2003). Partially opened triple helix is the biologically active conformation of 13-beta-glucans that induces pulmonary inflammation in rats. *J. Toxicol. Environ. Health Part A* **66**, 551–563.

Yu, L., Goldman, R., Sullivan, P., Walker, G. F., and Fesik, S. W. (1993). Heteronuclear NMR studies of ^{13}C-labeled yeast cell wall β-glucan oligosaccharides. *J. Biomol. NMR* **3**, 429–441.

Zimmerman, J. W., Lindermuth, J., Fish, P. A., Palace, G. P., Stevenson, T. T., and DeMong, D. E. (1998). A novel carbohydrate-glycosphingolipid interaction between a beta-(1-3)-glucan immunomodulator, PGG-glucan, and lactosylceramide of human leukocytes. *J. Biol. Chem.* **273**, 22014–22020.

2 Health Effects of (1→3)-β-*D*-Glucans: The Epidemiological Evidence

Jeroen Douwes

CONTENTS

2.1 Introduction 35
2.2 Field Studies..36
    2.2.1 Indoor Environment 36
    2.2.2 Occupational Environment...39
    2.2.3 Case Studies 42
2.3 Human Challenge Studies...42
2.4 The Epidemiological Evidence..44
2.5 Control of (1→3)-β-*D*-Glucan Exposure in the Home and
    Work Environment 47
2.6 Research Needs 48
2.7 Conclusions 49
Acknowledgments..49
References..49

2.1 INTRODUCTION

A remarkably consistent association between home dampness and respiratory symptoms or asthma has been observed in a large number of population studies conducted across many geographical regions (Peat et al., 1998; Bornehag et al., 2001). Positive associations have been shown in infants (Nafstad et al., 1998; Øie et al., 1999), children (Brunekreef et al., 1989; Andriessen et al., 1998) and adults (Norback et al., 1999; Kilpelainen et al., 2001). Some evidence for dose-response relationships has also been demonstrated (Engvall et al., 2001). Although the evidence is not entirely consistent (Douwes and Pearce, 2003a), it has often been suggested that mold exposure plays a causal role in the association between indoor dampness and respiratory morbidity (Peat et al., 1998; Zock et al., 2002).

However, if there is a causal association, it is not clear which specific fungal components cause symptoms and through which mechanisms. Some studies have shown associations between fungal exposure, sensitization, and asthma (Black et al., 2000; Zureik et al., 2002; Halonen et al., 1997), but the evidence that fungal allergens and IgE allergic responses play a major role in indoor-related respiratory symptoms is still very limited (Douwes and Pearce, 2003a). In addition to allergic mechanisms, nonallergic responses to fungal exposures have been reported mainly in relation to fungal (1→3)-β-D-glucans, the nonallergenic cell wall components of most fungi. The first reports suggesting a potential role for (1→3)-β-D-glucans in the development of indoor air-related health effects appeared in the late 1980s. Since then, a limited number of studies have been published, mainly due to the fact that commercially available methods to analyze (1→3)-β-D-glucan were not available until some years ago (see Chapter 10). Most of the published field studies in the indoor environment were conducted in Sweden and only a small number of studies were conducted in other countries, such as the Netherlands, Germany, Switzerland, Taiwan, Australia, Norway, Denmark, and the United Kingdom.

This chapter will give an overview of the epidemiological literature on health effects of (1→3)-β-D-glucan, focusing on respiratory health, i.e., atopy, airway inflammation and symptoms, asthma, and decreased lung function. In addition to population studies and one case study, studies in human volunteers experimentally exposed to (1→3)-β-D-glucan will be described. Studies will be discussed individually as only a small number of epidemiological and human challenge studies are available. This will be followed by a more general interpretation of the currently available evidence for a role of (1→3)-β-D-glucan in the development of adverse respiratory health effects. The chapter will further discuss the strengths and weaknesses of these studies, will include a discussion on potential confounding factors (e.g., other microbial exposures that may cause similar health effects), will briefly discuss susceptibility issues and options to control exposure, and will finally identify research needs.

2.2 FIELD STUDIES

2.2.1 INDOOR ENVIRONMENT

A population study in Sweden suggested a relationship between airborne (1→3)-β-D-glucan levels measured in various environments and questionnaire-assessed symptoms (Rylander et al., 1992). Two schools, a post office, and a day-care center were selected based on complaints about indoor air quality. In addition, one control office building was selected. All persons working full-time in those buildings (n = 8, 11, 14, 6, and 405, respectively) participated in the study. Mean glucan levels determined by an Limulus amebocyte lysate (LAL) assay ranged from 0.2–0.55 ng/m^3 in buildings with indoor air problems (n = 46) to a mean concentration of < 0.1 ng/m^3 (n = 36) in the control building. During the sampling process, settled dust was agitated using a machine specifically designed to make settled dust airborne. A significant correlation was found between glucan exposure and dry cough or skin rashes. No significant associations with any of the other studied symptoms (nose and eye irritations, chest tightness, headache and tiredness, joint pains, etc.) were found. The reported associations were not adjusted for other potential risk factors.

Another small survey in Sweden studied a day care center with dampness and mold problems before and after renovation (Rylander, 1997). The study population consisted of 11 female employees who worked at the day care center. Airway responsiveness and symptoms were measured before, two years after, and three years after renovation. Airborne glucan levels were determined by using an LAL assay. During sampling, floor dust was agitated to make it airborne. The mean $(1\rightarrow 3)$-β-D-glucan levels were 11.4 ng/m^3 (n = 24) before and 1.2 ng/m^3 (n = 13) after renovation. No significant difference in baseline lung function was observed after renovation. However, a significant decrease in airway responsiveness two years after renovation was found, although this was not confirmed three years after renovation. In addition, the number of subjects with a drop in FEV_1 of more than 4% after methacholine challenge (to determine airway responsiveness) was significantly reduced after renovation. This was consistent at two and three years. It is not clear why a cut-off of 4% was chosen. Finally, no significant improvement in other symptoms was recorded after renovation. The analyses were not adjusted for other exposures and involved a comparison of only two exposure situations. Therefore, no direct association between glucan exposure and symptoms could be assessed.

A larger field study in Sweden was conducted in 75 row houses, several of which had dampness problems (Thorn and Rylander, 1998a). A total of 129 occupants participated in the study. Airborne $(1\rightarrow 3)$-β-D-glucan levels were determined using an LAL assay. During sampling, floor dust was agitated to make it airborne. The $(1\rightarrow 3)$-β-D-glucan levels ranged from 0–19 ng/m^3. No significant association was found between exposure and atopy for the whole group, comparing subjects living in homes with levels below 3 ng/m^3 with those living in homes above this level. However, among subjects older than 65 years, the proportion of atopics was significantly larger in the group with exposures greater than 3 ng/m^3. Analyses in which the use of three exposure categories was explored showed that atopy was significantly increased in the high exposure group (> 4 ng/m^3; n = 52) compared to the low exposure group (0–2 ng/m^3; n = 45). No increased risk was found for the middle exposure group (>2–4 ng/m^3; n = 32). Analyses were adjusted for age, gender, cigarette smoking, and pets. The basis for choosing the cut-off points for the various exposure groups is not clear. Baseline FEV_1 values were unrelated to $(1\rightarrow 3)$-β-D-glucans levels for the whole group. However, when limiting the analyses to those subjects younger than 65 years and with glucan levels above 1 ng/m^3 an inverse association was found between glucan exposures and FEV_1. The analysis was adjusted for age, gender, cigarette smoking, asthma, atopy, and pets. It is not clear why subjects living in houses with glucan levels below 1 ng/m^3 were excluded and why 1 ng/m^3 was chosen as the cut-off for exclusion. The inverse association was shown only for males. When analyzing the data using three exposure groups (as described above), no association with FEV_1 was found. Also, no significant associations with airway responsiveness were found. Myeloperoxidase (MPO) levels in serum were significantly higher in subjects living in houses with glucan levels greater than 1 ng/m^3 compared to those living in houses with glucan levels less than 1 ng/m^3. When analyzing the data using three exposure groups, an increase was observed in the middle and high exposure group, but this was not statistically significant. Neither was there an indication of a dose-response trend. No associations with eosinophilic cationic protein (ECP) and C-reactive proteins (CRP) were found. Finally, using three exposure groups some significant associations with symptoms were found (joint pains and chest tightness), but these were only found for the middle and not for the high exposure group. Analyses were not adjusted for other potentially relevant exposures (endotoxin levels measured in a proportion of the samples were, however, low).

Rylander and colleagues (1998) conducted another study in Sweden comparing a school with mold problems with a control school. The study population comprised 65 pupils in the problem school and 141 in the control school. Airborne glucan levels were determined by using an LAL assay. Samples were taken in 6 classrooms in the problem school and 11 classrooms in the control school. During sampling, floor dust was agitated to make it airborne. The mean glucan concentration was 2.9 ng/m^3 (range 0–6.9 ng/m^3) and 15.3 ng/m3 (range 9.2–27.4 ng/m^3) in the control and problem school, respectively. Respiratory symptoms were significantly more prevalent among the pupils of the problem school. The differences in symptom prevalence (dry cough, cough with phlegm and hoarseness) were most pronounced in atopic children. The prevalence of atopy was not different between the schools. This study comprised a comparison of only two schools, and, therefore, no direct association between glucan exposure and symptoms could be assessed.

A study in Taiwan investigated the health effects of bioaerosol exposure in 40 day care center employees, 69 office employees, and 22 residents from the same area (Wan and Li, 1999). Day care centers (n = 8), office buildings (n = 8), and homes (n = 8) were selected randomly. Airborne glucan levels were measured using a glucan-specific LAL assay. Median (1→3)-β-D-glucan levels were 5.7, 3.2, and 3.7 ng/m^3 in the day care centers, office buildings and the home environment, respectively. It is not clear from the publication how many samples were taken. Elevated glucan levels were associated with an increased risk for lethargy and fatigue. No significant associations for any of the other symptoms (eye and nose irritations, skin and respiratory symptoms) were found. All analyses were adjusted for sex, ventilation, and type of building. None of the other measured exposures in this study (fungi, bacteria, and endotoxins) showed a positive association with lethargy and fatigue.

A study in the Netherlands investigated the effect of (1→3)-β-D-glucan exposure on peak flow (PEF) variability (a measure of asthma severity) in 148 children 7 to 11 years of age of whom 50% had self- or parent-reported chronic respiratory symptoms (Douwes et al., 2000a). (1→3)-β-D-glucan in settled house dust was analyzed using an inhibition enzyme-linked immunosorbent assay (ELISA) method. Geometric mean concentrations on living room floors of homes with nonsymptomatic and symptomatic children were 126 µg/m^2 (n = 69) versus 169 µg/m^2 (n = 74), respectively. In unadjusted analyses, the level of (1→3)-β-D-glucan was significantly associated with PEF variability in symptomatic children (n = 72), particularly in atopic children with asthma symptoms (n = 21). No association was shown with glucan levels measured in the bedroom (floor and mattress). Living room floor levels of endotoxin were also associated with PEF variability. Analyses adjusted for dust mite allergens, pets in the home, and type of floor cover showed the same association for (1→3)-β-D-glucans, but not for endotoxin.

Finally, an investigation in Sweden was performed to assess the effects of (1→3)-β-D-glucan exposures on inflammatory markers in blood. The study comprised 17 high (> 4.0 ng/m^3; median 6.0 ng/m^3) and 18 low exposed subjects (< 2.0 ng/m^3; median 0.9 ng/m^3) (Beijer et al., 2003). The study population was selected from a previous field study (Thorn and Rylander, 1998a; see above) and was the same as in the challenge study described a year before by the same authors (Beijer et al., 2002; see below). Blood samples were analyzed for granulocytic enzymes (ECP and MPO), T-cell subsets, and the secretion of cytokines (IFN-γ, IL-4, TNF-α, IL-1β, and IL-10) from *in vitro* incubated mononuclear cells with or without phytohemagglutinin (PHA) or endotoxin. CD8+ cells were lower in the high exposure group; this was statistically significant only for the cytotoxic CD8+ T-cells (CD8+ S6F1+). No difference in CD4+ T cells was found. TNF-α secretion from nonstimulated blood mononuclear cells (BMCs) was significantly greater in the high exposure group. However, no difference in TNF-α secretion was found after endotoxin stimulation. The IFN-γ/IL-4 ratio (after PHA stimulation) was higher in the high exposure group but this was only significant in the nonatopics. In addition, a significant correlation was shown in the nonatopics between airborne (1→3)-β-D-glucan levels in the home and the IFN-γ/IL-4 ratio (the IFN-γ/IL-4 ratio being used as a measure of the balance of the T-helper cell 1 [Th$_1$] and Th$_2$ response). One subject with a high amount of glucan in the home was excluded from the analysis. The reason for exclusion is not clear, particularly since an *a priori* decision to include this subject in the study was made based on the previously measured high exposure (i.e., 51.7 ng/m^3). No significant associations for IFN-γ and IL-4 alone were noted, although IFN-γ levels were somewhat elevated and IL-4 levels reduced in the high exposure group. Also, no significant differences were found between both exposure groups regarding secretion of IL-10 or IL-1β from BMCs, ECP or MPO levels in serum, differential counts in blood, or the prevalence of symptoms. Analyses were not adjusted for other exposures or any other confounding factors.

2.2.2 Occupational Environment

A few studies have measured glucan exposures in the wood industry (Mandryk et al., 1999; Mandryk et al., 2000; Ronald et al., 2003), two of which also reported health effects. The first study, in Australian sawmill workers (n = 87), showed a strong association between personal airborne (1→3)-β-D-glucan concentrations and a decrease in percentage predicted as well as cross-shift lung function (Mandryk et al., 2000). (1→3)-β-D-glucan exposures were also associated with an increase in respiratory symptoms, particularly chronic bronchitis and nasal irritations. This was demonstrated for both inhalable and respirable glucan but only for the workers in the green mills. (1→3)-β-D-glucan concentrations were determined by using an LAL assay. Geometric mean inhalable levels of approximately 3 ng/m^3 were reported in green mills based on 36 exposure measurements; levels in dry mills were approximately 10 times lower (n = 18 exposure measurements). Glucan levels were strongly correlated with endotoxin (r = 0.85) and moderately with dust levels (r = 0.31). The associations between glucan and lung function or symptoms remained after adjusting for endotoxin exposure. The models were not adjusted for dust exposure and were based on only 36 observations. A previous study in the wood processing industry (sawmill, wood chipping, and joinery; n = 168) conducted by the same research group showed that personal respirable glucan exposure was associated with an *increase* in lung function (expressed as percentage predicted) rather than a *decrease* (Mandryk et al., 1999). In addition, both positive and negative associations were found for cross-shift changes in lung function. Analyses were based on 53 observations, and analyses were not adjusted for other exposures. The geometric mean respirable $(1\rightarrow3)$-β-D-glucan exposure in sawmills was 0.26 ng/m^3 (n = 28) compared to 0.14 in the joineries (n = 25); the inhalable levels were 1.4 ng/m^3 (n = 54) and 0.6 ng/m^3 (n = 39), respectively.

One study in the paper industry in Sweden was performed investigating $(1\rightarrow3)$-β-D-glucan (and endotoxin) exposure and its potential effects on the airways (symptoms, airway responsiveness, and spirometry) (Rylander et al., 1999). Levels of ECP, MPO, and CRP in serum were also studied. A total of 83 workers and 44 controls were included in the study. Twenty stationary airborne samples were collected in the three main units of the plant, and another eight samples were collected to determine the exposure of the controls. $(1\rightarrow3)$-β-D-glucan was analyzed using an LAL assay. Workers had a significantly increased prevalence of symptoms, such as unusual tiredness, joint pains, nasal irritation, and short-lived flu-like symptoms. In addition, they had a decreased base-line lung function (FEV$_1$), increased airway responsiveness, and increased levels of ECP and MPO in the serum. Mean $(1\rightarrow3)$-β-D-glucan levels in the three units of the paper mill ranged from 2.0 to 97.7 ng/m^3; the mean $(1\rightarrow3)$-β-D-glucan level for controls was 0.1 ng/m^3. In order to assess a dose-response association, three exposure levels were defined. It is not clear how the cut-off points were determined as the numbers of subjects were unequal in the three exposure groups. Indications of a dose-response association were found for airway responsiveness and ECP in blood (data were, however, not shown). Positive associations were also found for symptoms, including throat irritation non-significant (ns) cough with phlegm (ns), congested nose, joint pains, unusual tiredness, and flu-like symptoms. Controls were included in the regression analyses, and no attempt was made to study dose-response relations within the group of workers only. Also, very similar results were found for endotoxin exposure, which was correlated very strongly with glucan exposure (r = 0.93). Due to this high correlation analyses could not be adjusted for endotoxin exposure. Therefore, it remains unclear which exposure caused the effects.

A series of relatively small studies of workers in household waste collection, recycling, and composting have been conducted in which airway inflammation and symptoms were studied in relation to $(1\rightarrow3)$-β-D-glucan exposure (as well as several other exposures, such as dust and endotoxin). A Swedish study in 25 household waste collectors and 24 controls in Sweden investigated the association between $(1\rightarrow3)$-β-D-glucan or endotoxin exposure and a variety of respiratory or other symptoms, as well as lung function, airway responsiveness, and inflammatory markers in induced sputum and serum/blood (Thorn and Rylander, 1998b). In total, 20 airborne samples were collected. Mean exposures of 19.1, 9.2, and 1.1 ng/m3 were found for collectors of compostable waste, collectors of unsorted waste and controls, respectively. (1→3)-β-D-glucan was analysed using an LAL assay. Some differences in symptoms and inflammatory markers in blood were found between waste collectors and controls, but only lymphocytes in blood were positively associated with (1→3)-β-D-glucan levels ($p < 0.01$). Endotoxin and glucan were significantly correlated; however, since endotoxin was not associated with any of the studied health outcomes, it was unlikely to account for the association between (1→3)-β-D-glucan and blood lymphocytes.

Two small surveys (n = 29) in the Netherlands in compost workers exposed to high levels of microorganisms indicated work-related acute and (sub-)chronic inflammatory reactions in the upper airways (assessed by using nasal lavage). However, no clear association with (1→3)-β-D-glucan exposure was found (Douwes et al., 2000b). (1→3)-β-D-glucan concentrations were assessed by using a glucan-specific inhibition ELISA, and personal inhalable geometric mean (1→3)-β-D-glucan levels ranged from 0.54 to 4.85 µg/m³. A subsequent study in 47 waste collectors showed a significant increase in total cells (mainly neutrophils) and IL-8 in the upper airways after a 5-day work week (Wouters et al., 2002). (1→3)-β-D-glucan glometric mean (GM = 1.3 µg/m³) was positively associated with IL-8 and cell counts, but the association did not reach statistical significance, in contrast to endotoxin and total dust levels that were significantly associated with both parameters.

Heldal et al. (2003a) studied upper airway inflammation by using nasal lavage in 31 Norwegian waste handlers. A significant increase in neutrophils and ECP was observed over a four-day work period. (1→3)-β-D-glucan exposure (median level 40 ng/m³) was significantly associated with an increase in IL-8 but not with any of the other studied inflammatory markers (neutrophils, MPO, or ECP). It was also significantly associated with nasal congestion assessed by acoustic rhinometry. (1→3)-β-D-glucan levels were assessed, using both the LAL assay and the inhibition ELISA. However, in the analyses only the LAL data were used since most of the samples were below the detection limit for the ELISA method. The same researchers also studied airway inflammation of the lower airways in 25 waste collectors selected from their initial study, using sputum induction (Heldal et al., 2003b). They found a significant increase in neutrophils and IL-8 over a four-day work period. Exposure to (1→3)-β-D-glucans (median level 52 ng/m³) was correlated with the increase in IL-8 (n = 21; $p < 0.05$) but not with the increase in neutrophils and other inflammatory mediators. The analyses were not adjusted for endotoxin exposure, although endotoxin was shown to be associated with airway inflammation in these two studies. No significant associations between glucan exposure and lung function or symptoms were found.

Another study in 159 recycling workers from the UK showed that (1→3)-β-D-glucans exposure was significantly associated with an increase in symptoms, such as cough with phlegm, hoarse/parched throat, and stomach problems (Gladding et al., 2003). (1→3)-β-D-glucan exposures were also positively associated with other respiratory symptoms, skin rash, and nausea; however, these associations were not statistically significant. (1→3)-β-D-glucan concentrations were determined using an LAL assay and median levels ranged from 4.8–40.1 ng/m³. In order to assess the association with health effects, the authors dichotomized exposure based on a cut-off of 12 ng/m³ (95 observations were above this value and 64 below). The reasons for choosing this arbitrary cut-off are not clear. The analyses were not adjusted for endotoxin and dust exposures despite the fact that these exposures were associated with respiratory symptoms. In addition, one of these exposures, i.e., endotoxin, was highly correlated with glucan levels.

Finally, Eduard et al. (2001) studied respiratory symptoms and eye and nose irritation in 106 farmers in Norway. Airborne levels of (1→3)-β-D-glucan and other bioaerosols were measured during specific tasks, and eight hour time-weighted average exposures were computed based on type and duration of the task for each farmer. (1→3)-β-D-glucan was assayed using an ELISA method. The personal geometric mean (1→3)-β-D-glucan exposure level was 0.82 μg/m^3. Although fungal spores were strongly associated with an increased risk for cough or eye and nose symptoms, no significant associations were found for (1→3)-β-D-glucan. Odds ratios were, however, elevated for (1→3)-β-D-glucan (ns).

2.2.3 CASE STUDIES

One case study, comprising a clinical investigation of two boys living in a mold-infested house in Switzerland, has been reported by Rylander et al., (1994). The mean (1→3)-β-D-glucan concentration was 41.9 ng/m^3 (determined using a LAL assay). Both children developed airway inflammation and symptoms, such as coughing, wheezing, and tiredness, after 6 months of living in the house. After moving out of the house, the symptoms disappeared. The parents developed airway inflammation about a year later. Symptoms may have been caused by (1→3)-β-D-glucans or any other agents associated with indoor mold.

2.3 HUMAN CHALLENGE STUDIES

In addition to the field studies described above, several experimental exposure studies have been described in the literature, most of which were conducted by the same research group. The first study showed a small increase in the severity of symptoms of nose and throat irritation in 26 subjects experimentally exposed for 4 hours to aerosolized (1→3)-β-D-glucan (curdlan) at a concentration of approximately 210 ng/m^3 (Rylander, 1996). No effects on FEV$_1$ or airway responsiveness were found for the whole group, nor for subgroups that were defined based on atopic status (n = 12). In the same study, it was shown that in 16 nonatopic and nonsymptomatic individuals exposed to particulate (1→3)-β-D-glucan (approximately 210 ng/m^3 for 4 hours) a statistically significant but very small decrease of FEV$_1$ was found immediately and 3 days after exposure, whereas no significant association was found with airway responsiveness.

In a subsequent study, 21 healthy students were exposed to an aerosol of (1→3)-β-D-glucan (grifolan suspended in saline) amounting to a dose of 125 ng (Thorne et al., 2001). Each student was subjected to two randomized inhalation challenges, one to (1→3)-β-D-glucan and one to saline. Spirometry, blood sampling, and sputum induction was conducted 24 and 72 hours after the challenge. No significant differences in sputum ECP, MPO, TNF-α, IL-8, IL-10, eosinophil, lymphocyte, macrophage, and neutrophil levels were observed between saline and glucan challenges either at 24 or 72 hours after the challenge. Also, no differences in ECP, TNF-α, IL-10, eosinophil, lymphocyte, monocyte, and neutrophil levels in blood were detected 24 hours after the challenge. Seventy-two hours after the challenge, eosinophil levels were lower (borderline significant, $p < 0.06$). In addition, TNF-α secretion by PHA-stimulated BMCs was significantly lower 72 hours after the glucan challenge compared to the saline challenge. No significant differences in lung function values were observed at 24 hours after the challenges. However, the FEV_1 value was significantly higher 72 hours after glucan challenge compared to the saline challenge.

A third study investigated whether an inhalation challenge to (1→3)-β-D-glucan (grifolan suspended in saline) for a period of 3 hours could induce effects on inflammatory markers in blood of nonasthmatic subjects (Beijer et al., 2002). The study also aimed to assess whether effects were related to recent home exposures to (1→3)-β-D-glucans. Therefore, 17 subjects with high (G-high; median exposure 6 ng/m^3) and 18 with low home exposures (G-low; 0.9 ng/m^3) were selected from a larger survey (Thorn and Rylander, 1998a; see above) to undergo two randomized inhalation challenges, one to (1→3)-β-D-glucan and another to saline. The mean exposure concentration was approximately 28 ng/m^3 (range 15.3–44.9 ng/m^3). Twenty-four hours after the challenge, blood was collected and BMCs were stimulated with LPS. (1→3)-β-D-glucan exposure decreased TNF-α production by BMC in both the G-low and G-high group, which was statistically significant only for the G-high group. In the G-high group, a significant increase was seen in blood lymphocytes compared with a decrease in the G-low group; however, this decrease was not statistically significant. The authors also suggested that IL-10 production was significantly affected; however, no clear evidence of this was shown in the tables. No significant differences were found in BMC secretion of the other studied cytokines, i.e., IL-1β, IL-4, and IFN-γ. Also no differences in ECP and MPO in serum were observed.

Finally, Sigsgaard et al. (2000) demonstrated an increase in albumin and a slight increase in IL-1β in the nasal mucosa in a small group of Danish garbage workers (n = 5) and controls (n = 5) who were exposed to solubilized (1→3)-β-D-glucan (curdlan) by nasal instillation (1 mg/ml). However, (1→3)-β-D-glucan exposure was not associated with an increase in the other tested cytokines, i.e., TNF-α, IL-6, and IL8. In a whole blood assay measuring cytokine release after *in vitro* exposure to high concentrations of (1→3)-β-D-glucan (250 µg/ml), a significant increase in all measured cytokines (TNF-α, IL-1β, IL-6, and IL8) was found. This was confirmed in two other studies in which blood was collected from healthy volunteers (n = 14) (Wouters et al., 2002) or municipality (n = 22) and fish factory workers (n = 20) (Bønløkke et al., 2004). The (1→3)-β-D-glucan concentration (curdlan) used to stimulate whole blood in these two studies was 12.5–25 µg/ml and 2 mg/ml, respectively. A subsequent experiment by the Danish group aimed to study (1→3)-β-D-glucan mediated cytokine production and related mRNA induction in a whole blood assay of 10 nonatopic and 10 atopic subjects. This study showed a glucan-dependent increase in both IL-1β and IL-8 mRNA production (Kruger et al., 2004). The steady-state level of IL-8 mRNA 3 hours after stimulation with glucan was lower in atopics than in nonatopics (in contrast to stimulation with LPS where steady-state levels were higher in the nonatopics, both for IL-8 mRNA and IL-1β mRNA). No significant difference between atopics and nonatopics was found for IL-1β mRNA after glucan stimulation.

2.4 THE EPIDEMIOLOGICAL EVIDENCE

How should we interpret these studies? A large number of health effects has been evaluated, including lung function (FEV_1, PEF variability), nasal congestion, bronchial hyperreactivity, atopy, symptoms (upper and lower respiratory symptoms, eye irritations, headache, fatigue/tiredness, joint pains, skin symptoms, flu-like symptoms, nausea, and gastro-intestinal symptoms), inflammatory cells (T-lymphocytes, neutrophils, eosinophils, and macrophages), and cytokines or other inflammatory markers (IL-1β, IL-4, IL-6, IL-8, IL-10, IFN-γ, TNF-α, ECP, MPO, CRP, and albumin) in blood, sputum, and nasal lavage. Most studies included a range of these outcomes, generally resulting in only a few significant associations. Although the focus has been on these positive findings, it is clear that the results were not always consistent (see below). Also, some of the positive associations described above were only observed when specific subgroup analyses were conducted. The criteria for stratified analyses, however, were not always clear, precluding a straightforward interpretation.

Several studies reported associations between glucan exposure and upper airway irritations and fatigue/tiredness (Wan and Li, 1999; Mandryk et al., 2000; Heldal et al., 2003a; Gladding et al., 2003). However, these associations were not confirmed in other studies (Rylander et al., 1992; Rylander, 1997; Thorn and Rylander, 1998b; Heldal et al., 2003b). The evidence with regard to lower airway symptoms is perhaps even more mixed. Also, no clear associations with lung function were found, i.e., some studies reported adverse effects on lung function (Rylander, 1996; Douwes et al., 2000a; Mandryk et al., 2000), whereas others found no association (Thorn and Rylander., 1998a; 1998b) or even a reversed association (Mandryk et al., 1999; Thorn et al., 2001). Similarly, some studies reported significant effects on airway responsiveness (Rylander, 1997), whereas others failed to demonstrate such an association (Thorn and Rylander, 1998a, 1998b). One study suggested that (1→3)-β-D-glucan was associated with an increased risk of atopy. This was not confirmed in a smaller study (Rylander et al., 1998). Also, an association between (1→3)-β-D-glucan exposure and an increased Th_1 immune response was suggested (Beijer et al., 2003), which appears to be contradictive with the finding of a higher prevalence of atopy in subjects with high (1→3)-β-D-glucan exposure (NB atopy is a Th_2 driven immune response). The latter, however, was only shown in nonatopic subjects. Further confirmation of the findings for atopy is thus essential.

The data with regard to the potential effects of glucan on airway inflammation is also mixed. *In vitro* studies indicated that glucan (at very high levels) can induce cytokine production (IL-8, IL-1β, and TNF-α) in blood mononuclear cells. However, this was not confirmed consistently in *in vivo* experiments in which subjects were exposed to specific glucans (e.g., curdlan, grifolan) (see above). Some observational studies in the occupational environment (Heldal et al., 2003a, 2003b; Wouters et al., 2002) showed neutrophilic airway inflammation in (1→3)-β-D-glucan-exposed subjects. However, the few human challenge studies conducted to date (see above) do not appear to support this finding. Therefore, the association between (1→3)-β-*D*-glucan exposure and neutrophilic airway inflammation may be more likely due to correlating endotoxin exposures (endotoxin induced airway inflammation is characterized by a rapid increase in IL-1, TNF-α, and IL-8 levels in the airways followed by an influx of neutrophils). On the other hand, results of the challenge studies should be interpreted with caution, since only a few are available and experiments were conducted with only two types of (1→3)-β-*D*-glucan (curdlan and grifolan) which may not necessarily be the most biologically active forms.

Some observational studies also focused on inflammatory mediators in blood. One study showed an association between (1→3)-β-*D*-glucan and MPO (Thorn and Rylander, 1998a), but this was not confirmed in another study (Rylander et al., 1999). Also, no association between (1→3)-β-*D*-glucan and MPO in sputum was found (Thorne et al., 2001; Heldal et al., 2003a). A positive association was demonstrated between (1→3)-β-*D*-glucan exposure and lymphocytes in blood (Thorn and Rylander, 1998b). Another study showed an inverse association between (1→3)-β-*D*-glucan exposure and the number of cytotoxic CD8+ T-cells (Beijer et al., 2003) and PHA- or LPS-induced TNF-α production by BMCs (Thorne et al., 2001; Beijer et al., 2002). The meaning of these findings is not clear, and results require further confirmation from both human and animal studies. The evidence regarding (1→3)-β-*D*-glucan airway inflammation based on animal models will be discussed in Chapter 4.

The lack of consistency between studies as indicated above may largely be due to the relatively small sample size of most of the reported studies. Another reason may be the weaknesses in exposure assessment. In most published studies the methodology to measure (1→3)-β-*D*-glucans in the indoor environment was not comparable; some studies used a Limulus Amebocyte Lysate test (Aketagawa et al., 1993), whereas others used an enzyme immunoassay (Douwes et al., 1996) (see Chapter 10). Currently it is not clear which method best captures the biological/toxicological properties of environmental (1→3)-β-*D*-glucan exposure. Also, methods to dissolve (1→3)-β-*D*-glucan were different — i.e., some studies used heat extraction whereas others applied alkaline treatment. Studies comparing these analytical procedures have not been conducted. Due to these differences in method, environmental exposures cannot be directly compared between studies (both quantitatively and qualitatively).

The exposure assessment strategy is another issue to consider. Some studies measured (1→3)-β-*D*-glucan concentrations in settled dust, and others measured it in the air, sometimes after rigorous disturbance of settled dust. Results are, therefore, not directly comparable between studies. In the occupational environment, it is usually best to measure personal airborne concentrations (Heederik et al., 2003). In the home environment, the preferred method is strongly dependent on the specific aims of the study. Considering the high temporal variation in airborne concentrations of bioaerosols in general, it is highly unlikely that the collection of a single airborne sample in the indoor environment (as is common practice in most studies) will capture a person's mean long-term (1→3)-β-*D*-glucan exposure. Therefore, in most studies reservoir dust from carpets or mattresses is collected. The advantage of this method is the presumed time-integration that occurs in the deposition of bioaerosols on surfaces over time (IOM, 2000). (1→3)-β-*D*-glucan concentrations measured in settled dust are, therefore, more likely to accurately reflect chronic exposure, particularly if only one or few samples per subject are taken. However, although surface sampling has advantages in many situations, airborne measurements may be more desirable in others. Airborne measurements allow fluctuations in exposures to be assessed over the course of a week, a day, or even hours, which can be essential when studying acute adverse effects, such as daily lung function changes. Airborne levels of (1→3)-β-*D*-glucan are, however, generally low in the residential indoor environment, and analytical sensitivity, thus, may not be sufficient (particularly for the ELISA method), precluding the option of short-term airborne sampling. Airborne sampling after agitation of settled dust, as has been employed in a number of studies discussed above (Rylander et al., 1992, 1998; Rylander, 1997; Thorn and Rylander, 1998a), may overcome this problem. In addition, it may have the advantage (over settled dust sampling) that the more appropriate dust fraction — i.e., airborne or inhalable particles — will be sampled, and that the temporal variation may be less compared with regular air sampling. However, these assumptions have not been confirmed, and procedures have not been standardized or validated.

In addition to sampling methods (which in itself may lead to substantial exposure misclassification), there is also the issue of the number of samples taken. In general, precision can be gained by increasing the number of samples, particularly when temporal variation in exposure is high. In most studies discussed above, only relatively few samples were taken, and, with exception of a few studies, no repeated sampling has been conducted. Therefore, uncertainty in exposure assessment was large, which may have resulted in obscured exposure-response relationships.

Another aspect that should be considered is the criteria for exposure grouping. In a number of publications (Thorn and Rylander, 1998a; Rylander et al., 1999; Gladding et al., 2003), exposure groups were constructed using cut-off points that were not based on objective criteria, such as the median exposure level, tertiles, quartiles, and so on. Instead the reasons for choosing cut-off points were not clear and did not appear to be driven by an *a priori* hypothesis. One study clearly indicated that the outcome of the analyses was highly dependent on the specific cut-off points chosen (Thorn and Rylander, 1998a), emphasizing that results from these studies should be interpreted with caution.

Finally, another weakness in many of the reported field studies was the lack of control for potential confounders, such as other common indoor or occupational exposures. Many of the studied health effects can also be caused by other bioaerosol exposures, such as allergens or bacterial endotoxin (Douwes et al., 2003). A number of studies showed a strong correlation between endotoxin and (1→3)-β-*D*-glucan levels (Rylander et al., 1999; Douwes et al., 2000a, 2000b; Gladding et al., 2003) suggesting that some of the findings may potentially have been caused by correlating exposures, such as endotoxin. In some of the studies where analyses were adjusted for other exposures no major changes in the relationship between (1→3)-β-*D*-glucan and health effects were observed after adjustment (Douwes et al., 2000a; Mandryk et al., 2000).

In summary, the observational and experimental studies described above suggest some association between (1→3)-β-*D*-glucan exposure and symptoms. However, results are mixed and specific symptoms associated with exposure cannot at this stage be identified. Based on subgroup analyses, it has been speculated that atopics and/or subjects with pre-existing symptoms may be more susceptible (Rylander et al., 1998; Douwes et al., 2000a), but this requires further study. *In vitro* observations indicate that (1→3)-β-*D*-glucan at high concentrations has the potential to initiate the secretion of various inflammatory cytokines. However, in inhalation experiments (at much lower doses), these effects could not consistently be reproduced. Currently, a clear interpretation of the potential health effects of (1→3)-β-*D*-glucan is hampered for the following reasons: (1) only a relatively small number of field studies are available, most of which lacked statistical power; (2) some of these studies have not appropriately controlled the analyses for other potential causal exposures and/or had weaknesses in the design and/or statistical analyses; and (3) different methods to assess exposure were used and only relatively few samples were collected, potentially resulting in substantial exposure misclassification obscuring exposure-response relationships. Thus, the currently available epidemiological data do not permit conclusions to be drawn regarding the presence (or absence) of an association between environmental glucan exposure and specific adverse health effects, nor is it clear from the currently available evidence which specific inflammatory mechanisms underlie the presumed health effects. For a more extensive discussion on risk evaluation of (1→3)-β-*D*-glucan, the reader is referred to Chapter 3.

2.5 CONTROL OF (1→3)-β-*D*-GLUCAN EXPOSURE IN THE HOME AND WORK ENVIRONMENT

If, as suggested, environmental exposures to (1→3)-β-*D*-glucans cause health effects, then control is essential. Specific measures to control exposure have not been developed, but some studies are available that investigated the association between exposure and housing characteristics or occupant behaviour. Findings from these studies may help to develop procedures to manage exposure in the home environment.

At present, preventive measures in the home environment are mainly directed at reduction of house dust mite allergen exposure. However, determinants of mite allergen concentrations in house dust are not necessarily the same as for (1→3)-β-*D*-glucan and other microbial contaminants, such as endotoxin. Only a few studies are available in which the relation between (1→3)-β-*D*-glucan levels in house dust and housing characteristics or occupant behavior were studied. The first study was a very small study in 25 German homes, showing that (1→3)-β-*D*-glucan levels on living room floors were higher in centrally heated houses built after 1970 compared to older individually heated houses (Douwes et al., 1998). No clear associations with any of the other home characteristics were found. Another much larger study in Germany (Gehring et al., 2001) showed that (1→3)-β-*D*-glucan levels in house dust were mainly determined by the amount of dust on the floor, suggesting that measures to lower the dust levels in the home or building will lower the exposure to (1→3)-β-*D*-glucan. Presence of carpets and pets (particularly dogs) in the home, intensive use of the home or apartment, and low frequency of cleaning were also associated with higher levels of dust and (1→3)-β-*D*-glucan, thus potentially providing measures to lower or control (1→3)-β-*D*-glucan exposure. Mold spots in the home were positively associated with (1→3)-β-*D*-glucans; therefore, any measure to control mold growth in the home is likely to also control (1→3)-β-*D*-glucan exposure. An association between the presence of mold infestation and high (1→3)-β-*D*-glucan levels has been demonstrated in other studies as well (Rylander et al., 1994; Rylander et al., 1998). Moreover, a (weak) positive association between indoor relative humidity and (1→3)-β-*D*-glucan levels in the home was demonstrated indicating that measures to reduce dampness in the home may lower the exposure to (1→3)-β-*D*-glucan. Finally, one study reported that homes in which separated organic waste was stored indoors (as is becoming increasingly common in many European countries) for 1 week or more had significantly (almost five times) higher levels of (1→3)-β-*D*-glucan (as well as endotoxin) indoors (Wouters et al., 2000). This suggests that storage of waste outdoors (and particularly the organic waste fraction) may substantially lower (1→3)-β-*D*-glucans in homes.

Source control is the most desirable option in the occupational environment as well. However, this is often not practical, since in many cases the source — i.e., molds or their substrate — is an integral part of the production process (e.g., in the composting, paper, and sawmill industries). In those cases where source control is not feasible, modification to the general work environment is the next level of control, followed by separating the exposed worker from the exposure either by the application of isolation environments or by employing personal protection equipment. These are general principles that apply to many other biological and chemical exposures and will, therefore, not be further discussed here. For a more extensive overview of control options for biological agents, the reader is referred to Martinez and Thorne (2003).

2.6 RESEARCH NEEDS

Currently, it is of highest importance to further study the health effects of (1→3)-β-*D*-glucan in large field studies to accomplish the following: (1) identify which specific effects are associated with (1→3)-β-*D*-glucan exposure in the home and work environment; and (2) assess at which levels these effects occur so that a safe level can be determined. In addition to more and larger observational studies, there is a need for human challenge studies testing various conformations of (1→3)-β-*D*-glucans (triple or single helix or random coil) as well as soluble and nonsoluble forms of glucan. This is essential, since conformation and solubility may be important factors determining the biological activity of (1→3)-β-*D*-glucans (Young et al., 2003a, 2003b; see also Chapter 4). Challenge studies should be conducted, using a wide range of doses, to gain more insight into which levels various glucans may cause symptoms and/or airway inflammation. Moreover, there is a clear need for validation of existing methods to measure environmental (1→3)-β-*D*-glucan and the subsequent further development of these assays to make them more suitable for large-scale epidemiological studies allowing a more valid risk assessment. Method validation should include a comparison of extraction methods as well as a comparison between the currently commonly applied assays to measure (1→3)-β-*D*-glucan;

i.e., the LAL and the ELISA assay. Other important areas that require further research include the following: (1) the issue of individual susceptibility for (1→3)-β-*D*-glucans, including the identification of potential genetic polymorphisms; (2) the potential interaction effects between (1→3)-β-*D*-glucans and other exposures, such as allergens, endotoxin, and infectious agents; (3) more research into other health effects, such as skin reactions, neurological and gastro-intestinal symptoms, and susceptibility to infectious diseases, all of which have been speculated to be related to (1→3)-β-*D*-glucan exposure; (4) the assessment of the biological properties of nonfungal (1→3)-β-*D*-glucan; and (5) the assessment of determinants of exposure to allow more specific control measures to be developed.

2.7 CONCLUSIONS

Some epidemiological evidence exists to suggest a role for (1→3)-β-*D*-glucan in the development of airway diseases. However, most studies were very small and inconclusive; thus, further and larger studies are needed to assess whether (1→3)-β-*D*-glucan is involved in respiratory health problems in the indoor and occupational environment.

ACKNOWLEDGMENTS

Jeroen Douwes is supported by a Sir Charles Hercus Research Fellowship from the Health Research Council (HRC) of New Zealand. The Centre for Public Health Research is supported by a Program Grant from the HRC of New Zealand. The author thanks Professor Neil Pearce for his comments on the draft manuscript.

REFERENCES

Andriessen, J.W., Brunekreef, B., and Roemer, W. (1998). Home dampness and respiratory health status in European children. *Clin. Exp. Allergy* 28,1191–1200.

Aketagawa, J., Tanaka, S., Tamura, H., Shibata, Y., and Saitô, H. (1993). Activation of limulus coagulation factor G by several (1→3)-β-D-glucans: comparison of the potency of glucans with identical degree of polymerization but different conformations. *J. Biochem.* 113, 683–686.

Beijer, L., Thorn, J., and Rylander, R. (2002). Effects after inhalation of (1→3)-β-D-glucans and relation to mould exposure in the home. *Mediators of Inflam.* 11, 149–153.

Beijer, L., Thorn, J., and Rylander, R. (2003). Mould exposure at homes relates to inflammatory markers in blood. *Eur. Respir. J.* 21, 317–322.

Black, P.N., Udy, A.A., and Brodie, S.M. (2000). Sensitivity to fungal allergens is a risk factor for life-threatening asthma. *Allergy* 55, 501–504.

Bønløkke, J.H., Thomassen, M., Viskum, S., Omland, O., Bonefeld-Jorgensen, E., Sigsgaard, T. (2004). Respiratory symptoms and *ex vivo* cytokine release are associated in workers processing herring. *Int. Arch. Occup. Environ. Health* 77,136–141.

Bornehag, C.G., Blomquist, G., Gyntelberg, F., Jarvholm, B., Malmberg, P., Nordvall, L., Nielsen, A., Pershagen, G., and Sundell, J. (2001). Dampness in buildings and health. Nordic interdisciplinary review of the scientific evidence on associations between exposure to "dampness" in buildings and health effects (NORDDAMP). *Indoor Air* 11, 72–86.

Brunekreef, B., Dockery, D., Speizer, F., Ware, J.H., Spengler, J.D., and Ferris, B.G., (1989). Home dampness and respiratory morbidity in children. *Am. Rev. Resp. Dis.* 140, 1363–1367.

Douwes, J., Doekes, G., Montijn, R., Heederik, D., and Brunekreef, B. (1996). Measurement of β(1→3)-glucans in the occupational and home environment with an inhibition enzyme immunoassay. *Appl. Environ. Microbiol.* 62, 3176–3182.

Douwes, J., Doekes, G., Heinrich, J., Koch, A., Bischof, W., and Brunekreef, B. (1998). Endotoxin and β(1→3)-glucan in house dust and the relation with home characteristics: a pilot study in 25 German houses. *Indoor Air* 8, 255–263.

Douwes, J., Zuidhof, A., Doekes, G., van der Zee, S., Wouters, I., Boezen, H.M., and Brunekreef, B. (2000a). (1→3)-β-D-glucan and endotoxin in house dust and peak flow variability in children. *Am. J. Respir. Crit. Care Med.* 162, 1348–1354.

Douwes, J., Wouters, I., Dubbeld, H., Zwieten, L., Steerenberg, P., Doekes, G., Heederik, D. (2000b). Upper airway inflammation assessed by nasal lavage in compost workers: a relation with bio-aerosol exposure. *Am. J. Indust. Med.* 37, 459–468.

Douwes, J., and Pearce, N. (2003). Is indoor mold exposure a risk factor for asthma? *Am. J. Epidemiol.* 58, 203–206.

Douwes, J., Thorne, P., Pearce, N., and Heederik, D. (2003). Bioaerosol health effects and exposure assessment: progress and prospects. *Annals Occup. Hyg.* 47, 87–200.

Eduard, W., Douwes, J., Mehl, R., Heederik, D., Melbostad, E. (2001). Short term exposure to airborne microbial agents during farm work: exposure-response relations with eye and respiratory symptoms. *Occup. Environ. Med.* 58, 113–118.

Engvall, K., Norrby, C., and Norback, D. (2001). Asthma symptoms in relation to building dampness and odour in older multifamily houses in Stockholm. *Int. J. Tuberc. Lung Dis.* 5, 468–477.

Gehring, U., Douwes, J., Doekes, G., Koch, A., Bischof, W., Wichmann, H.E., and Heinrich, J. (2001). β(1→3)-glucan in house dust of German homes related to culturable mold spore counts, housing and occupant characteristics. *Environ. Health Perspect.* 109, 139–144.

Gladding, T., Thorn, J., and Stott, D. (2003). Organic dust exposure and work-related effects among recycling workers. *Am. J. Indust. Med.* 43, 584–591.

Halonen, M., Stern, D.A., Wright, A.L., Taussig, L.M., and Martinez, F.D. (1997). *Alternaria* as a major allergen for asthma in children raised in a desert environment. *Am. J. Resp. Crit. Care Med.* 155, 1356–1361.

Heederik, D., Thorne, P., and Douwes, J. (2003). Biological agents – monitoring and evaluation of bioaerosol exposure. In *Modern Industrial Hygiene*, Vol. 2, J. Perkins, ed., ACGIH, Cincinnati, OH, pp. 293–327.

Heldal, K.K., Halstensen, A.S., Thorn, J., Eduard, W., and Halstensen, T.S. (2003a). Airway inflammation in waste handlers exposed to bioaerosols assessed by induced sputum. *Eur. Respir. J.* 21, 641–645.

Heldal, K.K., Halstensen, A.S., Thorn, J., Djupesland, P., Wouters, I., Eduard, W., and Halstensen, T.S. (2003b). Upper airway inflammation in waste handlers exposed to bioaerosols. *Occup. Eviron. Med.* 60, 444–450.

IOM (Institute of Medicine) (2000). *Clearing the Air: Asthma and Indoor Air Exposures*, National Academy Press, Washington DC.

Health Effects of (1→3)-β-D-Glucans: The Epidemiological Evidence 51

Kilpeläinen, M., Terho, E.O., Helenius, H., and Koskenvuo, M. (2001). Home dampness, current allergic diseases, and respiratory infections among young adults. *Thorax* 56, 462–467.

Kruger, T., Sigsgaard, T., and Bonefeld-Jorgensen, E.C. (2004). Ex vivo induction of cytokines by mould components in whole blood of atopic and non-atopic volunteers. *Cytokine* 25, 73–84.

Mandryk, J., Alwis, K.U., and Hocking, A.D. (1999). Work-related symptoms and dose-response relationships for personal exposures and pulmonary function among woodworkers. *Am. J. Indust. Med.* 35, 481–490.

Mandryk, J., Alwis, K.U., and Hocking, A.D. (2000). Effects of personal exposures on pulmonary function and work-related symptoms among sawmill workers. *Ann. Occup. Hyg.* 44, 281–289.

Martinez, K.F., and Thorne, P.S. (2003). Biological agents – Control in the occupational environment. In *Modern Industrial Hygiene*, Vol. 2, J. Perkins, ed., ACGIH, Cincinnati, OH, pp. 329–381.

Nafstad, P., Øie, L., Mehl, R., Gaarder, P.I., Lodrup-Carlsen, K.C., Botten, G., Magnus, P., and Jaakkola, J.J. (1998). Residential dampness problems and symptoms and signs of bronchial obstruction in young Norwegian children. *Am. J. Respir. Crit. Care Med.* 157, 410–414.

Norback, D., Bjornsson, E., Janson, C., Palmgren, U., and Boman, G. (1999). Current asthma and biochemical signs of inflammation in relation to building dampness in dwellings. *Int. J. Tuberc. Lung Dis.* 3, 368–376.

Øie, L., Nafstad, P., Botten, G., Magnus, P., and Jaakkola, J.K. (1999). Ventilation in homes and bronchial obstruction in young children. *Epidemiol.* 10, 294–299.

Peat, J.K., Dickerson, J., and Li, J. (1998). Effects of damp and mould in the home on respiratory health: a review of the literature. *Allergy* 53,120–128.

Ronald, L.A., Davies, H.W., Bartlett, K.H., Kennedy, S.M., Teschke, K., Spithoven, J., Dennekamp, M., Demers, P. (2003). (1→3)-glucan exposure levels among workers in four British Columbia sawmill workers. *Ann. Agric. Environ. Med.* 10, 21–29.

Rylander, R., Persson, K., Goto, H., Yuasa K., and Tanaka S. (1992). Airborne β-1,3-glucan may be related to symptoms in sick buildings. *Indoor Environ.* 1, 263–267.

Rylander, R., Hsieh V., and Courteheuse, C. (1994). The first case of sick building syndrome in Switzerland. *Indoor Environ.* 3,159–162.

Rylander, R. (1996). Airway responsiveness and chest symptoms after inhalation of endotoxin or (1→3)-β-D-glucan. *Indoor Built. Environ.* 5, 106–111.

Rylander, R. (1997). Airborne (1→3)-β-D-glucan and airway disease in a day-care centre before and after renovation. *Arch. Environ. Health* 52, 281–285.

Rylander, R., Norhall, M., Engdahl, U., Tunsäter, A., and Holt, P.G. (1998). Airways inflammation, atopy, and (1→3)-β-D-glucan exposure in two schools. *Am. J. Resp. Crit. Care Med.* 158, 1685–1687.

Rylander, R., Thorn, J., and Attefors, R. (1999). Airways inflammation among workers in a paper industry. *Eur. Resp. J.* 13, 1151–1157.

Sigsgaard, T., Bonefeld-Jørgensen, E.C., Kjaergaard, S.K., Mamas, S., and Pedersen, O.F. (2000). Cytokine release from the nasal mucosa and whole blood after experimental exposures to organic dusts. *Eur. Respir. J.* 16, 140–145.

Thorn, J., and Rylander, R. (1998a). Airway inflammation and glucan in damp rowhouses. *Am. J. Resp. Crit. Care Med.* 157, 1798–1803.

Thorn, J., and Rylander, R. (1998b). Airways inflammation and glucan exposure among household waste collectors. *Am. J. Indust. Med.* 33, 463–470.

Thorn, J., Beijer, L., and Rylander, R. (2001). Effects after inhalation of (1→3)-β-D-glucan in healthy humans. *Mediators Inflam.* 10, 173–178.

Wan, G.H., and Li, C.S. (1999). Indoor endotoxin and glucan in association with airway inflammation and systemic symptoms. *Arch. Environ. Health* 54, 172–179.

Wouters, I.M., Douwes, J., Doekes, G., Thorne, P.S., and Heederik, D.J. (2000). Increased levels of bacterial endotoxin and fungal antigens in homes with indoor storage of organic household waste. *Appl. Environ. Microbiol.* 66, 627–631.

Wouters, I.M., Hilhorst, S.K.M., Kleppe, P., Doekes, G., Douwes, J., Peretz, C., and Heederik, D. (2002). Upper airway inflammation and respiratory symptoms in domestic waste collectors. *Occup. Environ. Med.* 59, 106–112.

Young, S.H., Robinson, V.A., Barger, M., Frazer, D.G., Castranova, V., and Jacob, R.R. (2003a). Partially opened triple helix is the biologically active conformation of (1→3)-β-*D*-glucans that induces pulmonary inflammation in rats. *J. Toxicol. Environ. Health* 66, 551–563.

Young, S.H., Robinson, V.A., Barger, M., Whitmer, M., Porter, D.W., Frazer, D.G., and Castranova V. (2003b). Exposure to particulate (1→3)-β-*D*-glucans induces greater pulmonary toxicity than soluble (1→3)-β-*D*-glucans in rats. *J. Toxicol. Environ. Health* 66, 25–38.

Zock, J.P., Jarvis, D., Luczynska, C., Sunyer, J., and Burney, P. (2002). Housing characteristics, reported mold exposure, and asthma in the European Community Respiratory Health Survey. *J. Allergy Clin. Immunol.* 110, 285–292.

Zureik, M., Neukirch, C., Leynaert, B., Liard, R., Bousquet, J., and Neukirch, F. (2002). Sensitisation to airborne moulds and severity of asthma: cross sectional study from European Community respiratory health survey. *Brit. Med. J.* 325, 411–414.

3 (1→3)-β-*D*-Glucan in the Environment: A Risk Assessment

Ragnar Rylander, M.D.

CONTENTS

3.1 Introduction 53
3.2 General Considerations 53
3.3 Animal Inhalation Studies 55
3.4 Human Inhalation Studies 56
3.5 Synthesis 57
3.6 Application to Field Studies 59
3.7 Environmental Risk Evaluation 60
3.8 Conclusion 61
References 61

3.1 INTRODUCTION

While risk evaluations of environmental exposures to microbes have traditionally focused on infection, there is increasing evidence for other effects, particularly inflammation. Many microorganisms contain specific substances in or on their cell walls (microbial cell wall agents [MCWAs]), and a number of such agents have been found to have inflammagenic properties. Regarding (1→3)-β-*D*-glucan, an MCWA on fungi and some bacteria, much research has been devoted to its biological activities and possible relationship to diseases. This chapter will focus in the effects of (1→3)-β-*D*-glucan after inhalation, and extrapolate this information to evaluate the risks involved in occupational and general environments.

3.2 GENERAL CONSIDERATIONS

The dose is an important parameter in the risk assessment. There is thus a need to know the levels of airborne (1→3)-β-*D*-glucan, preferably obtained from measures in different environments. In experimental studies, such as when administering the agent by injection or intratracheal instillation, the doses on the cellular level exceed

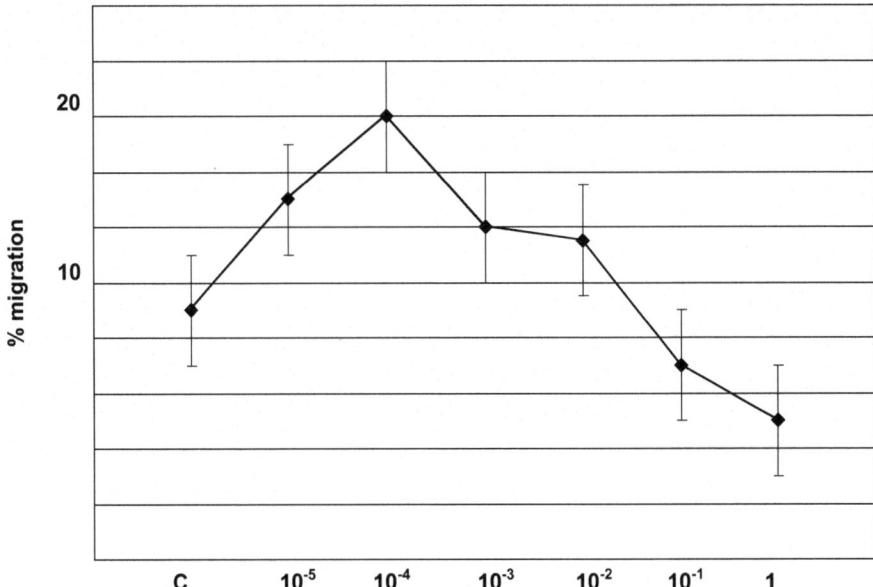

FIGURE 3.1 Neutrophil migration toward alveolar macrophages incubated in different concentrations of lipopolysaccharide (after Snella, 1986).

those normally present in the environment by as much as several orders of magnitude. It is important in this context that cellular effects seen at low exposure levels can be different than those observed at higher levels, as illustrated in Figure 3.1. It is seen that a low dose of endotoxin stimulated neutrophil migration toward macrophages, whereas a higher dose had an inhibition effect on the migration.

Regarding exposure to airborne (1→3)-β-D-glucan, data are available from occupational and general environments. In a study on garbage collectors, personal exposure levels ranged from 2.0 to 36.4 ng/m^3 (Thorn et al., 1998). In a wood processing industry, personal exposure levels ranged from 4 to 366 ng/m^3 (Rylander et al., 1999a). In a number of studies indoors, floor dust was agitated and collected from the air using filters. In a study comparing two schools, one of which had a mold problem, the levels were 15.3 and 2.9 ng/m^3 (Rylander et al., 1998). In studies on houses with mold problems, values ranging from 0.2 to 51.7 ng/m^3 were found (Beijer et al., 2002). In similar studies, airborne (1→3)-β-D-glucan was collected during 8 hours, giving values up to 7.6 ng/m^3 (Beijer et al., 2004). In individual homes, levels up to 22.0 ng/m^3 have been found.

Another important point in the risk assessment is the exposure specificity. In experimental work, it is possible to work with pure (1→3)-β-D-glucan, and the effects found thus imply causality. In field studies, however, measurements of (1→3)-β-D-glucan represent determinations of mold cell mass exposure. As the mold cell wall contains several agents with a range of biological potentials, a conclusion regarding causality for (1→3)-β-D-glucan cannot be drawn. Furthermore, samples of environmental air may also contain other agents, such as bacteria and chemical agents. Data on (1→3)-β-D-glucan exposure levels under field conditions can thus only be considered as an "index" of a multiple-agent environmental exposure. Conclusions concerning causality could, however, be drawn if the effects studied were specific for (1→3)-β-D-glucan.

An inhalation risk assessment of (1→3)-β-D-glucan is complicated by the many different configurations that exist for the molecule (Stone and Clark, 1994; Williams, 1997). An important characteristic is whether or not the molecule is soluble in water. In nature, (1→3)-β-D-glucan is present both in soluble and nonsoluble forms. Experience from some industries with organic dust exposure has shown that the water-soluble fraction was about 10% of the total. Studies using administration by injection or *in vitro* investigation have demonstrated important differences in effects between configurations, particularly between soluble and nonsoluble forms (Sherwood et al., 1987, Ishibashi et al., 2001 and 2002, Young et al., 2003a). It is likely that the configuration also influences effects after inhalation.

The effects after exposure to (1→3)-β-D-glucan are initiated through receptor binding to different cells. Regarding the lung, macrophages play a pivotal role mediated by the dectin-1 and Mac^{-1} receptors (Brown et al., 2003). Vertebrates do not possess specific β-glucan hydrolases and (1→3)-β-D-glucan is broken down by slow oxidative degradation; it may thus be retained intracellularly for prolonged periods of time (Ohno et al., 1999).

3.3 ANIMAL INHALATION STUDIES

A number of inhalation studies have been performed using a guinea pig model. One experiment evaluated the effects of an acute inhalation of several different forms of (1→3)-β-D-glucan (Fogelmark et al., 1992). After exposure to the nonsoluble form curdlan (3 μg/m^3, 1–4 hours), there was a decrease in the number of lymphocytes in the airways but no effect on neutrophils. In the lung tissue, the number of macrophages and lymphocytes was decreased at 4 to 5 days after exposure. Of particular interest is that exposure to (1→3)-β-D-glucan blunted the neutrophil increase induced by a simultaneous inhalation of endotoxin. When curdlan was suspended in NaOH, dissolving the triple helix and rendering it water soluble, there was a significant increase in number of neutrophils in the lung lavage after inhalation. In a subchronic 5 week exposure study using endotoxin and (1→3)-β-D-glucan (curdlan in distilled water, 300 μg/m^3, 4 hours), the (1→3)-β-D-glucan exposure did not alter the number of lung lavage cells, although there was a tendency to exhibit a decrease in lymphocytes (Fogelmark et al., 1994). When (1→3)-β-D-glucan was administered together with endotoxin, numbers of all cell types were significantly increased above values found after exposure to endotoxin only. This is in contrast to the inhibitory effect on endotoxin-induced neutrophilia in an acute exposure as described above. The production of N-acetyl-β-D-glucosaminidase (NAG) and cathepsin D in lysed lavage or lung wall cells was not affected by the inhalation of (1→3)-β-D-glucan, but the increase in activity observed after endotoxin was lowered by simultaneous exposure to (1→3)-β-D-glucan.

A similar approach with combined subchronic exposures involved (1→3)-β-D-glucan and cigarette smoke (Sjöstrand and Rylander, 1997). Exposure to (1→3)-β-D-glucan (curdlan in distilled water, 30 μg/m^3, 4 hours) or cigarette smoke did not influence the number of inflammatory cells, except for neutrophils which were slightly increased. No effect was seen on the enzyme production in lung tissue cells. However, the combination of the two agents caused an increase in lung tissue macrophages, lymphocytes, neutrophils, and eosinophils.

In another subchronic experiment on guinea pigs, a 5 week inhalation of (1→3)-β-D-glucan (grifolan in NaOH, 30 µg/m^3, 4 hours) caused an increase in the number of eosinophils in lung lavage, lung interstitium, and airway epithelium (Fogelmark et al., 2001). There was also an increase in lymphocytes in the lung interstitium.

The development of antibodies to an inhaled antigen was evaluated in a 5 week exposure study (Rylander and Holt, 1998). Guinea pigs inhaled ovalbumin (OVA), endotoxin, (1→3)-β-D-glucan (curdlan in saline, 300 µg/m^3, 4 hours), or a mixture of these. Endotoxin caused an increase in IgG serum antibodies to OVA, an effect that was abolished by a simultaneous inhalation of (1→3)-β-D-glucan. The OVA-induced increase in eosinophils was also depressed by (1→3)-β-D-glucan.

Mice have also been used in inhalation studies. After inhalation of (1→3)-β-D-glucan (grifolan in NaOH, 37–1189 µg/m^3, 20 minutes) mice were sensitized with OVA (Wan et al., 1999). The OVA-specific IgE response was higher after this exposure, and these animals also produced lower amounts of IgG$_{2a}$ antibodies to OVA. In the same experiment, it was shown that (1→3)-β-D-glucan suppressed the antigen-induced IgE-specific tolerance.

In another experiment using mice, the sensory irritation in the airways was studied after inhalation of (1→3)-β-D-glucan (grifolan in NaOH) using respiratory function (Korpi et al., 2003a). Although a slight irritation effect was found, it was small in comparison to that induced by other agents such as suspensions of Aspergillus (Korpi et al., 2003b) and not dose-related. In this experiment, no effect of (1→3)-β-D-glucan exposure on total IgE levels was found.

3.4 HUMAN INHALATION STUDIES

In an initial study, subjects were exposed to (1→3)-β-D-glucan (curdlan in particulate form, 0.21 µg/m^3, 4 hours) in an inhalation chamber (Rylander, 1996). No effect was found on forced expiratory volume in 1 second (FEV$_1$) or on airway reactivity, measured with a methacholine challenge test. There were only small and nonsignificant effects on throat irritation and cough.

Another experimental study comprised 21 healthy subjects (Thorn et al., 2001). Using a nebulizer, subjects inhaled saline or (1→3)-β-D-glucan (grifolan in NaOH, dose 0.125 µg) in a double-blind, crossover fashion. No (1→3)-β-D-glucan-related effects were found on eosinophilic cationic protein (ECP), myeloperoxidase (MPO), tumour necrosis factor alpha (TNFα), IL-8, IL-10, or inflammatory cells in induced sputum. Peripheral blood monocytic cells (PBMCs) were collected 72 hours after inhalation and stimulated *in vitro* with lipopolysaccharide (LPS). The amounts of TNFα secreted were significantly lower after inhalation of (1→3)-β-D-glucan.

As (1→3)-β-D-glucan is present in pollen (Rylander et al., 1999b), a study was undertaken where persons sensitive to pollen received a nasal dose of 50 or 500 µg -glucan. No effects on the number of eosinophils or the amount of eotaxin in nasal lavage fluid were seen at 6 and 24 hours after application (Beijer and Rylander, 2005).

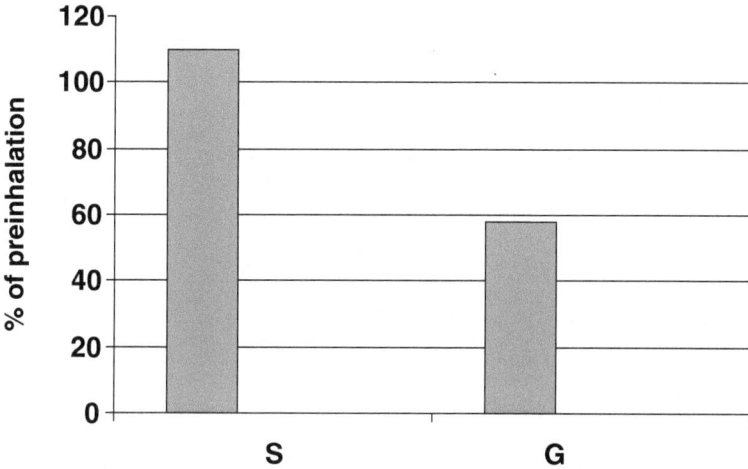

FIGURE 3.2 Secretion of TNFα from blood monocytic cells after inhalation of (1→3)-β-D-glucan among persons living in houses with high levels of molds. Values after inhalation as a percent of the pre-exposure level. S = saline inhalation, G = (1→3)-β-D-glucan inhalation (after Beijer et al., 2002).

Another inhalation study comprised subjects recruited from a field investigation. They were living in houses with different degrees of contamination with molds, measured as the amount of (1→3)-β-D-glucan in agitated floor dust, sampled in the air (Beijer et al., 2002). Among persons living in houses with higher levels of molds, there was a decrease in the secretion of TNFα and IL-10 from PBMC that had been stimulated with LPS *in vitro* after inhalation of (1→3)-β-D-glucan (grifolan in NaOH, 0.03 μg/m^3, 3 hours), as compared to conditions after inhalation of saline (Figure 3.2). In this group, the inhalation of (1→3)-β-D-glucan caused an increase in the number of lymphocytes in the blood.

3.5 SYNTHESIS

The number of inhalation studies on (1→3)-β-D-glucan is as yet limited and conclusions have to be drawn with care, particularly in view of the many forms of (1→3)-β-D-glucan that exist. Regarding dose levels, most experiments on animals have employed atmospheres with amounts up to 300 μg/m^3. These levels are above those so far measured in general and occupational environments. In inhalation experiments on humans, the doses have been more similar to those found in occupational and indoor environments.

Differences in effects have been found between acute and subchronic exposures and between nonsoluble and soluble (1→3)-β-D-glucan. The classic neutrophil invasion in the lung tissue and the airways after inhalation of endotoxin cannot be reproduced with nonsoluble (1→3)-β-D-glucan. On the other hand, studies where (1→3)-β-D-glucan has been given as intratracheal instillations or in *in vitro* experiments demonstrate a clear inflammagenic effect (Adachi et al., 1997; Ishibashi et

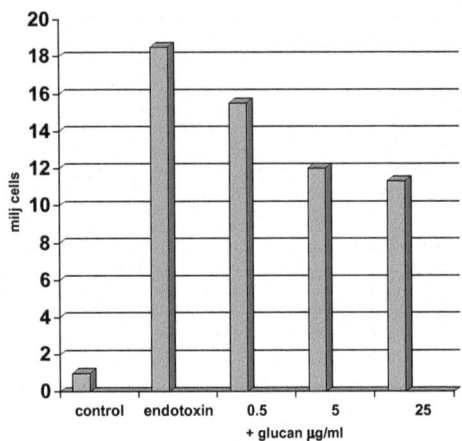

FIGURE 3.3 Neutrophil migration into airways after inhalation of lipopolysaccharide with and without previous inhalation of different concentrations of (1→3)-β-D-glucan.

al., 2002; Young et al., 2003a and 2003b). The reason for the discrepancy in findings from these studies and from inhalation experiments is unclear but the most likely explanation is the large difference in exposure levels.

Data from several experiments demonstrate a depressing effect on the inflammatory reactions induced by endotoxin, both *in vivo* and *ex vivo*. This effect is summarized in Figure 3.3 in terms of the neutrophil response after inhalation of endotoxin and how this is suppressed in a dose-related manner by a previous exposure to (1→3)-β-D-glucan.

This blunting effect has also been found outside of the lungs in studies on the secretion of TNFα from PBMC.

With reference to the inflammatory response in terms of an acute neutrophil invasion in the lung and airways, there is little evidence that (1→3)-β-D-glucan can induce this type of response. After a longer exposure times, however, (1→3)-β-D-glucan may elicit a neutrophil reaction, but only if other inflammagenic agents are given simultaneously. It is hypothesized that this effect is due to a disturbance of the adaptation to the inflammagenic agent, probably by interfering with the secretion of IL-10 or some other antiinflammagenic agent.

After a subchronic inhalation of (1→3)-β-D-glucan, one has observed an accumulation of lymphocytes together with a suppression of IgG formation and an increased IgE formation. The relevance of these findings is supported by model experiments *in vitro* where (1→3)-β-D-glucan from *Candida albicans* was found to suppress the endotoxin induced IL-6 production in cultures of PBMC (Nakagawa et al., 2003). In a mouse model where (1→3)-β-D-glucan was given as a subcutaneous injection in the right hind footpath, (1→3)-β-D-glucan administered with OVA caused elevated levels of IgE and IgG1 but not IgG2 (Ormstad et al., 2000).

The mechanisms behind the above inflammatory responses are open for speculation. An appealing hypothesis is that the effects are regulated through the influence of (1→3)-β-D-glucan on macrophage function and secondarily on the Th1/Th2

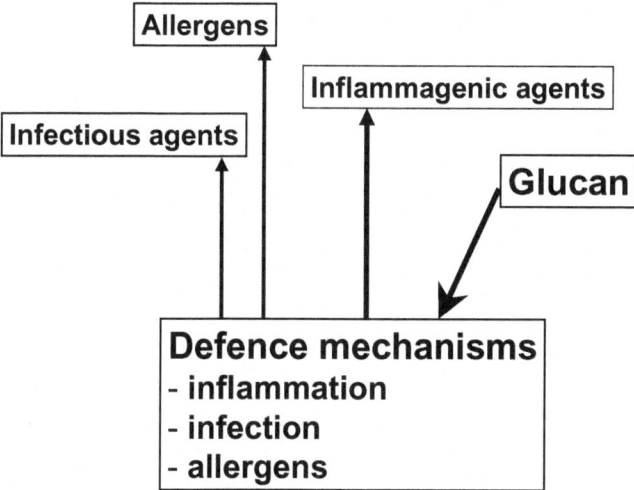

FIGURE 3.4 Working model of the effects of (1→3)-β-*D*-glucan on body defense mechanisms.

lymphocyte reaction patterns. The Th2 up-regulation is probably induced by (1→3)-β-*D*-glucan itself and orchestrated by macrophages. This reaction represents a suppression of the host defences, and it is possible that this renders the host more susceptible not only to inflammation but also to infection and allergens, resulting in atopy. This concept needs to be further tested in experimental models and epidemiological studies. A summary of this working model is given in Figure 3.4.

3.6 APPLICATION TO FIELD STUDIES

As stated in the introduction, a relation between exposure to (1→3)-β-*D*-glucan and an effect in field studies does not imply causality in view of the number of other agents present on the mold cell wall and the concomitant presence of other microbes and agents. An attempt can be made to evaluate the relationships found in view of the particular toxicological properties of (1→3)-β-*D*-glucan as summarized above.

Several studies have found a relation between symptoms of inflammation and irritation and increased levels of (1→3)-β-*D*-glucan (Rylander et al., 1992, Douwes et al., 2000). The data presented earlier do not support a conclusion that (1→3)-β-*D*-glucan causes this response. It is likely that it is caused by a concomitant presence of endotoxin or some other inflammagenic agent. There is thus evidence that indoor air dust can induce production of Th1 type as well as Th2 type cytokine secretion. The final outcome in a specific case would depend on the exposure balance between different agents in the dust.

It is also likely that exposure to molds or to (1→3)-β-*D*-glucan in connection with other inflammagenic agents may worsen the airway inflammation already present among atopic or allergic persons (Rylander et al., 1998). Evidence for such a mechanism is also present from studies on asthmatic persons who deteriorate in the clinical status when they are exposed to endotoxin in their homes (Michel et al., 1996). The consequences for children living in moldy homes is obvious in view of the potential influence on the maturation of their immune system (Holt, 1999).

Some studies have found an increased incidence of infections among children exposed to molds in their home or school environment (Pirhonen et al., 1996; Savilahti et al., 2000; Taskinen et al., 1999). Against the model as suggested in Figure 3.4, it can be hypothesized that the increased risk for infection, when present in a moldy environment, is caused by the $(1\rightarrow3)$-β-D-glucan component of the mold through a depression of the defence mechanisms to infectious agents. Contradictory data are, however, present from animal and *in vitro* experiments. In a mouse model, protective effects were reported against infections with *Mycobacterium bovis* (Hetland et al., 1998) and *Streptococcus pneumoniae* (Hetland et al., 2000). Another study reported protection against viral infections by oral administration of $(1\rightarrow3)$-β-D-glucan (Itoh, 1997). The dose levels in the latter experiments were, however, much larger that those encountered in the environment. Conclusions regarding the role of $(1\rightarrow3)$-β-D-glucan for the increased risk for infection after exposure to molds thus cannot be drawn before the importance of the dose level has been further elucidated.

Several reports suggest that lymphocytes are affected by exposure to pure $(1\rightarrow3)$-β-D-glucan (Fogelmark et al., 1992; Beijer et al., 2002). In occupational settings an exposure to high levels of molds causes lymphocytosis in the airways and at later stages hypersensitivity pneumonitis (Richerson, 1994). This reversible granulomatous pneumonitis is initially characterized by an accumulation of mononuclear cells in the interstitium with the formation of granulomas. In asymptomatic persons, there may be a lymphocytosis in the airways (Larsson et al., 1988). In view of the effects by pure $(1\rightarrow3)$-β-D-glucan exposure on lymphocytes and the indications of macrophage and mononuclear cell dysfunction after such exposure, it is tempting to speculate that $(1\rightarrow3)$-β-D-glucan is the causative agent. Support for this hypothesis is found in animal experiments where injection of $(1\rightarrow3)$-β-D-glucan was found to induce granulomas in the lung (Johnson et al., 1984).

An extrapolation of the effects of pure $(1\rightarrow3)$-β-D-glucan on antibody formation suggests an alteration of the reaction pattern of the antibody response and specific as well as nonspecific IgE (Rylander and Holt, 1998; Wan et al., 1999). In field studies an increased proportion of persons with atopic sensitization has been reported from environments with high levels of mold exposure (Beijer et al., 2005; Garrett et al., 1998; Savilahti et al., 2001). Several studies show a relation between total IgE levels in serum and exposure to molds (Müller et al., 2002; Beijer et al., 2005; Su et al., 2004). These effects could reflect a lowered resistance to inhaled environmental antigens by the effect of $(1\rightarrow3)$-β-D-glucan on the immune system as evidenced in the animal experiments described previously and summarized in Figure 3.4.

3.7 ENVIRONMENTAL RISK EVALUATION

Risk assessments ultimately aim to identify exposure levels that can be used to eliminate risk or to monitor preventive measures. Regarding $(1\rightarrow3)$-β-D-glucan, it has previously been underlined that some of the effects related to levels of this agent in the environment do not reflect causality. This does not exclude that (1→3)-β-*D*-glucan could be used as an indicator of exposure and thus fulfill an important purpose for risk assessment and prevention.

In an occupational setting, a dose-response relationship was found between the number of blood lymphocytes among garbage collectors and levels of (1→3)-β-*D*-glucan (Thorn et al., 1998). In indoor environments, a study on flats and offices found an increased prevalence of symptoms of cough, irritation in the nose, and itching skin at airborne levels of (1→3)-β-*D*-glucan of 0.49 ng/m^3 (Rylander et al., 1992). In a day care center, airway responsiveness decreased when the building was repaired, with (1→3)-β-*D*-glucans levels decreasing from 11.4 to 1.2 ng/m^3 (Rylander 1997). In a study of persons living in house with suspected mold problems, there was a higher number of cytotoxic T-cells in blood among persons living in houses with (1→3)-β-*D*-glucan levels higher than 4 ng/m^3 (Beijer et al., 2004).

In summary, the few studies available suggest that future studies of humid or moldy buildings should focus on (1→3)-β-*D*-glucan in the range of 1–2 ng/m^3 for a possible establishment of a threshold value for effects. A prerequisite for a further advance in this area is the use of uniform collection and analysis methods for (1→3)-β-*D*-glucan.

3.8 CONCLUSION

This review of responses to inhaled (1→3)-β-*D*-glucan demonstrates effect characteristics that are different from those induced by classic inflammagenic agents, such as endotoxin. Prolonged exposures and the simultaneous presence of inflammagenic agents seem to be prerequisites for several effects that could involve an increased risk for hypersensitivity pneumonitis, infection, and atopic sensitization.

REFERENCES

Adachi, Y., Okazaki, M., Ohno, M., Ohno, N., and Yadomae, T. (1997). Leukocyte activation by (1→3)-β-D-glucans. *Mediators Inflam.* 6, 251–256.

Beijer, L., Thorn, J., and Rylander, R. (2002). Effects after inhalation of (1→3)-β-D-glucan and relation to mold exposure in the home. *Mediators Inflam.* 11, 149–153.

Beijer, L., Thorn, J., and Rylander, R. (2004). Mold exposure at home relates to inflammatory markers in blood. *Eur. Resp. J.* 21, 317–322.

Beijer, L. and Rylander, R. (2004). (1→3)-β-D-Glucan does not induce acute inflammation after nasal deposition in pollen sensitive persons. *Mediators Inflam.* 14, in print.

Beijer, L., Larsson, L., Eduard, W., Szponar, B., and Rylander, R. (2004). Microbial cell wall agents indoors – measurements and relations to inflammatory markers. *Eur. Resp. J.* (submitted).

Brown, G.D., Herre, J., Williams, D.L., Willment, J.A., Marshall, S.J., and Gordon, S. (2003). Dectin-1 mediates the biological effects of β-glucans. *J. Exp. Med.* 197, 1119–1124.

Douwes, J., Zuidhof, A., Doekes, G., van deer Zee, S., Wouters, I., Boezen, H.M., and Brunekreef, B. (2000). (1→3)-β-D-Glucan and endotoxin in house dust and peak flow variability in children. *Am. J. Respir. Crit. Care Med.* 162, 1348–1354.

Fogelmark, B., Lacey, J., and Rylander, R. (1991). Experimental allergic alveolitis after exposure to different microorganisms. *Int. J. Exp. Path.* 72, 387–395.

Fogelmark, B., Goto, H., Yuasa, K., Marchat, B., and Rylander, R. (1992). Acute pulmonary toxicity of inhaled (1→3)-β-D-glucan and endotoxin. *Agents Actions* 35, 50–56.

Fogelmark, B., Sjöstrand, M., and Rylander, R. (1994). Pulmonary inflammation induced by repeated inhalations of (1→3)-β-D-glucan and endotoxin. *Int. J. Exp. Path.* 75, 85–90.

Fogelmark, B., Thorn, J., and Rylander, R. (2001). Inhalation of (1→3)-β-D-glucan causes airway eosinophilia. *Mediators Inflamm.* 10, 13–19.

Garrett, M.H., Rayment, P.R., Hooper, M.A., Abramson, M.J., and Hooper, B.M. (1998). Indoor airborne fungal spores, house dampness and association with environmental factors and respiratory health in children. *Clin. Exp. Allergy* 28, 459–467.

Hetland, G., Lovik, M., and Wiker, H.G. (1998). Protective effect of beta-glucan against *Mycobacterium bovis*, BCG infection in BALB/c Mice. *Scand. J. Immunol.* 47, 584–553.

Hetland, G., Ohno, N., Aaberge, I.S., and Lovik, M. (2000). Protective effect of beta-glucan against systemic *Streptococcus pneumoniae* infection in mice. *FEMS Immunol. Med. Microbiol.* 27, 111–116.

Holt, P.G. (1999). Potential role of environmental factors in the etiology and pathogenesis of atopy: a working model. *Env. Health Persp.* 107, 485–487.

Ishibashi, K.I., Miura, N.N., Adachi, Y., Ohno, N., and Yadomae, T. (2001). Relationship between solubility of grifolan, a fungal 1,3-β-D-glucan, and production of tumor necrosis factor by macrophages *in vitro*. *Biosci. Biotechnol. Biochem.* 65, 1993–2000.

Ishibashi, K.I., Miura, N.N., Adachi, Y., Ogura, N., Tamura, H., Tanaka, S., and Ohno, N. (2002). Relationship between the physical properties on *Candida albicans* cell wall beta-glucan and activation of leukocytes *in vitro*. *Int. Immunopharmacol.* 2, 1109–1122.

Itoh, W. (1997). Augmentation of protective immune responses against viral infection by oral administration of schizophyllan. *Mediators Inflamm.* 6, 267–269.

Johnson, K.J., Glovsky, M., and Schrier, D. (1984). Pulmonary granulomatous vasculitis induced in rats by treatment with glucan. *Am. J. Path.* 114, 515–516.

Korpi, A., Kasanen, J.P., Kosma, V.M., Rylander, R., and Pasanen, A.L. (2003a). Slight respiratory irritation but not inflammation in mice exposed to (1→3)-β-D-glucan aerosols. *Mediators Inflamm.* 12, 139–146.

Korpi, A., Kasanen, J.P., Raunio, P., and Pasanen, A.L. (2003b). Acute effects of *Aspergillus versicolor* aerosols on murine airways. *Indoor Air* 13, 260–266.

Larsson, K., Malmberg, P., Eklund, A., Belin, L., and Blaschke, E. (1988). Exposure to microorganisms, airway inflammatory changes and immune reactions in asymptomatic dairy farmers. *Int. Arch. Allergy Immunol.* 87, 127–133.

Michel, O., Kips, J., Duchateau, J., Vertongen, F., Robert, L., Collet, H., Pauwels, R., and Sergysels, R. (1996). Severity of asthma is related to endotoxin in house dust. *Am. J. Respir. Crit. Care Med.* 154, 1641–1646.

Müller, A., Lehmann, I., Seiffart, A., Diez, U., Wetzig, H., Borte, M., and Herbarth, O. (2002). Increased incidence of allergic sensitisation and respiratory diseases due to mold exposure; results from the Leipzig allergy risk study (LARS). *Int. J. Hyg. Environ. Health* 204, 363–365.

Nakagawa, Y., Ohno, N., and Murai, T. (2003). Suppression by *Candida albicans* β-glucan of cytokine release from activated human monocytes and from T cells in the presence of monocytes. *J. Inf. Dis.* 187, 710–713.

Ohno, N., Uchiyama, M., Tsuzuki, A., Tokunaka, K., Miura, N.N., Adachi, Y., Aizawa, W., Tamura, H., Tanaka, S., and Yadome, T. (1999). Solubilisation of yeast cell-wall (1→3)-β-D-glucan by sodium hypochlorite oxidation and dimethyl sulfoxide extraction. *Carbohydr. Res.* 316, 161–172.

Ormstad, H., Groeng, E.C., Lovik, M., and Hetland, G. (2000). The fungal wall component beta.1,3-glucan has an adjuvant effect on the allergic response to ovalbumin in mice. *J. Toxicol. Environ. Health* 61, 55–67.

Pirhonen, I., Nevalainen, A., Husman, T., and Pekkanen, J. (1996). Home dampness, molds and their influence on respiratory infections and symptoms in adults in Finland. *Eur. Respir. J.* 9, 2618–2622.

Richerson, H. (1994). Hypersensitivity penumonitis. In *Organic Dusts: Exposure, Effects and Prevention*, R. Rylander and R.R. Jacobs, eds., Lewis Publishing, Boca Raton, FL, pp. 139–160.

Rylander, R., Persson, K., Goto, H., Yuasa, K., and Tanaka, S. (1992). Airborne beta-1,3-glucan may be related to symptoms in sick buildings. *Indoor Environ.* 1, 263–267.

Rylander, R. (1996). Airway responsiveness and chest symptoms after inhalation of endotoxin or (1→3)-β-D-glucan. *Indoor Built. Environ.* 5, 106–111.

Rylander, R. (1997). Airborne (1→3)-β-D-glucan and airway disease in a day care center before and after renovation. *Arch. Environ. Health* 52, 281–285.

Rylander, R., Norrhall, M., Engdahl, U., Tunsäter, A., and Holt, P.G. (1998). Airways inflammation, atopy, and (1→3)-β-D-glucan exposures in two schools. *Am. J. Respir. Crit. Care Med.* 158, 1685–1687.

Rylander, R. and Holt, P.G. (1998). (1→3)-β-D-glucan and endotoxin modulate immune response to inhaled allergen. *Mediators Inflam.* 7, 105–110.

Rylander, R., Thorn, J., and Attefors, R. (1999a). Airways inflammation among workers in a paper industry. *Eur. Resp. J.* 13, 1151–1157.

Rylander, R., Fogelmark, B., McWilliam, A., and Currie, A. (1999b). (1→3)-β-D-glucan may contribute to pollen sensitivity. *Clin. Exp. Immunol.* 115, 383–384.

Savilahti, R., Uitti, J., Laippala, P., Husman, T., and Roto, P. (2000). Respiratory morbidity among children following renovation of a water-damaged school. *Arch. Environ. Health* 55, 405–410.

Savilahti, R., Uitti, J., Roto, P., Laippala, P., and Husman, T. (2001). Increased prevalence of atopy among children exposed to mold in a school building. *Allergy* 56, 175–179.

Shahan, T.A., Sorenson, W.G., Paulauskis, J.D., Morey, R., and Lewis, D.M. (1998). Concentration and time dependent upregulation and release of the cytokines MIP-2, KC, TNF and MIP-1α in rat alveolar macrophages by fungal spores implicated in airway inflammation. *Am. J. Respir. Cell Mol. Biol.* 18, 435–440.

Sherwood, E.R., Williams, D.L., McNamee, R.B., Jones, E.L., Browdes, I.W., and Di Luzio, N.R. (1987). Enchantment of interleukin-1 and interleukin-2 production by soluble glucan. *Int. J. Immunopharmac.* 9, 261–267.

Sjöstrand, M. and Rylander, R. (1997). Pulmonary cell infiltration after chronic exposure to (1→3)-β-D-glucan and cigarette smoke. *Inflamm. Res.* 46, 93–97.

Snella, M-C. (1986). Production of a neutrophil chemotactic factor by endotoxin stimulated alveolar macrophages *in vitro*. *Br. J. Exp. Path.* 67, 801–807.

Stone, B.A. and Clarke, A.E. (1994). *Chemistry and Biology of (1→3)-β-D-Glucans*, La Trobe University Press, Victoria, Australia.

Su, H-J., Wu, P-C., Lei, H-Y., and Wang, J-Y. (2004). The allergic sensitisation of asthmatic children with exposure to domestic fungi and dust mite allergen. Personal communication.

Tada, H., Nemoto, E., Shimauchi, H., Watanabe, T., Mikami, T., Matsumota, T., Ohno, N., Tamura, H., Shibata, K.I., Akashi, S., Miyake, K., Sugawara, S., and Takada, H. (2002). *Saccharomyces cerevisiae* and *Candida albicans* derived mannan induced production of tumor necrosis factor alpha by human monocytes in a CD14 and toll like receptor 4-dependent manner. *Microbiol. Immunol.* 46, 503–512.

Taskinen, T., Hyvärinen, A., Meklin, T., Husman, T., Nevalainen, A., and Korppi, M. (1999). Asthma and respiratory infections in school children with special reference to moisture and mold problems in the school. *Acta. Paediatrica.* 88, 1373–1379.

Thorn, J., Beijer, L., and Rylander, R. (1998). Airways inflammation and glucan exposure among household waste collectors. *Am. J. Ind. Med.* 33, 463–470.

Thorn, J., Beijer, L., and Rylander, R. (2001). Effects after inhalation of (1→3)-β-D-glucan in healthy humans. *Mediators Inflam.* 11, 173–178.

Wan, G-H., Li, C-S., Guo, S-P., Rylander, R., and Lin, R-H. (1999). An airborne mold-derived product, (1→3)-β-D-glucan, potentiates airway allergic responses. *Eur. J. Immunol.* 9, 2491–2497.

Williams, D.L. (1997). Overview of (1→3)-β-D-glucan immunobiology. *Mediators Inflam.* 6, 247–250.

Young, S.H., Robinson, V.A., Barger, M., Frazer, D.G., and Castranova, V. (2003a). Partially opened triple helix is the biologically active conformation of (1→3)-β-glucans that induces pulmonary inflammation in rats. *J. Toxicol. Environ. Health* 55, 551–563.

Young, S.H., Robinson, V.A., Barger, M., Whitmer, M., Porter, D.W., Frazer, D.G., and Castranova, V. (2003b). Exposure to particulate (1→3)-β-glucans induces greater pulmonary toxicity than soluble (1→3)-β-glucan in rats. *J. Toxicol. Environ. Health* 66, 25–38.

4 Animal Model of (1→3)-β-Glucan-Induced Pulmonary Inflammation in Rats

Shih-Houng Young and Vincent Castranova

CONTENTS

4.1 Introduction: Why Study (1→3)-β-Glucans?...66
4.2 What is the Cause of Controversy Regarding Glucan-Induced Pulmonary Inflammation Studies?..67
4.3 Important Factors Determining the Biological Activity of (1→3)-β-Glucans 67
    4.3.1 Type of Bond Linkage 67
    4.3.2 Higher Molecular Weight..68
    4.3.3 Degree of Branching (DB) 68
    4.3.4 Conformation 69
    4.3.5 Solubility of Polysaccharides..69
4.4 Why Choose Zymosan as the Test Glucan in Animal Studies? 70
4.5 Similarities between Symptoms Observed in Workers and Responses in an Animal Model 70
4.6 Parameters Monitored in the Animal Model...71
    4.6.1 Breathing Frequency 71
    4.6.2 Cell Differentials in Bronchoalveolar Lavage Fluid 71
    4.6.3 Lung Damage Indicators in Bronchoalveolar Lavage Fluid...........71
        4.6.3.1 Lactate Dehydrogenase (LDH)..71
        4.6.3.2 Albumin Content 72
    4.6.4 Oxidant Production 72
        4.6.4.1 Alveolar Macrophage Chemiluminescence....................72
        4.6.4.2 Nitric Oxide Release from AMs 72
4.7 Dose-Response Relationship of Zymosan-A-Induced Pulmonary Inflammation...73
4.8 Time Course of Recovery from Zymosan A Exposure.................................74
4.9 Which Form of Zymosan, Soluble or Insoluble, Causes Greater Inflammation? 75

4.10 Which Conformation of Particulate Zymosan A, Partially Open Triple Helix or Closed Triple Helix, Induces Greater Pulmonary Inflammation in Rats?...78
    4.10.1 NaOH-Treated Zymosan Annealed for 9 Days is Less Inflammatory than Fresh NaOH-Treated Zymosan..........................80
    4.10.2 Annealing of Zymosan Does Not Induce Changes in Pulmonary Responses 80
    4.10.3 Inhibition of Annealing by Freezing Retains the Potency of Fresh NaOH-Treated Zymosan 81
    4.10.4 Summary of Findings from Conformation Studies.........................83
4.11 Conclusions from Zymosan-Induced Pulmonary Inflammation Studies......83
4.12 Pretreatment with (1→3)-β-Glucans Modifies Endotoxin Response...........84
    4.12.1 Regimen Determination Following Combined Exposure to Zymosan and LPS 85
    4.12.2 Comparison of Results from Combined Treatment Groups to Calculated Expected Values..86
4.13 Conclusion and Need for Developing Methods for Analyzing Insoluble Glucans..87
References..89

4.1 INTRODUCTION: WHY STUDY (1→3)-β-GLUCANS?

Exposure to organic dust has been associated with pulmonary inflammatory symptoms (Rylander and Jacobs, 1994). Two of the most common etiological agents that have been identified as possible respiratory biohazards are endotoxin (lipopolysaccharide [LPS]) and (1→3)-β-glucans (Young et al., 1998; Zejda and Dosman, 1993). Endotoxin is a known potent stimulant of inflammatory responses. High endotoxin levels, measured by the Limulus amebocyte lysate (LAL) assay, are often reported for organic dusts exhibiting inflammatory potential. However, measured endotoxin content in organic dusts does not always correlate well with the magnitude of pulmonary inflammation measured in animal models (Castranova et al., 1991). Thus other components of organic dusts appear to contribute to the observed symptoms. There are reasons to consider (1→3)-β-glucans as a possible etiological candidate beside endotoxin. For example, (1→3)-β-glucans often coexist with endotoxin in organic dusts. They share many similarities: (1) both are cell wall constituents, (2) they have polysaccharides in their structure, and (3) both are positive in the LAL assay. In 1989, Rylander was the first to hypothesize that a correlation between (1→3)-β-glucans and pulmonary inflammation in workers exposed to organic dusts may exist. This hypothesis has since been supported by several studies by his group (Fogelmark and Rylander, 1993; Fogelmark et al., 1997; Rylander et al., 1989a and 1989b; Rylander and Goto, 1991; Rylander and Peterson, 1993; Rylander and Fogelmark, 1994; Rylander and Lin, 2000). However, there are controversies in the literature concerning the inflammatory potency of glucans. For example, inhalation of curdlan, a (1→3)-β-glucan from soil bacteria, by guinea pigs did not result in pulmonary inflammation in either acute (Fogelmark et al., 1992) or chronic

(Fogelmark et al., 1994) experiments. On the other hand, inhalation of a purified (1→3)-β-glucan product, zymosan A, resulted in an acute inflammatory response (Robinson et al., 1996). The differences between these two results maybe not be simply due to the use of different types of glucans. Since intratracheal exposure to curdlan resulted in a dose-dependent pulmonary reaction (Schuyler et al., 1998). In light of these conflicting results, there is a strong need to further investigate the role of (1→3)-β-glucans in pulmonary inflammatory symptoms associated with exposure to organic dusts.

4.2 WHAT IS THE CAUSE OF CONTROVERSY REGARDING GLUCAN-INDUCED PULMONARY INFLAMMATION STUDIES?

A major source of the controversy concerning the inflammatory potency of glucans may come from the difficulty in characterizing the physicochemical properties of (1→3)-β-glucans. (Please refer to Chapter 1.) Without a well-characterized glucan sample, there is no valid comparison between studies. This is because the biological activity of glucans is dependent on several factors: types of bond linkages, molecular weight, degree of branching, and conformation. Beside the above factors, solubility may influence the potency of (1→3)-β-glucans in biological systems. Thus, in order to estimate correctly the health hazards of glucans, it is important to investigate one with a representative biological activity. Results can be quite different for different glucans. Even if two experiments used glucans from the same source, results still could be quite different if the glucans studied were different in molecular weight or conformation. These important factors are discussed in more detail in the following section.

4.3 IMPORTANT FACTORS DETERMINING THE BIOLOGICAL ACTIVITY OF (1→3)-β-GLUCANS

4.3.1 TYPE OF BOND LINKAGE

D-glucans are polymers of D-glucopyranose joined by glucosidic linkages between the hemiacetal oxygen at C-1 on one monosaccharide residue and one of the four hydroxyls at C-2, C-3, C-4, or C-6 on the next glucose residue (Figure 4.1). Theoretically, eight different glucans, with respect to linkage type, can be formed by a combination of the four possible linkage positions and two possible configurations (α and β) of the hemiacetal (anomeric) hydroxyl groups. Seven of the eight possible glucans, homogeneous with respect to linkage type, occur naturally; the exception is the (1→2)-β-glucans. Glucans containing both α- and β-linkages have only occasionally been reported. Glucans can have several different types of bond linkages, such as (1→2) or (1→3) or (1→4) or (1→6), and those with linkages of more than one positional type are quite common. Seljelid et al. (1981) have studied 42 different glucans *in vitro* with mouse macrophages. They found that (1→3)-β types of bond linkages stimulate macrophages most effectively; although, other types of bond

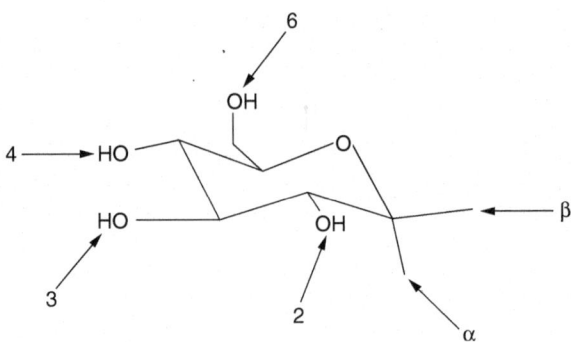

FIGURE 4.1 Possible intermolecular linkage position in D-glucose.

linkage could also cause macrophage stimulation, such as a (1→4), (1→6) linkage. Lichenan, a (1→3), (1→4)-glucan, demonstrated no significant antitumor activity except when modified with a β-glucopyranosyl unit (degree of branching 0.08) (Demleitner et al., 1992a, 1992b). This result supports the importance of the (1→3)-linked backbone.

4.3.2 Higher Molecular Weight

Glucans are polysaccharides with different chemical structures, but all exhibit immunomodulating activity. This fact suggests that the immune response is in part determined by size rather than by chemical structure (Whistler et al., 1976). The antitumor activity could be completely lost when degree of polymerization (DP) of a linear (1→3)-β-*D*-glucan is less than 38, as in the case of laminarin (Saito et al., 1991). Tanaka et al. (1991) found that, if the number average molecular weight (MW) exceeds a critical value of 6800 Da, glucans can activate factor-G. This activating ability increases with the increasing number average MW of linear glucans. Indeed, schizophyllan (SPG) exhibits antitumor activity against Sacroma 180 only when its MW is higher than 5×10^4 Da (Kojima et al., 1986).

4.3.3 Degree of Branching (DB)

Glucans isolated from fungus or yeast often contain varying degrees of side-chain branching. Most of these side chains are (1→6) bond linkages. An increased degree of (1→6) side chain substitution not only increases the solubility in water but also affects the biological activity of glucans. Bohn and BeMiller (1995) found that the most active polymers for antitumor activity have a DB between 0.20 and 0.33. For example, SPG and lentinan have single β-D-glucopyranosyl unit 1→6 branches with a DB of 0.33 and are antitumorgenic. In contrast, highly branched glucans such as *Auricularia auricula judae* (DB 0.75) and *Pesstalotia* sp. 815 (DB 0.67) have limited activity against tumors. Likewise, the much less branched glucan extracted from *Ganoderm lucidum* with an alkaline solution without heating (DB 0.06) is only weakly active (Misaki et al., 1993).

The structure of the branch substituents has only a slight influence on the antitumor activity of the polysaccharide derivatives (Kraus and Franz, 1992). The influence of glycosyl units attached to O-6 has been investigated (Demleitner et al., 1992a). Curdlan, a linear (1→3)-β-glucan, and lichenan, a linear mixed-linkage (1→3), (1→4)-β-D-glucan with 33% (1→3) and 66% (1→4) linkage, were modified with β-D-glucopyrqanosyl, α-L-arabinofurosyl, α-L-rhamnosyl, and β-gentiobiosyl units added to O-6. Curdlan modified with D-glucosyl, L-rhamnosyl, L-arabinosyl, and, to a lesser extent, gentiobiosy branch units significantly inhibited sarcoma 180 tumors. However, only D-glucosyl-modified lichenan showed significant antitumor activity. The other lichenan derivatives showed little or no antitumor activity and, in some cases, stimulated tumor growth. For curdlan, the specific glucosyl substitution was not important for anti-tumor activity, since all derivatives showed a significant antitumor effect. Furthermore, when curdlan was etherified with sulfonylethyl, sulfonylpropyl, or sulfonylbutyl groups, it had significant antitumor activity against allogeneic sarcoma 180 (Demleitner et al., 1992b).

4.3.4 CONFORMATION

Three conformations of glucans in solution have been reported in the literature: single helix, triple helix, and random coil (Yadomae and Ohno, 1996). Recently, a closed triple helix and an opened triple helix conformation was reported by Young et al. (2000). Aketagawa et al. (1993) concluded that the MW is not the dominant factor inducing antitumor activity, provided that the MW is large enough to allow formation of the single helix conformation. Different conformers may have different biological activities toward different assays. Therefore, a comparison of the biological activity of glucans will also be dependent on the assay of choice. So far, there is no general agreement as to which conformer is the potent form for assays monitoring either antitumorgenic or pulmonary inflammatory potential. Some have suggested that the single helix conformation is the most effective for antitumor activity (Aketagawa et al., 1993; Yoshioka et al., 1992). Yoshioka et al. (1992) found that the random coil and triple helix of linear glucans have limited antitumor activity. Other studies have supported the triple helix as the important conformer (Bohn and BeMiller, 1995). These discrepancies may arise from difficulties in identifying the conformation in solution.

4.3.5 SOLUBILITY OF POLYSACCHARIDES

Solubility of (1→3)-β-glucans may determine the availability of (1→3)-β-glucans in the study system. Therefore, it is important to understand the general concept of solubility of polysaccharides (Whistler, 1973). Polysaccharides consisting of one type of sugar unit uniformly linked in linear chains are usually water insoluble even when the molecules have low MW (DP 20–30). Exceptions are the (1→6)-linked homoglycans, which, because of the extra degrees of freedom provided by the rotation about the C-5 to C-6 bonds (Figure 4.1), give higher solution entropy values. Almost all low MW polysaccharides (DP less than 15–20) are soluble in water. Homoglycans with two types of sugar linkages or heteroglycans composed of two types of sugars are more soluble than purely homogeneous polymers. Highly branched polysaccharides are almost always soluble in water. (1→3)-β-glucan generally consists of only one type (1→3) of backbone bond linkage. Therefore, it is water-insoluble. The (1→6) branching increases the solubility of glucans.

4.4 WHY CHOOSE ZYMOSAN AS THE TEST GLUCAN IN ANIMAL STUDIES?

Zymosan has a long history associated with glucan research. The study of glucans originated from studies of zymosan. The term refers to crude yeast cell wall preparations, consisting chiefly of protein–carbohydrate complexes. Zymosan was first described in 1941 after Pillemer and Ecker found that the active component of yeast was an insoluble wall fraction and named it zymosan (Pillemer and Ecker, 1941). The same fraction interacted with properdin, another serum protein, which was also involved in killing infectious organisms (Pillemer and Ross, 1955). Benacerraf and Sebestyen (1957) presented a report on the influence of intravenously administered zymosan on stimulation of the reticulo-endothelial system (RES). Since then a range of immunological and physiological effects, including tumor regression, has been attributed to zymosan (Bradner et al., 1958). The history and status of the experimental work with zymosan up to 1964 was presented in a detailed review by Fizpatrick and DiCarlo (1964). In 1961, Riggi and Di Luzio (1961) reported that the active fraction of zymosan, which induced proliferation and activation of the RES, was (1→3)-β-D-glucan. Each component of zymosan has been tested for activity in stimulating the RES. Mannan and a $CHCl_3$:MeOH soluble fraction were reported to be inactive (Riggi and Di Luzio, 1961). On the other hand, alkali-insoluble yeast glucan preparations are consistently active; e.g., glucans prepared by the methods of Hassid et al. (1941) and Peat et al. (1958) are both active.

Considering the important factors for biological activity discussed in Section 4.3, zymosan has several advantages over other (1→3)-β-D-glucans. Zymosan is a purified form of (1→3)-β-D-glucan from *Saccharomyces cervisiae*. It has a high MW, and has some degree of branching (not being a linear glucan like curdlan) (DiCarlo and Fiore, 1957). The conformation of zymosan is generally considered to be a triple helix structure. The zymosan A sample used in experiments for our laboratory was obtained from Sigma Chemical Company (St. Louis, MO). The majority of particles were 2–5 μm in diameter, which is in the range of respirable particles. The particle size distribution in suspension was about the same as in inhalation studies (Robinson et al., 1996). Zymosan A was freshly prepared as an aqueous suspension before every experiment.

4.5 SIMILARITIES BETWEEN SYMPTOMS OBSERVED IN WORKERS AND RESPONSES IN AN ANIMAL MODEL

We have developed an animal model for studying pulmonary reactions to organic dust exposures (Castranova et al., 1996). We have demonstrated that guinea pigs and rats exhibit numerous pulmonary responses to cotton dust and other organic particles. Symptoms, observed both in human and animal models, included an increased breathing rate, pulmonary inflammation (measured as an increase in bronchoalveolar lavage [BAL] polymorphonuclear leukocytes), alveolar septal leakage (measured as an increase in BAL RBCs), airway obstruction, and activation of alveolar macrophages.

4.6 PARAMETERS MONITORED IN THE ANIMAL MODEL

4.6.1 BREATHING FREQUENCY

Organic toxic dust symptoms in workers are often associated with chest tightness and reduced forced expiratory volume in 1 second (FEV_1). Exposure of guinea pigs to aerosolized cotton dust has been shown to cause airway closure (Frazer et al., 1984). In addition, exposure of rats or guinea pigs to cotton dusts results in an increase in CO_2-challenged breathing frequencies (Ellakkani et al., 1984; Robinson et al., 1988). Such changes in breathing patterns have been associated with airway irritation (Castranova et al., 2002). Breathing frequencies were determined using a flow plethysmograph (Frazer et al., 1997). The plethysmograph chamber was constructed from an acrylic tube enclosed at both ends. One end of the chamber had a circular port that contained four 400-mesh stainless steel screens. Pressure variations across the screens generated by flow into and out of the chamber were measured with a pressure transducer (Setra, Inc., Foxborough, MA). A digital oscilloscope (Tektronix, Inc., Wilsonville, OR) was used to record flow signals, which were transferred to a digital computer for analysis. To measure breathing frequencies, a rat was placed in the flow plethysmograph, equilibrated with 10% CO_2, and its average breathing frequency calculated based on the time between zero crossings of the flow signal at the beginning and end of an inhalation–exhalation cycle.

4.6.2 CELL DIFFERENTIALS IN BRONCHOALVEOLAR LAVAGE FLUID

Polymorphonuclear (PMN) leukocyte infiltration, measured as an increase in neutrophil count in BAL samples, was observed in patients with organic dust toxic syndrome (ODTS) (Lecours et al., 1986). Similar results were noted in guinea pigs and rats exposed to various organic dust (Castranova et al., 1996). In animal studies, aliquots of BAL cell suspensions were taken for determination of total cell and differential cell counts using a Coulter Multisizer II and AccuComp software (Coulter Electronics, Hialeah, FL).

4.6.3 LUNG DAMAGE INDICATORS IN BRONCHOALVEOLAR LAVAGE FLUID

4.6.3.1 Lactate Dehydrogenase (LDH)

The activity of LDH, a cytosolic enzyme, in the first acellular BAL fluid was assayed to evaluate general lung cell damage (Sendelbach and Witschi, 1987).

4.6.3.2 Albumin Content

Albumin content of the acellular BAL fluid from the first lavage was used to evaluate the damage/permeability of the alveolar–capillary barrier.

The albumin content and LDH activity were measured using an automated Cobas Fara II Analyzer (Roche Diagnostic Systems, Montclair, NJ). The albumin content was determined colorimetrically at 628 nm based on albumin binding to bromcresol green (albumin BCG diagnostic kit, Sigma Chemical Company, St. Louis, MO) and expressed as mg/ml BAL fluid. LDH activity was measured by the formation of NADH and expressed as U/L bronchoalveolar lavage fluid (BALF), using Roche Diagnostic reagents and procedures (Roche Diagnostic Systems).

4.6.4 OXIDANT PRODUCTION

4.6.4.1 Alveolar Macrophage Chemiluminescence

The alveolar macrophage (AM) chemiluminescence (CL) assay was conducted in a 0.25 ml reaction mixture of HEPES-buffered solution. Resting AM–CL was determined by incubating 0.25×10^6 AMs at 37°C for 20 min, then adding 0.008 mg% (w/v) luminol (Sigma Chemical Company, St. Louis, MO) followed by the measurement of CL for 15 min. To determine zymosan-stimulated AM–CL, the reaction mixture was modified to include 0.5 mg of unopsonized zymosan (Sigma Chemical Company, St. Louis, MO), which was added immediately prior to measure of CL. Measurement of AM–CL was conducted with an automated luminometer (Berthold Autolumat LB 953, Wallace, Inc., Gaithersburg, MD) at 390–620 nm for 15 min, and the integral of counts per minute (cpm) verses time was calculated. Zymosan-stimulated CL was calculated as the cpm in the zymosan-stimulated assay minus the cpm in the resting assay. The nitric oxide (NO)-dependent CL was determined by adding an inhibitor of nitric oxide synthase, 1 mM N-nitro-L-arginine methyl ester HCl (L-NAME, Sigma Chemical Company, St. Louis, MO), to the cells prior to preincubation. NO-dependent CL was calculated as the difference between zymosan-stimulated CL measurements in the absence and presence of L-NAME.

4.6.4.2 Nitric Oxide Release from AMs

Nitric oxide was determined as nitrite (NO^{2-}) with Griess reagent as described by Green et al. (1982). Specifically, AM-conditioned media (100 µl) were collected, mixed with 100 µl of Griess reagent (0.5% sulfanilamide, 0.05% naphthylethylenediamine dihydrochloride, 2.5% H_3PO_4) and analyzed spectrophotometrically at 546 nm. Nitrite concentrations were determined using a standard curve prepared from sodium nitrite with a linear range from 1.6 µM to 206 µM. Three replicates were done for each sample in the experiment.

4.7 DOSE-RESPONSE RELATIONSHIP OF ZYMOSAN-A-INDUCED PULMONARY INFLAMMATION

Dose-response of zymosan-A-induced pulmonary inflammation in rats exposed to intratracheal instillation (IT) was reported by Young et al. (2001). The dose range of zymosan A tested was from 0–5 mg/kg body weight. Upon challenge with zymosan A, rats exhibited a dose-dependent pulmonary response at 1 day post IT. All the parameters measured were significantly elevated above saline control levels and had a good correlation with zymosan A dose. For example, post-IT enhancement of breathing frequencies and polymorphonuclear leukocytes harvested from lavage fluid strongly correlated with zymosan A concentration (r^2 = 0.95 and 0.99, respectively) (Figures 4.2 and 4.3). Albumin concentration and LDH activity also increased with increasing dose of zymosan A (r^2 = 0.80) (Table 4.1). Alveolar macrophage chemiluminescence increased with increasing dose of zymosan A, except for the highest dose (5 mg/kg) where cytotoxicity was observed (Table 4.1). The above results demonstrated that acute exposure of rats to zymosan A results in significant pulmonary inflammation, damage, and AM reaction.

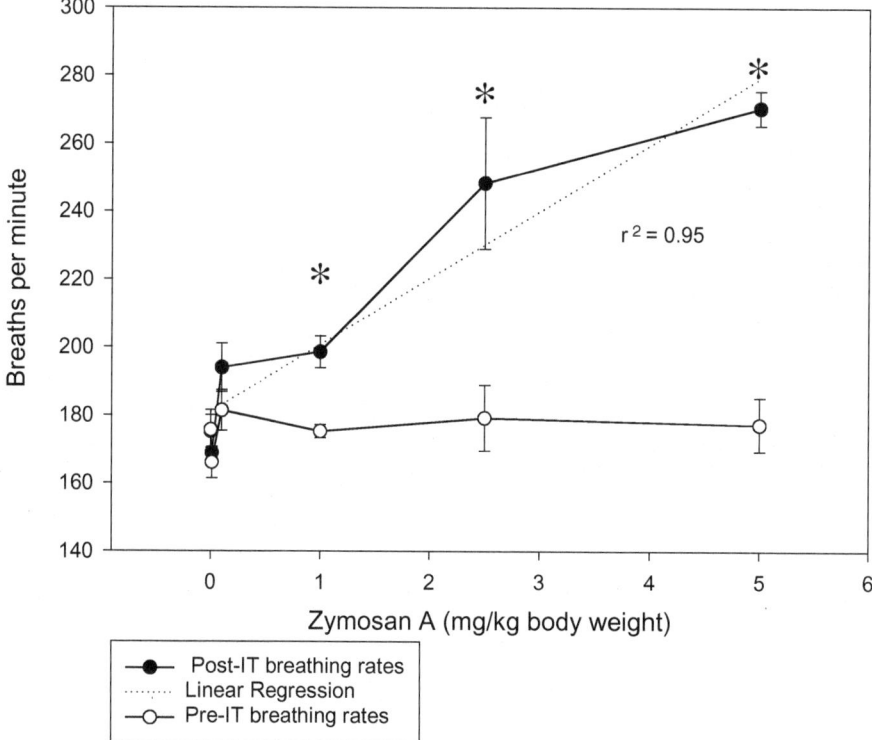

FIGURE 4.2 Breathing frequencies of rats in 10% CO_2 1 day after IT instillation with various doses of zymosan. "*" indicates a significant increase above the pre-IT level. The correlation coefficient (r^2) equals 0.95 between zymosan and 1 day post-IT breathing frequencies. Values are means ± SEM of 3 to 6 rats (Young et al., 2001).

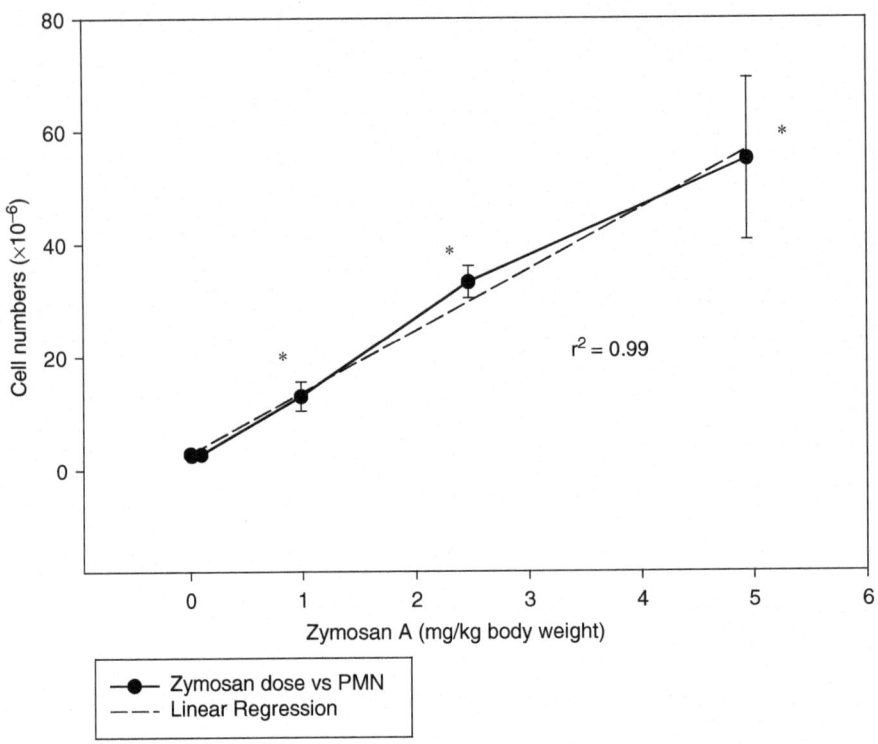

FIGURE 4.3 Polymorphonuclear (PMN) leukocytes obtained by BAL of rats one day after IT instillation with various doses of zymosan A. "*" indicates a significant increase above the control PMN level. Values are means ± SEM of 3 to 6 rats. $r^2 = 0.99$ between zymosan A and 1 day post-IT PMN cells (Young et al., 2001).

4.8 TIME COURSE OF RECOVERY FROM ZYMOSAN A EXPOSURE

The recovery from a single intratracheal administration of zymosan A (2.5 mg/kg) was monitored for 1 week (Young et al., 2001). A decrease in all parameters was observed from day 1 to day 4 post IT. By day 7 post IT, all parameters monitored were no different than the saline control levels. These decreases were correlated well with days post IT (Table 4.2). For example, the decrease in post-IT breathing frequencies had a correlation coefficient of 0.97 with days post-IT (Figure 4.4). PMN harvested by BAL was decreased with increasing time after a single IT exposure, which suggested recovery from inflammation (Figure 4.5). A good correlation ($r^2 = 0.8$) between recovery of PMN in BAL, CL, or nitric oxide production and the days after exposure was observed (Figure 4.5 and Table 4.2). Albumin level and LDH activity were still elevated for the first three days after exposure. Only at day 7 after instillation did the albumin level and LDH activity significantly decrease from the day 1 level, indicating a delayed recovery from injury.

TABLE 4.1
Pulmonary Responses to Zymosan A Exposure

Zymosan (mg/kg)	Albumin (mg/ml)	LDH Activity (U/L)	Chemiluminescence (cpm x 10^{-5}/0.25 x 10^6 AM/15 min)
Control	0.22 ± 0.03	56.6 ± 4.4	2.6 ± 1.1
0.01	0.20 ± 0.02	47.3 ± 4.2	6.7 ± 4.9
0.1	0.24 ± 0.06	47.0 ± 4.2	3.6 ± 1.4
1	0.32 ± 0.06	82.0 ± 13.8	10.7 ± 2.6*
2.5	0.41 ± 0.03*	95.0 ± 6.5*	12.7 ± 1.2*
5	0.66 ± 0.17*	108 ± 13.3*	8.2 ± 4.0

Values are expressed as means ± SEM; n = 3–6 rats.

*Indicates a significant increase from control level ($p < 0.05$).

Data modified from Young et al. (2001).

TABLE 4.2
Recovery of Pulmonary Responses after Zymosan A Exposure (2.5 mg/kg IT)

Time (Days Post-Exposure)	Albumin (mg/ml)	LDH Activity (U/L)	Chemiluminescence (cpm x 10^{-5}/0.25 x 10^6 AM/15 min)	Nitric Oxide (μM)
Control	0.13 ± 0.01	26.7 ± 2.2	2.4 ± 0.6	4.5 ± 0.9
1	0.42 ± 0.05	112.2 ± 15.8	17.7 ± 5.4	166.4 ± 18.4
2	0.38 ± 0.09	187.2 ± 24.1	17.3 ± 2.7	164.6 ± 17.1
3	0.57 ± 0.03	177.7 ± 39.8	8.1 ± 1.7*	142.6 ± 30.0
4	0.41 ± 0.05	109.6 ± 15.6	2.9 ± 0.3*	102.9 ± 13.1*
7	0.15 ± 0.01*	31.8 ± 5.5*	0.6 ± 0.1*	5.0 ± 3.2*

Values are expressed as means ± SEM; n = 8–10 rats.

*Significantly decreased from day 1 level ($p < 0.05$).

Data modified from Young et al. (2001).

4.9 WHICH FORM OF ZYMOSAN, SOLUBLE OR INSOLUBLE, CAUSES GREATER INFLAMMATION?

Although the particulate form of (1→3)-β-glucans is found in nature, many labs have reported the effects of solubilized (1→3)-β-glucans in the literature. However, the toxicological characteristic of soluble (1→3)-β-glucans may not be comparable

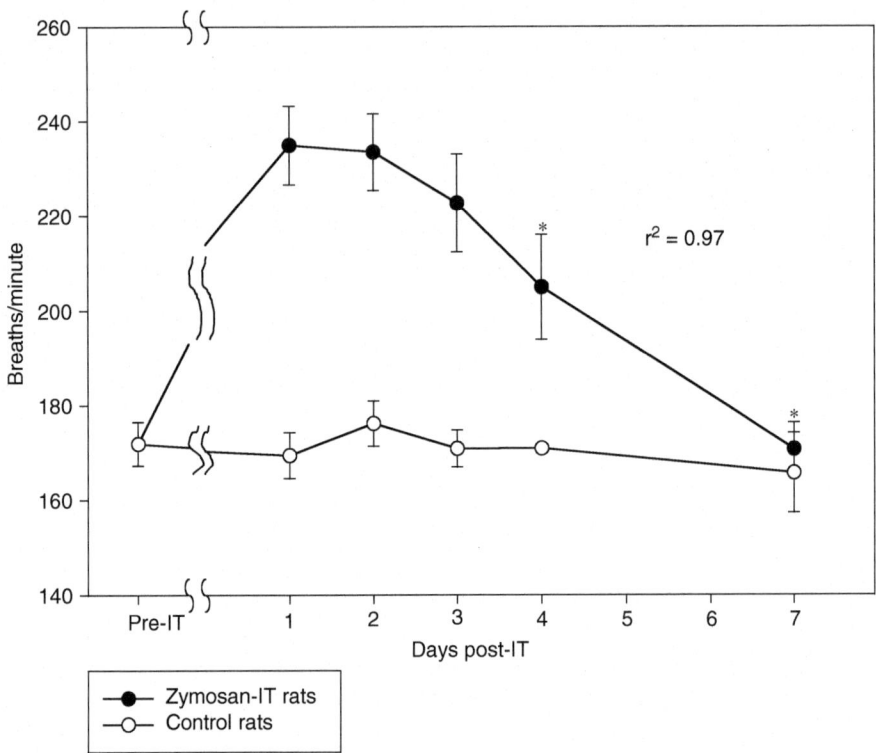

FIGURE 4.4 Recovery of breathing frequency after zymosan A exposure. Breathing frequencies of rats were measured at 1–7 days after a single IT exposure to zymosan A (2.5 mg/kg). "*" indicates a significant decrease from the 1 day post level. $r^2 = 0.972$ between breathing frequencies and days after zymosan A exposure. Values are means ± SEM of 8 to 28 rats (Young et al., 2001).

with the original particulate form of (1→3)-β-glucans. Four major types of methodologies have been used to solublize particulate (1→3)-β-glucans. These methods include: heating, solublization in acid, solublization in base, or use of organic solvents. While solubilizing (1→3)-β-glucans, each method also may modify the toxicological properties of (1→3)-β-glucans to some degree. Therefore, caution is recommended when making comparisons with the original particulate form of (1→3)-β-glucans.

Among these four methods of solublization, heating and the use of acid are less desirable methods, because they both can reduce the molecular weight of (1→3)-β-glucans. Decreased molecular weight significantly reduces the immunological stimulating ability of glucans (Adachi et al., 1990; Nono et al., 1991; Ohno et al., 1995). Linear glucans, such as curdlan (no branching) and pachyman (few branches), were easily degraded to oligosaccharides by treatment for 20 min at 100°C with 90% formic acid (Ohno et al., 1995). In contrast, the highly-branched glucans were more resistant to degradation (Ohno et al., 1995). In comparison to acid and heating methods, a weak alkaline solution normally will not degrade (1→3)-β-glucans.

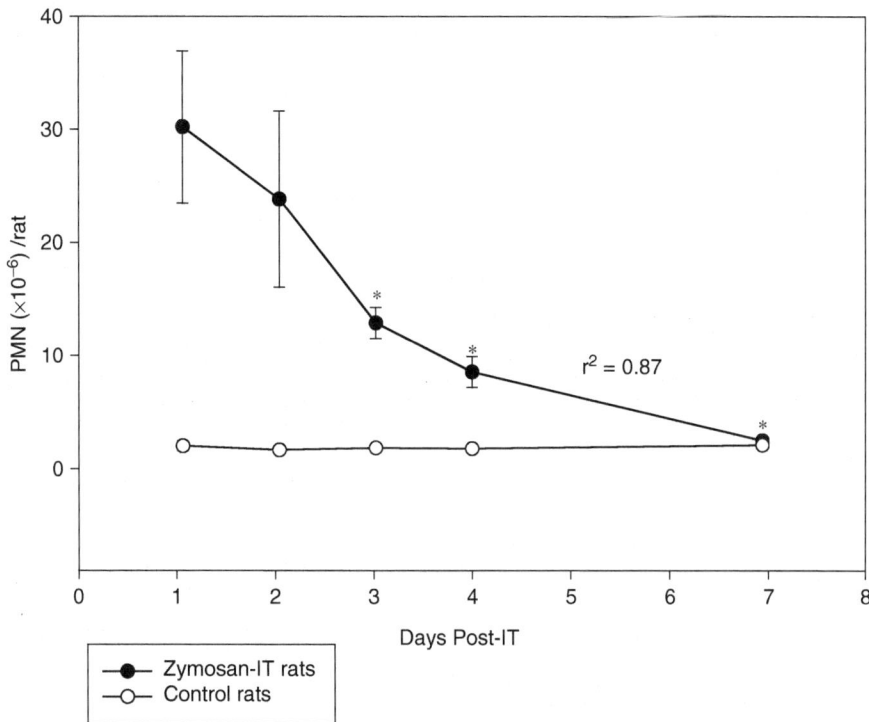

FIGURE 4.5 Recovery of PMN levels after zymosan A exposure. PMN harvested by BAL of rats 1–7 days after a single IT exposure to zymosan A (2.5 mg/kg). "*" indicates a significant decrease from the 1 day post level. $r^2 = 0.869$ between PMN and days after zymosan A exposure. Values are means ± SEM of 8 to 10 rats (Young et al., 2001).

Furthermore, the (1→6) branched glucans are more stable in alkaline solution compared to unbranched glucans (Whistler and BeMiller, 1958). However, a strong alkaline solution will change the conformation of glucans from a helix to a random coil. Dependent on the reannealing condition, it may or may not return to the original triple helix state. The fourth method of solublization is to use of organic solvents, such as 8 M urea or dimethyl sulfoxide. This method has been used for preparing different conformers of glucans. However, data indicate that glucans lyophilized from organic solvent can have a very different biological activity than the parent glucans (Maeda et al., 1988; Stone and Clarke, 1992). In addition, the nature of the resulting conformer is not always clear. In light of the above discussion, we choose NaOH as the least destructive method for solubilization in our studies.

We have examined the pulmonary inflammatory potential from the two fractions of NaOH-treated zymosan A (Young et al., 2003b). The effect of NaOH alone on inflammation was removed by neutralization and dialysis. A 30 min of 0.25 N NaOH treatment followed by neutralization and dialyzsis for 2 days using deionized water was chosen as the method to obtain two fractions of zymosan A. Because there is no direct method of measuring zymosan A concentration in the solution, an estimation of soluble zymosan A concentration was done in a separate but parallel experiment.

TABLE 4.3
Pulmonary Responses to Soluble or Insoluble NaOH-treated Zymosan

	Breathing Frequencies (breaths/min)	Albumin (mg/ml)	LDH Activity (U/L)	NO (μM)	Chemiluminescence (cpm x 10^{-5}/0.25 x 10^6 AM/15 min)
Control	159 ± 3	0.2 ± 0.02	32.8 ± 4	0 ± 0	4.68 ± 0.84
Soluble zymosan	176 ± 4	0.22 ± 0.02	59.3 ± 2.1	0 ± 0	15.08 ± 4.93
Insoluble zymosan	198 ± 6*	0.36 ± 0.04*	82.8 ± 14.5	65.0 ± 13.1*	40.32 ± 10.54*

Values are expressed as means ± SEM; n = 5–4 rats.

*Significantly different from soluble zymosan group ($p < 0.05$).

Data modified from Young et al. (2003b).

The soluble fraction of zymosan A was collected, and the dry weight of zymosan A was recovered after heating at 160°C for 2 hr. The dosage of soluble zymosan A administered by IT instillation was calculated to be 1.6 mg/kg body weight and was comparable to the dose of particulate zymosan, 1.9 mg/kg body weight, used in the study.

The results (Table 4.3) show that exposure to the insoluble fraction of NaOH-treated zymosan induced a significant increase in lung damage (BAL fluid albumin levels and LDH activity), inflammation (yield of PMN in BAL, Figure 4.6), pulmonary irritation (increase in breathing frequency), and AM activity (nitric oxide production and CL). In contrast, rats exposed to the NaOH-soluble fraction did not exhibit a significant increase in breathing rate or NO production and showed a smaller degree of damage and inflammation than rats exposed to the insoluble fraction of NaOH-treated zymosan. The results demonstrate that insoluble zymosan A is more potent in inducing pulmonary inflammation and damage in rats than the soluble form of this zymosan A. Therefore, the insoluble fraction of glucan is the dominating factor determining the inflammatory potency of glucan. Measurement of soluble fraction of glucan would not reflect the inflammatory potency of glucan.

4.10 WHICH CONFORMATION OF PARTICULATE ZYMOSAN A, PARTIALLY OPEN TRIPLE HELIX OR CLOSED TRIPLE HELIX, INDUCES GREATER PULMONARY INFLAMMATION IN RATS?

It is known that the biological activity of $(1\rightarrow 3)$-β-glucans is dependent on three factors: MW, DB, and molecular conformation (Bohn and BeMiller, 1995). Studies have shown that high MW (100,000~200,000 g/mol) glucans with a degree of branching of 0.20~0.33 are most active (Bohn and BeMiller, 1995). However, the relationship between conformation and biological activity is less clear. While many

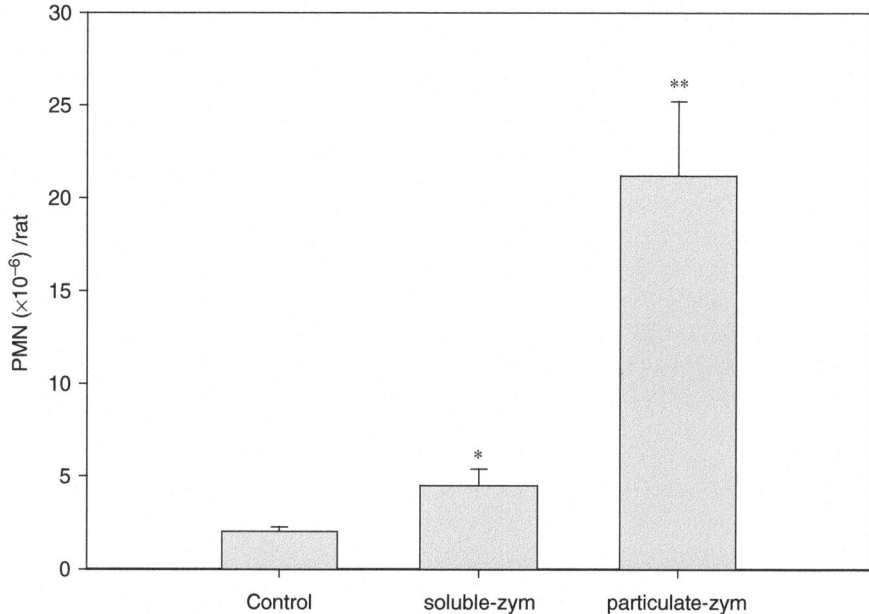

FIGURE 4.6 BAL PMN yield. Particulate zymosan induced a greater increase in PMN infiltration than soluble zymosan at 18 hr post-IT. "*" indicates a significant increase versus control ($p < 0.05$). "**" indicates a significant increase versus both control and soluble zymosan ($p < 0.05$). Values are means ± SEM of 4 to 5 rats (Young et al., 2003b).

different kinds of (1→3)-β-glucans have been examined for their biological activity (Ohno et al., 1998), their corresponding conformations were often unknown. Thus, a comparison of their biological activity based on conformation has been very difficult to make. In addition, not only might different conformations react differently for a given biological endpoint, but different conformations may exhibit different reactivities across biological endpoints. Therefore, we were interested in examining the role of different conformations of (1→3)-β-glucans in the induction of pulmonary inflammation in rats (Young et al., 2003a).

The native conformation of (1→3)-β-glucans is thought to be a triple helix conformation (Bluhm and Sarko, 1976; Chuah et al., 1983; Deslandes et al., 1980; Kashiwagi et al., 1981). The conversion between different conformations can be mediated by different chemical or physical treatments (Yadomae and Ohno, 1996). Sodium hydroxide (NaOH) treatment is a common method for conformation conversion of (1→3)-β-glucans. Previously, we reported that NaOH induced a conformational change from the closed triple helix to the partially opened triple helix in the glucan, laminarin, and that the biological activity of laminarin was related to the degree of strand opening of this (1→3)-β-glucan (Young et al., 2000). After treatment with NaOH, laminarin would gradually anneal back to the closed triple-helix conformation in about 8 days (Young et al., 2000). Therefore, by comparing the inflammatory potency of reannealing NaOH-treated zymosan A at different days posttreatment, we were able to investigate the role of conformation on biological activity.

The conformation status of reannealing zymosan was monitored by using a (1→3)-β-glucans-specific dye: aniline blue. The above methodology has an advantage that it only compares the effect of conformation by itself and not the effect of changes in MW or DB, because a mild NaOH treatment did not cause a change in MW (see our previous discussion).

Aniline blue was chosen to monitor the conformational status of (1→3)-β-glucans. Young et al. (2000) have reported that aniline blue can be used to detect the NaOH-induced conformational changes in schizophyllan, a closed triple helix glucan. Aniline blue contains an impurity, sirofluor (sodium carbonylbis[4-(phenyleneamino)benzenesulfonate]), that binds specifically to (1→3)-β-glucans (Evans et al., 1984). Aniline blue is a nonfluorescent dye; however, when bound to (1→3)-β-glucans, sirofluor fluoresces (ca. 140 times increase). This enhancement of fluorescence induced by dye binding to (1→3)-β-glucans is thought to occur via hydrogen binding rather than by hydrophobic binding (Thistlethwaite et al., 1986). The triple helix conformation of glucans has three hydrogen bonds with oxygen in the C-2 position (Chuah et al., 1983; Deslandes et al., 1980). Therefore, when the triple helix becomes partially opened, sirofluor can bind to hydrogen binding sites on the glucans and fluoresce. Thus, relative fluorescent intensity can serve as an indirect indicator of the conformational status of (1→3)-β-glucans. The fluorescent intensities of day 1 (fresh NaOH-treated zymosan) and day 9 (annealed NaOH-treated zymosan) zymosan–aniline blue were quite different (Figure 4.7).

Figure 4.7 shows that there is an increase of aniline blue binding sites after NaOH treatment (fresh NaOH-treated zymosan). In contrast, annealed NaOH-treated zymosan has fewer available binding sites for aniline blue. The corresponding pulmonary responses for these two preparations are shown in Table 4.4.

4.10.1 NaOH-Treated Zymosan Annealed for 9 Days is Less Inflammatory than Fresh NaOH-Treated Zymosan

Table 4.4 shows that fresh NaOH-treated zymosan induced significantly greater PMN infiltration and nitric oxide production than annealed NaOH-treated zymosan. These data support the hypothesis that an open triple helix is more potent in inducing pulmonary inflammation than a closed triple helix conformer of (1→3)-β-glucans. However, not every pulmonary index measured was sensitive to conformation changes in zymosan. Although breathing frequency and LDH or albumin (from first BAL fluid) were elevated above control after zymosan exposure, no significant difference in potency was found between the fresh and annealed NaOH-treated zymosan samples.

4.10.2 Annealing of Zymosan Does Not Induce Changes in Pulmonary Responses

To further support the hypothesis that the decrease in pulmonary response was due to the annealing of NaOH-treated zymosan but not by annealing storage alone, we compare the response to stored untreated zymosan to the response to fresh untreated

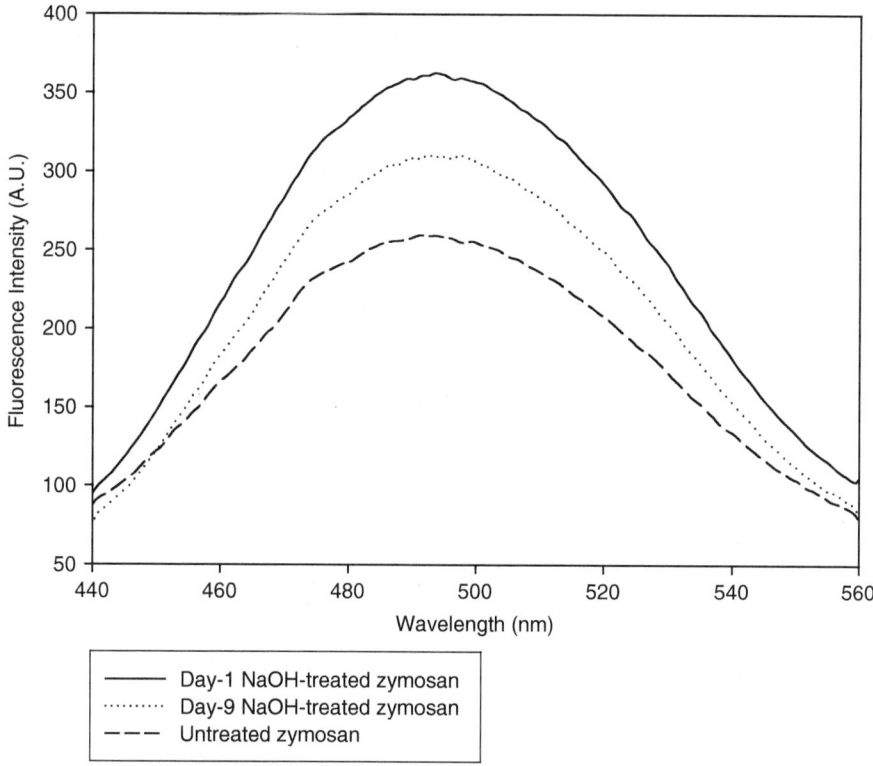

FIGURE 4.7 Fluorescent emission spectrum of zymosan stained with aniline blue. Zymosan was treated with 1 ml 0.25 N NaOH, then neutralized with 80 µl 3 N HCl, and dialyzed for 2 days. NaOH-treated zymosan or untreated zymosan (1 mg/ml PBS) was stained with aniline blue (3 µl, 4.221 mg/ml). Fresh NaOH-treated zymosan is the day-1 NaOH-treated zymosan; annealed NaOH-treated zymosan is the day-9 NaOH-treated zymosan (Young et al., 2003a).

zymosan (Table 4.5). The results demonstrate that without the NaOH treatment, there are no differences in pulmonary response observed from fresh untreated zymosan and annealed untreated zymosan.

4.10.3 INHIBITION OF ANNEALING BY FREEZING RETAINS THE POTENCY OF FRESH NaOH-TREATED ZYMOSAN

Removing thermal energy through freezing can inhibit the annealing process. Table 4.6 shows that there are no differences between the potency of fresh and 7-day frozen NaOH-treated zymosan for all the indices of pulmonary inflammation and damage measured in rats. These data further support the conclusion that the observed decrease in PMN infiltration and nitric oxide production in response to annealed NaOH-treated zymosan was due to a change in conformation.

TABLE 4.4
Pulmonary Responses to Fresh or Annealed NaOH-treated Zymosan

	Breathing Frequency Increase Post-IT (breaths/min)	PMN (x 10^6)	Albumin (mg/ml)	LDH Activity (U/L)	NO (μM)
Control	12.3 ± 2.8	1.5 ± 0.1	0.14 ± 0.02	41.0 ± 4.6	0.8 ± 0.1
Fresh NaOH-treated zymosan	67.1 ± 7.5*	18.0 ± 1.6**	0.36 ± 0.04*	89.5 ± 16.6*	104.9 ± 6.5**
Annealed NaOH-treated zymosan	60.9 ± 13.6*	11.4 ± 1.3*	0.37 ± 0.08*	83.5 ± 6.9*	65.3 ± 8.7*

Values are 18 h after IT exposure to 3 mg/kg body weight zymosan, expressed as means ± SEM; n = 5 rats.

* Significantly greater than control ($p < 0.05$).

** Significantly different from the annealed NaOH-treated zymosan group ($p < 0.05$).

Data modified from Young et al. (2003a).

TABLE 4.5
Comparison of Pulmonary Response Induced by Fresh Untreated Zymosan and Untreated Zymosan Annealed for 7 Days

	Breathing Frequency Increase Post-IT (breaths/min)	PMN (x 10^6)	Albumin (mg/ml)	LDH Activity (U/L)	NO (μM)
Control	6.9 ± 1.5	1.9 ± 0.4	0.24 ± 0.03	37.0 ± 8.0	0 ± 0
Fresh untreated zymosan	60 ± 1*	22.9 ± 3.8*	0.52 ± 0.06*	80.5 ± 7.2*	94.9 ± 3.5*
Untreated zymosan annealed for 7 days	50.6 ± 16.4*	19.5 ± 3.3*	0.34 ± 0.07*	69.4 ± 7.1*	83.2 ± 9.4*

Values are 18 h after IT exposure to 1.8 mg/kg body weight zymosan and expressed as means ± SEM; n = 5 rats.

*Significantly greater than control ($p < 0.05$).

Data modified from Young et al. (2003a).

TABLE 4.6
Comparison of Pulmonary Response Induced by Fresh NaOH-treated Zymosan and NaOH-treated Zymosan Frozen for 7 Days

	Breathing Frequency Increase Post-IT (breaths/min)	PMN (x 10^6)	Albumin (mg/ml)	LDH Activity (U/L)	NO (μM)
Control	5.8 ± 3.7	0.9 ± 0.1	0.12 ± 0.01	39.6 ± 4.8	0 ± 0
Fresh NaOH-treated zymosan	34.4 ± 3.7*	21.2 ± 4.0*	0.36 ± 0.04*	82.8 ± 14.4*	64.9 ± 13.1*
Frozen NaOH-treated zymosan (for 7 days)	32.0 ± 4.1*	23.1 ± 4.8*	0.41 ± 0.03*	118.3 ± 11.5*	75.4 ± 8.7*

Values are 18 h post IT exposure to 1.9 mg/kg body weight zymosan and are expressed as means ± SEM; n = 5 rats.

*Significantly greater than control ($p < 0.05$).

Data modified from Young et al. (2003a).

4.10.4 SUMMARY OF FINDINGS FROM CONFORMATION STUDIES

In summary, we compared the pulmonary potency of fresh NaOH-treated zymosan to annealed NaOH-treated zymosan. The result shows that fresh NaOH-treated zymosan was more potent than annealed NaOH-treated zymosan in inducing pulmonary inflammation in rats. The fact that these two samples were not subject to any further treatment except for the annealing period of time suggests that a difference in conformation (partially open triple helix versus closed triple helix) is most likely the contributing factor for the observed difference in pulmonary inflammatory responses. The difference in conformation was confirmed by the fluorescence measurements, which indicate that fresh NaOH-treated zymosan has a more partially opened triple helix than the annealed NaOH-treated zymosan. The above results strongly support the hypothesis that the partially opened triple helix is the more potent conformation for inducing pulmonary inflammation in rats.

4.11 CONCLUSIONS FROM ZYMOSAN-INDUCED PULMONARY INFLAMMATION STUDIES

Zymosan A is a representative example of (1→3)-β-glucans. Untreated zymosan induces pulmonary inflammation in rats following IT instillation in both a dose and time-dependent fashion (Young et al., 2001). An analysis of which form of zymosan is more inflammatory in rats indicates that NaOH-insoluble particulate zymosan was significantly more potent than NaOH-soluble zymosan (Young et al., 2003b). Further examination of the contribution of conformation to the inflammatory potency of NaOH-treated insoluble zymosan suggests that the partially opened triple helix is the more potent conformational form of particulate zymosan in some pulmonary toxicity assays (Young et al., 2003a).

4.12 PRETREATMENT WITH (1→3)-β-GLUCANS MODIFIES ENDOTOXIN RESPONSE

Endotoxin and (1→3)-β-glucans are both widely distributed in the environment. Therefore the interaction between endotoxin and (1→3)-β-glucans with respect to the pulmonary responsiveness in an animal model is of interest. (1→3)-β-glucans can function as an inflammatory reagent as shown above. However, they can also function as biological response modifiers (Augustin, 1998; Bohn and BeMiller, 1995; Di Luzio, 1983; 1985). This classification was given pharmacologically for immunotherapeutic purposes to those compounds that can stimulate immunity and increase resistance to microbial disease. (1→3)-β-Glucans have been shown to enhance the immune system systemically. Therefore, animals treated with (1→3)-β-glucans exhibit an enhanced ability to resist microbial invasion, such as bacterial infection. Since (1→3)-β-glucans exhibit these two distinct effects on hosts, it is of interest to investigate how hosts would respond to endotoxin in the presence of (1→3)-β-glucans; i.e., will (1→3)-β-glucans have an inhibitory or potentiating effect on pulmonary responses to endotoxin?

The literature suggests that, depending on experimental conditions, (1→3)-β-glucans can either enhance (Bower et al., 1986; Cook et al., 1980) or inhibit (Soltys and Quinn, 1999; Vereschagin et al., 1998) endotoxin effects on the host. In some studies, it has been suggested that (1→3)-β-glucans enhance the toxicity of endotoxin. Cook et al. (1980) pretreated rats with particulate glucans and reported marked increases in their sensitivity to endotoxic shock. Likewise, Bower et al. (1986) reported that rats pretreated with particulate glucans, but not soluble glucans, became sensitized to endotoxins. Recently, inhibitory effects of (1→3)-β-glucans on the potency of endotoxin have been reported. Soltys and Quinn (1999) found that *in vivo* treatment of mice with soluble glucans resulted in suppression of production of proinflammatory cytokines, such as IL-6 and TNF-α, in response to endotoxin. Vereschagin et al. (1998) found that pretreatment of mice with soluble carboxymethyl-β-1,3-glucan (CMG) protected against endotoxin shock. A trend was observed in that most of the reported inhibitory effects of (1→3)-β-glucans on endotoxin were seen with soluble (1→3)-β-glucans, while most of the studies reporting enhancement of endotoxin toxicity used particulate (1→3)-β-glucans. Because the particulate form of (1→3)-β-glucans is the form most likely found in environmental settings, our laboratory was interested in inhalation toxicity of particulate (1→3)-β-glucans, and how particulate glucan treatment modified the pulmonary response of rats to endotoxin (Young et al., 2002).

4.12.1 Regimen Determination Following Combined Exposure to Zymosan and LPS

The combined exposure to zymosan and LPS was evaluated after both simultaneous and sequential exposure. However, after analysis only Group 1 (first day zymosan, second day LPS) resulted in pulmonary response levels which were statistically different than the sum of the separate exposure to zymosan or LPS. Therefore, this results section will focus on the pulmonary response of Group 1 only, i.e., the effect of pretreatment with particulate $(1\rightarrow 3)$-β-glucans (zymosan) on the subsequent pulmonary response to endotoxin (Table 4.7).

TABLE 4.7
Pulmonary Responses to LPS Exposure Following Pretreatment with Zymosan

	PMN (x 10^6)	Net Breathing Frequency Increase (breaths/min)	Albumin (mg/mL)	LDH (U/L)	NO-Dependent Chemiluminescence (cpm x 10^{-5}/0.25 x 10^6 AM/15 min)
PBS control (day 1, PBS/day 2, PBS/day 3, euthanized)	2.7 ± 0.3	0.59 ± 0.59	0.14 ± 0.01	48.8 ± 1.7	0.9 ± 0.2
Zymosan (day 1, zymosan/day 2, PBS/day 3, euthanized)	17.7 ± 2.4	61.2 ± 10.3	0.27 ± 0.02	145.5 ± 8.2	4.5 ± 1.9
LPS (day 1, PBS/day 2, LPS/day 3, euthanized)	15.2 ± 3.9	4.9 ± 3.5	0.16 ± 0.01	102.7 ± 18.1	47.8 ± 9.2
Zym + LPS (day 1, zymosan/day 2, LPS/day 3, euthanized)	25.2 ± 4.9	89.9 ± 9.4	0.21 ± 0.02*	135.5 ± 17.2*	11.2 ± 5.7*
Expected additive value (zymosan + LPS – PBS)	30.8 ± 4.3	65.5 ± 14.4	0.30 ± 0.02	207.5 ± 17.3	51.4 ± 8.8

Values are expressed as means ± SEM; n = 4–6 rats.

*Significantly lower than the expected additive value ($p > 0.05$).

Data modified from Young et al. (2002).

4.12.2 COMPARISON OF RESULTS FROM COMBINED TREATMENT GROUPS TO CALCULATED EXPECTED VALUES

Three experimental groups were investigated to evaluate the interaction of $(1\rightarrow3)$-β-glucans and endotoxin: group 1 IT-zymosan at day 1 and IT-LPS at day 2; group 2, IT-LPS at day 1 and IT-zymosan at day 2; group 3, IT zymosan and LPS at day 1, IT PBS at day 2. Four additional groups were added as controls of zymosan or LPS alone at different days: group 4, IT zymosan at day 1, IT PBS at day 2; group 5, IT LPS at day 1, IT PBS at day 2; group 6, IT PBS at day 1, IT zymosan at day 2; group 7, IT PBS at day 1, IT LPS at day 2. Group 8 was the PBS-treated control. The expected values were calculated by Equation 4.1. The results from the combined zymosan and LPS treatment groups were compared to the expected value with their matched control groups. For example, the results from group 1 (first day zymosan, second day LPS) would be compared to the results from $(1\rightarrow3)$-β-glucans-IT at day 1 (group 4) plus results from LPS-IT at day 2 (group 7) minus the result from PBS-IT group (group 8).

$$\text{Expected value} = ([1\rightarrow3]\text{-}\beta\text{-glucans-IT at day x}) + (\text{endotoxin-IT at day y}) - (\text{PBS-IT}) \qquad (4.1)$$

At day 3, the following pulmonary responses were monitored: (1) breathing frequency, (2) differential cell counts of BAL cells, (3) AM–CL, (4) NO-dependent CL as a measure of alveolar macrophage activation, (5) nitric oxide production from alveolar macrophages, (6) TNF-α levels, (7) albumin levels, and (8) LDH activity in the first acellular lavage fluid. Interaction between zymosan and endotoxin exposures was determined as the deviation from the sum of the individual effects of these agents.

Table 4.7 shows that zymosan did not inhibit the inflammatory reaction to subsequent LPS exposure; that is, increases in breathing rate and PMN infiltration in response to zymosan and then LPS were no different than that expected from the additive effects of separate exposure to zymosan A and LPS. However, an inhibition of LPS-induced cytotoxicity (LDH activity and albumin level in first acellular BALF), macrophage activation (CL) and cytokine (TNF-α) production (Figure 4.8) were observed in the rats pretreated with zymosan (Young et al., 2002). This inhibition was not observed when rats were treated with LPS before zymosan exposure or when rats were exposed to LPS and zymosan simultaneously. Inhibition of LPS-induced macrophage activation (chemiluminescence) by pretreatment with zymosan was also reported in a guinea pig inhalation model (Robinson et al., 1996). Likewise, *in vivo* zymosan exposure followed by *ex vivo* LPS stimulation produced a lower than expected LPS induction of nitric oxide production from AM (Figure 4.9) (Young et al., 2002).

In summary, the interaction of zymosan and LPS on the pulmonary response of rats was investigated in three different exposure groups: zymosan treatment prior to LPS exposure, zymosan treatment after LPS exposure, and simultaneous zymosan and LPS exposure. The pulmonary responses to combined exposures did not differ significantly for the sum of the individual exposures to zymosan or endotoxin in the

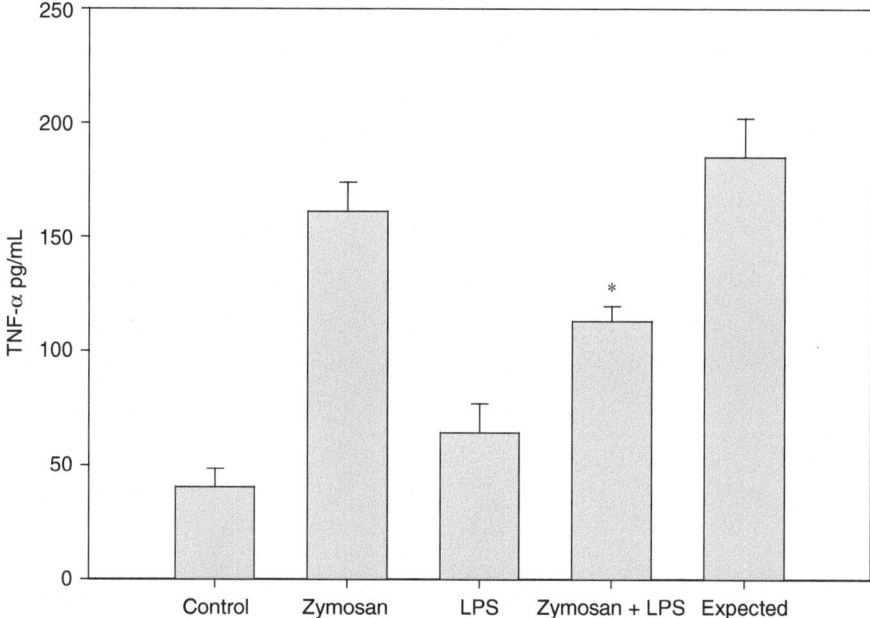

FIGURE 4.8 Inhibition of responsiveness to LPS by zymosan pretreatment. TNF-α levels were measured in the first acellular BALF. Sacrifice was on day 3 after administration of PBS or zymosan on day 1 or PBS or LPS on day 2. The expected value was calculated from the expected additive value; i.e., expected = zymosan + LPS - PBS. "*" indicates a significant lower stimulation than the expected additive effect from zymosan or LPS exposure alone ($p \leq 0.05$). Values are means ± SEM of 5 rats (Young et al., 2002).

last two exposure groups; i.e., LPS pretreatment followed by zymosan or simultaneous LPS plus zymosan treatments. A lower than expected LPS-induced pulmonary response was observed only in the group receiving zymosan pretreatment followed by LPS exposure (Table 4.7). Therefore, pretreatment of rats with a particulate form of (1→3)-β-glucans appears to inhibit the overall pulmonary response to LPS stimulation. These results suggest that complex interaction of components may exist in exposure to organic dusts. Therefore, hazard may not be defined by measuring endotoxin or (1→3)-β-glucans alone.

4.13 CONCLUSION AND NEED FOR DEVELOPING METHODS FOR ANALYZING INSOLUBLE GLUCANS

Results of studies from our laboratory indicate that zymosan A (a crude form of [1→3]-β-glucans) can induce pulmonary inflammation in rats when inhaled. There is a direct dose-response and time course relationship between zymosan A exposure and development of pulmonary inflammation. The NaOH-soluble forms of zymosan A cause mild inflammation in rats, while NaOH-insoluble zymosan A induces a

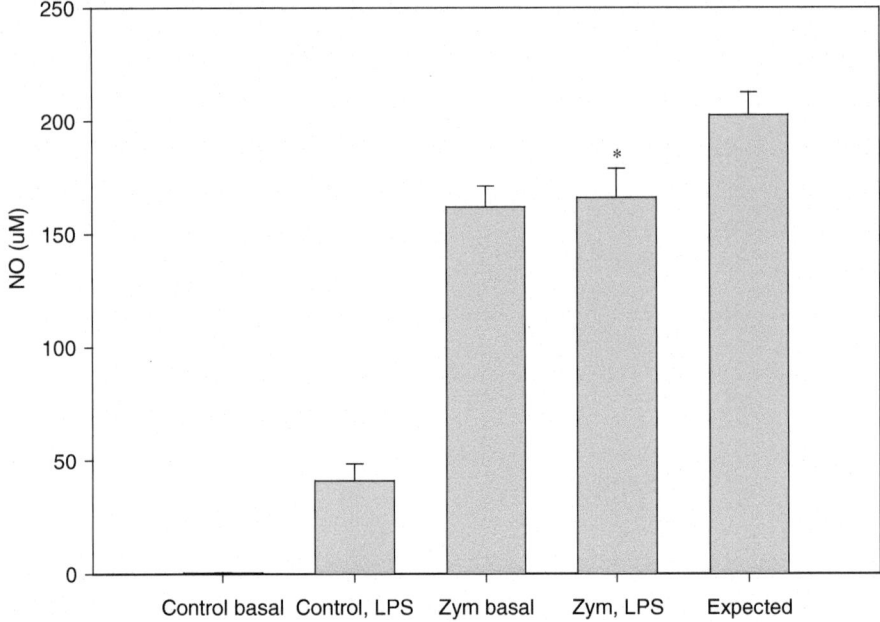

FIGURE 4.9 *In vivo* zymosan exposure decreased nitric oxide production from AM in response to *ex vivo* LPS exposure. AM (0.5 x 10^6 cells/well) were plated in a 24-well plate and cultured for 24 h before measurement of NO levels in the culture medium. The control basal was unstimulated cultured AM cells harvested from PBS-IT control rats. The control LPS was cultured LPS-stimulated (10 μg/ml LPS) AM cells harvested from PBS-IT control rats. The zymosan basal was cultured unstimulated AM cells harvested from day 1 PBS-IT, day 2 zymosan-IT rats. The zymosan LPS was LPS-stimulated cultured AM cells from day 1 PBS-IT, day 2 zymosan-IT rats. The expected value was calculated from the expected additive value, i.e., expected = zymosan basal + control LPS - control basal. "*" indicates a significant lower than expected additive effect ($p \leq 0.05$). Values are means ± SEM of 5–10 rats (Young et al., 2002).

significant greater degree of pulmonary inflammation in rats. An open triple helix conformation of zymosan is more potent than a closed triple helix toward some assays. Pretreatment with particulate (1→3)-β-glucans inhibits pulmonary responses to subsequent LPS exposure.

Most of the current (1→3)-β-glucan analysis methods can only analyze glucans in the soluble form (either by heat extraction or NaOH extraction or a combination of both). However, our studies indicate that the level of inflammation correlates well with the insoluble fraction of (1→3)-β-glucans but not with the soluble fraction of (1→3)-β-glucans. Therefore, there is a need to develop alternative methods for analysis the insoluble fraction of (1→3)-β-glucans.

REFERENCES

Adachi, Y., Ohno, N., Ohsawa, M., Oikawa, S., and Yadomae, T. (1990). Change of biological activities of (1→3)-beta-D-glucan form Grifola frondosa upon molecular weight reduction by heat treatment. *Chem. Pharm. Bull.* 38, 477–481.

Aketagawa, J., Tanaka, S., Tamura, H., Shibata, Y., and Saito, H. (1993). Activation of limulus coagulation factor G by several (1→3)-β-D-glucans: comparison of the potency of glucans with identical degree of polymerization but different conformations. *J. Biochem.* 113(6), 683–686.

Augustin, J. (1998). Glucans as modulating polysaccharides: their characteristics and isolation from microbiological sources. *Biologia* 53, 277–282.

Benacerraf, B. and Sebestyen, M.M. (1957). Effect of bacterial endotoxins on the reticuloendothelial system. *Fed. Proc.* 16, 860–867.

Bluhm, T.L. and Sarko, A. (1976). The triple-helical structure of lentinan, a linear -(1→3)-D-glucan. *Can. J. Chem.* 55, 293–299.

Bohn, J.A. and BeMiller, J.N. (1995). (1→3)-β-D-Glucans as biological response modifiers: a review of structure-functional activity relationships. *Carbohydr. Poly.* 28, 3–14.

Bower, G.J., Patchen, M.L., MacVittie, T.J., Hirsch, E.F., and Fink, M.P. (1986). A comparative evaluation of particulate and soluble glucan in an endotoxin model. *Int. J. Immunopharmc.* 8(3), 313–321.

Bradner, W.T., Clarke, D.A., and Stock, C.C. (1958). Stimulation of host defense against experimental cancer - I. Zymosan and Sarcoma 180 in mice. *Cancer Res.* 18, 347–351.

Castranova, V., Domelsmith, L.N., Barger, M., Ma, J.K.H., Olenchock, S.A., Jones, T.A., and Frazer, D.G. (1991). Efficacy of acid/base washes in removing the bioactive agent from cotton dust: response of the guinea pig model to inhalation of treated dust. In *Proceedings of the 15th Cotton Dust Research Conferences*, R.R. Jacobs, P.J. Wakelyn, and L.N. Domelsmith, eds., National Cotton Council, Memphis, TN, pp. 252-255.

Castranova, V., Robinson, V.A., and Frazer, D.G. (1996). Pulmonary reactions to organic dust exposures: development of an animal model. *Environ. Health Perspect.* 104 (suppl 1), 41–53.

Castranova, V., Frazer, D.G., Manley L.K., Dey, R.D. (2002). Pulmonary alterations associated with inhalation of occupational and environmental irritants. *Inter. Immunopharmarcol.* 2, 163–172.

Chuah, C.T., Sarko, A., Deslandes, Y., and Marchessault, R.H. (1983). Triple-helical crystalline structure of curdlan and paramylon hydrates. *Macromolecules* 16, 1375–1382.

Cook, J.A., Dougherty, W.J., and Holt, T.M. (1980). Enhanced sensitivity to endotoxin induced by the R E stimulatant, glucan. *Circulatory Shock* 7, 225–238.

Demleitner, S., Kraus, J., and Franz, G. (1992a). Synthesis and antitumour activity of derivatives of curdlan and lichenan branched at C-6. *Carbohydr. Res.* 226(2), 239–246.

Demleitner, S., Kraus, J., and Franz, G. (1992b). Synthesis and antitumour activity of sulfoalkyl derivatives of curdlan and lichenan branched. *Carbohydr. Res.* 226(2), 247–252.

Deslandes, Y., Marchessault, R.H., and Sarko, A. (1980). Triple-helical structure of (1-3)-beta-D-glucan. *Macromolecules* 13, 1466–1471.

Di Luzio, N.R. (1983). Immunopharmacology of glucan: a broad spectrum enhancer of host defense mechanisms. *Trends Pharmacol. Sci.* 4, 344–347.

Di Luzio, N.R. (1985). Update on the immunomodulating activities of glucans. *Springer Semin. Immunopathol.* 8, 387–400.

DiCarlo, F.J. and Fiore, J.V. (1957). On the composition of zymosan. *Science* 127, 756–757.

Ellakkani, M.A., Alarie, Y.C., Weyel, D.A., Mazumdar, S. and Karol, M.H. (1984). Pulmonary reactions to inhaled cotton dust: an animal model for byssinosis. *Toxicol. Appl. Pharmacol.* 94, 267–284.

Evans, N.A., Hoyne, P.A., and Stone, B.A. (1984). Characteristics and specificity of the interaciton of a fluorochrome from aniline blue (sirofluor) with polysaccharides. *Carbohydr. Poly.* 4, 215–230.

Fizpatrick, F.W. and DiCarlo, F.J. (1964). Zymosan. *Ann. N.Y. Acad. Sci.* 118, 235–261.

Fogelmark, B., Goto, H., Yuasa, K., Marchat, B., and Rylander, R. (1992). Acute pulmonary toxicity of inhaled beta-1,3-glucan and endotoxin. *Agents and Actions* 35(1–2), 50–56.

Fogelmark, B. and Rylander, R. (1993). Lung inflammatory cells after exposure to mouldy hay. *Agents and Actions* 39, 25–30.

Fogelmark, B., Sjostrand, M., and Rylander, R. (1994). Pulmonary inflammation induced by repeated inhalations of beta-(1,3)-D-glucan and endotoxin. *Int. J. Exp. Pathol.* 75, 85–90.

Fogelmark, B., Sjostrand, M., Williams, D.L., and Rylander, R. (1997). Inhalation toxicity of (1→3)-beta-D-glucan: recent advances. *Mediators of Inflammation* 6, 263–265.

Frazer, D.G., Afshari, A.A., Goldsmith, W.T., Phillips, N., and Robinson, V.A. (1997). Estimation of guinea pig specific airway resistance following exposure to cotton dust measured with a whole body flow plethysmograph. In *Proceedings of the 21st Cotton and Organic Dust Research Conferences*, P.J. Wakelyn, R.R. Jacobs, and R. Rylander, eds., National Cotton Council, Memphis, TN, pp. 175–180.

Frazer, D.G., Robinson, V.A., DeLong, D.S., Castranova, V., Jones, T.A., and Petsonk, E.L. (1989). Comparisons of breathing rate, cellular response, airway obstruction (determined by postmortem pulmonary hyperinflation) and the wet/dry weight ratios of guinea pigs exposed to cotton dust aerosol. In *Proceedings of the 13th Cotton Dust Research Conferences*, R.R. Jacobs and P.J. Wakelyn, eds. National Cotton Council, Memphis, TN, pp. 129–133.

Green, L.C., Wagner, D.A., Glogowski, J., Skipper, P.L., Wishnok, J.S., and Tannenbaum, S.R. (1982). Analysis of nitrate, nitrite, and [^{15}N]nitrate in biological fluids. *Anal. Biochem.* 126, 131–138.

Hassid, W.Z., Joslyn, M.A., and McCready, R.M. (1941). The molecular constitution of an insoluble polysaccharide from yeast, Saccharomyces cerevisiae. *J. Am. Chem. Soc.* 63, 295–298.

Kashiwagi, Y., Norisuye, Y., and Fujita, H. (1981). Triple helix of Schizophyllum commune polysaccharide in dilute solution. 4. Light scattering and viscosity in dilute aqueous sodium hydroxide. *Macromolecules* 14, 1220–1225.

Kojima, T., Tabata, K., Itoh, W., and Yanaki, T. (1986). Molecular weight dependence of the antitumor activity of schizophyllan. *Agric. Biol. Chem.* 50(1), 231–232.

Kraus, J. and Franz, G. (1992). Immunomodulating effects of polysaccharides from medicinal plants. *Adv. Exp. Med. Biol.* 319, 299–308.

Lecours, R., Laviolette, M., and Cormier, Y. (1986). Bronchoalveolar lavage in pulmonary mycotoxicosis (organic dust toxic syndrome). *Thorax* 41, 924–926.

Maeda, Y.Y., Watanabe, S.T., Chihara, C., and Rokutanda, M. (1988). Denaturation and renaturation of a beta-1,6;1,3-glucan, lentinan, associated with expression of T-cell-mediated responses. *Cancer Res.* 48, 671–675.

Misaki, A., Kishida, E., Kakuta, M., and Tabata, K. (1993). Antitumor fungal (1→3)-beta-D-glucans: structural diversity and effects of chemical modification. In *Carbohydrates and Carbohydrate Polymers— Analysis, Biotechnology, Modification, Antiviral, Biomedical and Other Applications*, Vol. 1, M. Yalpani, ed.. ATL Press, Inc., Mount Prospect, IL, pp. 116–129.

Nono, I., Ohno, N., Masuda, A., Oilawa, S., and Yadomae, T. (1991). Oxidative degradation of an antitumor (1→3)-beta-D-glucan, grifolan. *J. Pharmacobio-Dyn.* 14, 9–19.

Ohno, N., Terui, T., Chiba, N., Kurachi, K., Adachi, Y., and Yadomae, T. (1995). Resistance of highly branched (1-3)-beta-D-glucans to formolysis. *Chem. Pharm. Bull.* 43, 1057–1060.

Ohno, N., Miura, N.N., Adachi, Y., and Yadomae, T. (1998). Inflammatory and immunotoxicological activities of 1→3-beta-glucans. In *Proceedings of the 22th Cotton and Organic Dust Research Conferences*, R.R. Jacobs, P.J. Wakelyn, and R. Rylander, eds., Vol. 1, National Cotton Council, Memphis, TN, pp. 240–244.

Peat, S., Whelan, W.J., and Edwards, T.E. (1958). Polysaccharide of baker's yeast. Part II. Yeast glucan. *J. Chem. Soc.*, 3862–3868.

Pillemer, L. and Ecker, E.E. (1941). Anticomplementary factor in frest yeast. *J. Biol. Chem.* 137, 139–142.

Pillemer, L. and Ross, O.A. (1955). Alternations in serum properdin levels following injection of zymosan. *Science* 121, 732–733.

Riggi, S.J. and Di Luzio, N.R. (1961). Identification of a reticuloendothelial stimulating agent in zymosan. *Am. J. Physiol.* 200, 297–300.

Robinson, V.A., DeLong, D.S., Frazer, D.G., and Castranova, V. (1988). Dose-response relationships for pulmonary reations of guinea pigs to inhalation of cotton dust. In *Proceedings of the 12th Cotton Dust Research Conferences*, R.R. Jacobs and P.J. Wakelyn, eds., National Cotton Council, Memphis, TN, pp. 149–152.

Robinson, V.A., Frazer, D.G., Afshari, A.A., Goldsmith, W.T., Olenchock, S., Whitmer, M.P., and Castranova, V. (1996). Guinea pig response to zymosan and a serial exposure of zymosan and endotoxin. In *Proceedings of the 20th Cotton and Organic Dust Research Conferences*, R.R. Jacobs, P.J. Wakelyn, and R. Rylander, eds., National Cotton Council, Memphis, TN, pp. 356–360.

Rylander, R., Bergstrom, R., Goto, H., Yuasa, K., and Tanaka, S. (1989a). Studies on endotoxin and beta-1,3 glucan in cotton dust. In *Proceedings of the 13th Cotton Dust Research Conferences*, R.R. Jacobs and P.J. Wakelyn, eds., National Cotton Council, Memphis, TN, pp. 46–47.

Rylander, R., Goto, H., and Marchat, B. (1989b). Acute toxicity of inhaled beta, 1-3 glucan and endotoxin. In *Proceedings of the 13th Cotton Dust Research Conferences*, R.R. Jacobs and P.J. Wakelyn, eds., National Cotton Council, Memphis, TN, pp. 145–146.

Rylander, R. and Goto, H. (1991). *First Glucan Lung Toxicity Workshop*. Committee on Organic Dusts, ICOH, Tokyo.

Rylander, R. and Peterson, Y. (1993). *Second Glucan Inhalation Toxicity Workshop*. Committee on Organic Dusts, ICOH, Tokyo.

Rylander, R. and Fogelmark, B. (1994). Inflammatory responses by inhalation of endotoxin and (1-3)-beta-D-glucan. *Am. J. Ind. Med.* 25, 101–102.

Rylander, R. and Jacobs, R.R. (1994). *Organic Dust Exposure, Effects, and Prevention*. CRC Press, Inc., Boca Raton, FL.

Rylander, R. and Lin, R.H. (2000). (1→3)-beta-D-glucan — relationship to indoor air-related symptoms, allergy and asthma. *Toxicology* 152, 47–52.

Saito, H., Yoshioka, Y., and Uehara, N. (1991). Relationship between conformation and biological response for (1-3)-beta-d-glucans in the activation of coagulation factor G from limulus amebocyte lysate and host-mediated antitumor activity. Demonstration of single-helix conformation as a stimulant. *Carbohydr. Res.* 217, 181–190.

Schuyler, M., Gott, K., and Cherne, A. (1998). Effect of glucan on murine lungs. *J. Toxicol. Environ. Health Part A* 53, 493–505.

Seljelid, R., Bogwald, J., and Lundwall, A. (1981). Glycan stimulation of macrophages *in vitro*. *Exp. Cell Res.* 131, 121–129.

Sendelbach, L.E. and Witschi, H.P. (1987). Bronchoalveolar lavage in rats and mice following beryllium sulfate inhalation. *Toxicol. Appl. Pharmacol.* 90, 322–329.

Soltys, J. and Quinn, M.T. (1999). Modulation of endotoxin- and enterotoxin-induced cytokine release by *in vivo* treatment with beta-(1,6)-branched beta-(1,3)-glucan. *Infection Immunity* 67, 244–252.

Stone, B.A. and Clarke, A.E. (1992). (1→3)-beta-glucans and animal defense mechanisms. In *Chemistry and Biology of (1→3)-β-Glucans*, La Trobe University Press, Victoria, Australia, pp. 525–564.

Tanaka, S., Aketagawa, J., Takahashi, S., Shibata, Y., Tsumuraya, Y., and Hashimoto, Y. (1991). Activation of a limulus coagulation factor G by (1-3)-beta-D-glucans. *Carbohydr. Res.* 218, 167–174.

Thistlethwaite, P., Porter, I., and Evans, N. (1986). Photophysics of aniline blue fluorophore: a fluorescent probe showing specificity toward (1-3)-beta-D-glucans. *J. Phy. Chem.* 90, 5058–5063.

Vereschagin, E.I., van Lambalgen, A.A., Dushkin, M.I., Schwartz, Y.S., Polyakov, L., Heemskerk, A., Huisman, E., Thijs, L.G., and van den Bos, G.C. (1998). Soluble glucan protects against endotoxin shock in the rat: the role of the scavenger receptor. *Shock* 9, 193–198.

Whistler, R.L. and BeMiller, J.N. (1958). Alkaline degradation of polysaccharides. *Adv. Carbohydr. Chem.* 13, 289–329.

Whistler, R.L. (1973). Solubility of polysaccharides and their behavior in solution. In *Carbohydrates in Solution*, R.F. Gould, ed., Vol. 117, American Chemical Society, Washington, D.C., pp. 243–255.

Whistler, R.L., Bushway, A.A., and Singh, P.P. (1976). Noncytotoxic, antitumor polysaccharides. *Adv. Carbohydr. Chem. Biochem.* 51, 235–275.

Yadomae, T. and Ohno, N. (1996). Structure-activity relationship of immunomodulating (1→3)-beta-D-glucans. *Recent Res. Devel. Chem. Pharm. Sci.* 1, 23–33.

Yoshioka, Y., Uehara, N., and Saito, H. (1992). Conformation-dependent change in antitumor activity of linear and branched (1-3)-beta-D-glucans on the basis of conformational elucidation by carbon-13 nuclear magnetic resonance spectroscopy. *Chem. Pharm. Bull.* 40(5), 1221–1226.

Young, R.S., Jones, A.M., and Nicholls, P.J. (1998). Something in the air: endotoxins and glucans as environmental troublemakers. *J. Pharm. Pharmacol.* 50, 11–17.

Young, S.-H., Dong, W.J., and Jacobs, R.R. (2000). Observation of a partially opened triple-helix conformation in 1→3-beta-glucan by fluorescence resonance energy transfer spectroscopy. *J. Biol. Chem.* 275, 11874–11879.

Young, S.-H., Robinson, V.A., Barger, M., Porter, D.W., Frazer, D.G., and Castranova, V. (2001). Acute inflammation and recovery in rats after intratracheal instillation of a 1→3-beta-glucan (zymosan A). *J. Toxicol. Environ. Health Part A* 64, 311–325.

Young, S.-H., Robinson, V.A., Barger, M., Zeidler, P., Porter, D.W., Frazer, D.G., and Castranova, V. (2002). Modified endotoxin responses in rats pre-treated with 1→3-beta-glucan (zymosan A). *Toxicol. Appl. Pharmacol.* 178, 172–179.

Young, S.-H., Robinson, V.A., Barger, M., Frazer, D.G., Castranova, V., and Jacobs, R.R. (2003a). Partially opened triple helix is the biologically active conformation of 1→3-beta-glucans that induces pulmonary inflammation in rats. *J. Toxicol. Environ. Health A* 66, 551–563.

Young, S.-H., Robinson, V.A., Barger, M., Whitmer, M., Porter, D.W., Frazer, D.G., and Castranova, V. (2003b). Exposure to particulate 1→3-beta-glucans induces greater pulmonary toxicity than soluble 1→3-beta-glucans in rats. *J. Toxicol. Environ. Health A* 66, 25–38.

Zejda, J.E. and Dosman, J.A. (1993). Respiratory disorders in agriculture. *Tuber. Lung Dis.* 74(2), 74–86.

5 β-Glucan Receptor(s) and Their Signal Transduction

Yoshiyuki Adachi

CONTENTS

5.1 Introduction 95
5.2 Soluble β-Glucan Recognition Proteins 96
    5.2.1 Surfactant Protein D (SP-D) 96
    5.2.2 β-Glucan Recognition Proteins from Plants and Invertebrates.........96
5.3 1,3-β-Glucan Receptors on the Plasma Membrane of Leukocytes 97
    5.3.1 Monocyte 20-kDa Receptor...97
    5.3.2 CR3..98
    5.3.3 Lactosylceramide 98
    5.3.4 Scavenger Receptor...98
    5.3.5 Dectin-1...99
5.4 Signaling via 1,3-β-Glucan Receptors on Plasma Membranes 100
    5.4.1 Lactosylceramide-Mediated Leukocyte Activation 100
    5.4.2 Contribution of the Toll-Like Receptor to Zymosan-Mediated Inflammatory Cytokine Production 101
    5.4.3 Dectin-1 May Explain the Biological Activities of Fungal 1,3-β-Glucan 103
5.5 Concluding Remarks...103
References..104

5.1 INTRODUCTION

1,3-β-glucan is a natural abundant constitutive component in the cell wall of fungi or plants. 1,3-β-glucan has been shown to stimulate the immune system (Yadomae and Ohno, 1996), and 1,3-β-glucan obtained from edible mushrooms has been used as a biological response modifier in Japanese cancer patients (Chihara et al.,1970; Miyazaki et al., 1995). Yeast contain a significant amount of 1,6-β-glucan and 1,3-β-glucan complex, which are major components of cell walls (Lipke and Ovalle, 1998). Because these yeast-like fungi include *Candia*, or *Cryptocuccus*, which are representative opportunistic pathogens, especially in deep mycosis (Anaissie, 1992; Myskowski et al., 1997), it is important to characterize the recognition mechanism of fungal β-glucans to understand how fungal infections are caused. In this article, various molecules are described by which fungal β-glucans are recognized and generate an activation signal in the host defense to fungi.

5.2 SOLUBLE β-GLUCAN RECOGNITION PROTEINS

In general, it has been understood that specific antibodies against 1,6-β-monoglucosyl branched 1,3-β-glucans are difficult to develop in mice (Adachi et al., 1994). However, other 1,3-β-glucans having a long 1,6-β-glucosyl branch, CSBG, obtained from the *Candida* yeast cell wall, can induce a specific antibody in some mouse strains and humans (Masuzawa et al., 2003). This finding suggests that such fungal cells can stimulate the acquired immune system via 1,3-β-glucans in addition to proteinous substances. The physiological role of the β-glucan antibody in the host defense should be examined in future studies. Please refer to Chapter 9 in this book.

5.2.1 SURFACTANT PROTEIN D (SP-D)

Surfactant proteins function as carbohydrate recognition molecules with Ca^{2+} dependency, which belong to the C-type lectin family. Surfactant proteins are produced by lung alveolar type II epithelial cells and play important roles in modulating alveolar surface tension, regulating secretion, and mediating innate immunity (Kuroki and Voelker, 1994). Surfactant protein A (SP-A), a major surfactant protein, is similar to surfactant protein D (SP-D). Both proteins are composed of a short N-terminal region involved in covalent cross-linking, followed by a collagen-like domain, a neck region, and a C-terminal carbohydrate recognition domain (CRD) (Kuroki and Voelker, 1994). SP-D forms a predominantly cruciform-like dodecamer, whereas SP-A forms a bouquet-like octadecamer. The carbohydrate recognition of SP-D is demonstrated to be specific for 1,6-α and 1,6-β-glucosyl chains (Allen et al., 2001). The amino acid residues in CRD, Glu321/Asn323, and Glu329/Asn341, associate with C3-OH and C4-OH of the glucose residues via a hydrogen bond, respectively (Allen et al., 2001). The SP-D multimer accelerates to form multivalent ligand interaction, which may contribute to the host defense by aggregating fungal cells, such as *Aspergillus* and *Candida* (Kishor et al., 2002).

5.2.2 β-GLUCAN RECOGNITION PROTEINS FROM PLANTS AND INVERTEBRATES

β-glucan recognition proteins are present in plants or invertebrates. Plants have β-glucan elicitor receptors to recognize the 1,3-β-linked glucosyl linkage to sense invading phytopathogens and to produce host defense molecules named phytoalexins (Sharp et al., 1984). Soybeans possess such receptor proteins to recognize 1,6-β-linked and 1,3-β-branched heptaglucosides in pathogen cell walls as a signal compound eliciting the onset of defense reactions (Sharp et al., 1984; Umemoto et al., 1997). A specific and high affinity glucan-binding site is contained in the β-glucan-binding protein (GBP), which in turn is part of a proposed receptor complex (Fliegmann et al., 2004). The ability to perceive and respond to *Phytophthora* cell wall-derived β-glucan elicitors is exclusive to plants that belong to the *Fabaceae* family. However, it is proposed that the presence of the GBP is essential, but not sufficient, for β-glucan elicitor-dependent disease resistance, because genes encoding GBP-related proteins can be retrieved from many plant species. The GBP in soybeans is composed of two different carbohydrate-reactive protein domains, one containing the β-glucan-binding site, and the other related to glucan endoglucosidases of fungal origin (Fliegmann et al., 2004). The glucan hydrolase displays a most likely endo-specific mode of action, cleaving only 1,3-β-D-glucosidic linkages of oligoglucosides (Fliegmann et al., 2004). Glucanase domains have been detected in a different set of host defense proteins. The β-glucan recognition proteins (β-GRP) participate in the innate immune response of insects (Ochiai et al., 1992), shrimp (Sritunyalucksana et al., 2002; Roux et al., 2002), and horseshoe crabs (Tanaka et al., 1993, Takaki et al., 2002). 1,3-β-glucans from fungi lead to the activation of their specific defense systems; i.e., the prophenoloxidase pathway (Ochiai and Ashida, 1988; Beschin et al., 1998), the induction of anti-microbial peptide genes (Kim et al., 2000), or proteolysis for clot formation cascade (Tanaka et al., 1993; Takaki et al., 2002). β-GRP from *Drosophila melanogaster*, which was shown to be a carboxyl terminal anchored via a glycosylphosphatidylinositol modification (Kim et al., 2000), may also be the β-GRP from *Bombyx mori* (Ochiai et al., 1992). The membrane bound β-GRPs from soybeans, *B. mori* and *D. melanogaster*, also exist in soluble forms. However, neither the full-length recognition proteins nor the glucanase-like domains of plants and insects show any similarity in primary structure. Thus, it is not clear how similar β-GRPs recognize β-glucans from their protein ultrastructure. β-GRP from horseshoe crabs, factor G, which is a heterodimeric serine-protease zymogen, is activated by 1,3-β-glucan. The glucan-binding domain has a xylanase Z-like domain composed of two tandem-repeating units, each of which exhibits sequences similar to the cellulose-binding domains of bacterial xylanases. Each subunit weakly binds to laminaribiose, but the association constant increases by forming a tandem-repeating structure (Takaki et al., 2002). A multivalent binding site in factor G molecules seems to be involved in the stable and specific binding to 1,3-β-glucan.

5.3   1,3-β-GLUCAN RECEPTORS ON THE PLASMA MEMBRANE OF LEUKOCYTES

5.3.1   MONOCYTE 20-KDA RECEPTOR

The first report concerning the specific recognition of 1,3-β-glucan by leukocytes was presented by Czop et al. (1985). They reported that the ingestion of unopsonized yeast particles or zymosan, a cell wall preparation containing 1,3-β-glucan, was inhibited by pretreatment of the monocyte with soluble 1,3-β-glucan. The report was a trailblazing finding that opened the study into β-glucan receptors on immune cells. The β-glucan receptor on monocytes was characterized by using an anti-idiotype antibody to the receptor (Czop et al., 1990). The antibody was prepared by immunizing an anti-β-glucan monoclonal antibody that recognizes particulate yeast β-glucan and purified yeast heptaoligosaccharide in rabbits. The rabbit anti-idiotype antibody reacts to 160kDa and 180kDa proteins (Szabo et al., 1995). The 180-kDa receptor was composed of three disulfide-linked polypeptides of 95, 60, and 20kDa. The 160-kDa protein was composed of two polypeptide subunits, 27 kDa and 20 kDa, and was constitutively phosphorylated on the tyrosine residue (Szabo et al., 1995). However, the molecular cloning of these receptor proteins has not yet been achieved.

5.3.2 CR3

Ross et al. (1987) demonstrated that $β_2$-integrin, especially CR3 (CD11b/CD18), serves as the leukocyte receptor for β-glucan by using CR3-specific monoclonal antibodies. A molecular biology technique revealed that CR3 recombinant protein interacts with a soluble β-glucan preparation (Xia and Ross, 1999). The study of recombinant CD11b fragments demonstrated that a broad polypeptide, including part of the divalent cation binding site and C-terminal domain, is required for specific binding to β-glucan (Xia et al., 1999), whereas the I-domain relating to iC3b recognition is not required for interaction with β-glucan (Xia et al., 1999 and 2002). The physiological significance of CR3 ligation with β-glucan was that β-glucan stimulation of CR3 on natural killer (NK) cells augments its cytotoxic activity against tumor cells that express the iC3b complement fragment (Vetvicka et al., 1997). This suggests that dual ligation of CR3 via β-glucan and iC3b on target cells confers a strong signal to NK cells to kill tumor cells efficiently (Ross, 2000).

5.3.3 LACTOSYLCERAMIDE

Glycosylsphingolipids, such as galactosylceramide and lactosylceramide, have the ability to interact with β-glucan (Zimmerman et al., 1998). These glycolipids tend to form clusters of cholesterol-rich microdomains, so-called "lipid rafts" or "detergent-insoluble membranes," in a plasma membrane (Harder and Simons, 1997). The interaction between lactosylceramide (LacCer) and β-glucan requires 1,3-β-glucosyl linkage, a temperature above 30°C, and a higher molecular weight β-glucan (Zimmerman et al., 1998). Interestingly, a relatively small β-glucan, laminarin, which is used to demonstrate specific binding to the β-glucan receptor proteins, is insufficient for the interaction with LacCer. Furthermore, the binding of β-glucan to LacCer was not completely inhibited by lactose, galactose, glucose, or the corresponding methylglycoside at 100 mM (Zimmerman et al., 1998). This suggests that the multiple recognition of a 1,3-glucosyl linkage. The conformation of a sugar moiety formed by lipid conjugation, and the thermal sensitive molecular dynamics of LacCer are important for stable binding to higher molecular weight β-glucans.

5.3.4 SCAVENGER RECEPTOR

The acetylated low-density lipoprotein (AcLDL) receptor, the scavenger receptor (SR), is responsible for the uptake of modified LDL and the accumulation of cholesteryl esters in macrophages (Brown and Goldstein, 1983; Dejagar et al., 1993).

This scavenger receptor also binds negatively charged, high molecular weight polysaccharides, including dextran sulfate, carrageenan, lipopolysaccharide (LPS), lipoteichoic acid, and polyphosphates (Steinberg, 1995). The chemical modification of 1,3-β-glucan with carboxylmethylation or phosphorylation allows the β-glucan to associate with SRs (Rice et al., 2002). The binding is competitive with anionic polymers; therefore, the interaction between 1,3-β-glucan derivatives and SRs seems to be mediated by charged residues. However, the modification of 1,3-β-glucans confers their binding ability to not only SRs but also other recognition systems specific for 1,3-glucosyl linkage on macrophages (Rice et al., 2002). These receptors may also play significant roles in the pattern-recognition system in innate immunity (Pearson et al., 1995).

5.3.5 Dectin-1

Dectin-1 is a C-type lectin that was originally cloned from dendritic cells by Ariizumi et al. (2000). Although it has been found that Dectin-1 co-stimulates T-cell proliferation, the function of Dectin-1 as a lectin was unknown before Gordon's group elucidated the role of the sugar ligand (Brown and Gordon, 2001). The role of Dectin-1 in carbohydrate recognition was revealed by functional cDNA cloning from a RAW264.7 macrophage cell line. Dectin-1 can bind to zymosan and the yeast form of *Candida* cells in a soluble 1,3-β-glucans-inhibitable fashion (Brown and Gordon, 2001). Dectin-1 is a type II membrane protein with an immunoreceptor tyrosine-based activating motif (ITAM) in the N-terminal cytoplasmic domain and its C-terminal moiety, in which six cystein residues are linked to form a highly conserved carbohydrate recognition domain (Ariizumi et al., 2000; Yokota et al., 2001). C-type lectins are classified into at least 14 groups, and Dectin-1 belongs to the NK receptor group, which has the largest number of 20 species among the 71 human C-type lectins (Drickamer, 2002). The β-glucan binding site on the Dectin-1 molecule was found near the β-sheet strand number three of the typical NK group C-type lectin by amino acid mutation of the carbohydrate recognition domain and by blocking the monoclonal antibody (Adachi et al., 2004). This β-glucan binding site is on an apical region corresponding to an interface area for the binding of NK receptors, including Ly49A and NKG2D, to MHC class I molecules (Natarajan et al., 2002). The amino acid residues in the binding site were not similar to other representative sequences, Glu-Pro-Asn or Gln-Pro-Asp, which form a Ca^{2+}-chelating site to recognize mannose- or galactose-residues by the type II receptor group of C-type lectin (Taylor et al., 2004). In addition to the recognition of 1,3-β-glucan, Dectin-1 binds to T lymphocytes and augments their mitogenic response by cross-linking T-cell receptors (Ariizumi et al., 2000; Grunebach et al., 2002). The binding domain on Dectin-1 molecules for T-cell activation has not been clarified, although it has been confirmed that β-glucan binding does not interfere with T-cell interaction (Herre et al., 2004). Although the crystal structure of Dectin-1 protein has not been clarified, information on the β-glucan binding site may point the way for the exploration of the T-cell interaction moiety, a tantalizing future problem to be solved.

5.4 SIGNALING VIA 1,3-β-GLUCAN RECEPTORS ON PLASMA MEMBRANES

CR3 is a heterodimer molecule composed of a CD11b α-M chain and a CD18 β-chain. Previous studies of CR3 suggested a role for PKC-mediated phosphorylation in phagocytosis (Roubey et al., 1991). CR3-mediated phagocytosis of β-glucan particles was inhibited by the PKC inhibitor, staurosporine. Furthermore, it was demonstrated that the staurosporine-inhibited phosphorylation of CD18 occurred by phagocytosing particulate β-glucan (Roubey et al., 1991). The role of CR3 as a signaling receptor has been reported, since the incubation of phagocytes with unopsonized zymosan or particulate β-glucan resulted in the tyrosine phosphorylation of several molecules, including lyn in monocytes and fgr in neutrophils (Zaffran et al., 1995). However, it is unclear whether the phosphorylation of lyn and fgr proteins is mediated solely by the ligation of CR3, since other zymosan-reactive receptors — i.e., Dectin-1 or lactosylceramide-associated intracellular proteins — may be involved in the signaling. It has been reported that the stimulation of monocytes through the engagement of CD11b or CD11c by antibodies results in the phosphorylation and activation of ERK1, ERK2, and p38/SAPK2 MAP kinases (Rezzonico et al., 2000). Therefore, the mechanism underlying the phosphorylation of signaling molecules, including tyrosine- or serine/threonine-phosphorylation, by β-glucan-mediated CR3 cross-linking is still under discussion. Further examination may be required to determine the exact role of CR3.

5.4.1 LACTOSYLCERAMIDE-MEDIATED LEUKOCYTE ACTIVATION

Lactosylceramide (LacCer, CDw17) is a unique β-glucan binding molecule, which was demonstrated by Zimmerman et al. (1998). A β-glucan obtained from *Pneumocystis carinii* can stimulate alveolar epithelial cells (AECs) to release chemokines, MIP-2 (Hahn and Limper, 2001). MIP-2 production could be inhibited by pretreatment with an anti-CDw17 antibody, suggesting that LacCer participates in the β-glucan-induced inflammatory response in the lungs. The signaling mechanism has not been determined clearly. However, the characteristics of glycosphingolipids (GSLs), such as LacCer, on plasma membranes suggest the following possibility in signaling events. Briefly, GSLs are membrane components consisting of hydrophobic ceramide lipid and hydrophilic sugar moieties. They are characterized by clustering to form cholesterol-rich microdomains in plasma membranes (Harder and Simons, 1997). GSL clusters are essentially insoluble in detergent and constitute a "detergent-insoluble membrane" (DIM), which is separable from soluble components by density gradient centrifugation (Harder and Simons, 1997). A number of studies have shown that DIM contains several transducer molecules, such as *Src* family kinases, and that microdomains form functional units, termed lipid rafts, which mediate signal transduction and cell functions (Young et al., 2003). Recent studies on gangliosides have proposed the concept that GSL clusters form GSL signaling domains (GSDs), and extracellular stimuli are transmitted via GSL clusters to signal transducer molecules, initiating cellular functions (Young et al., 2003). It has been demonstrated that LacCer activates NADPH oxidase to modulate the intercellular adhesion of

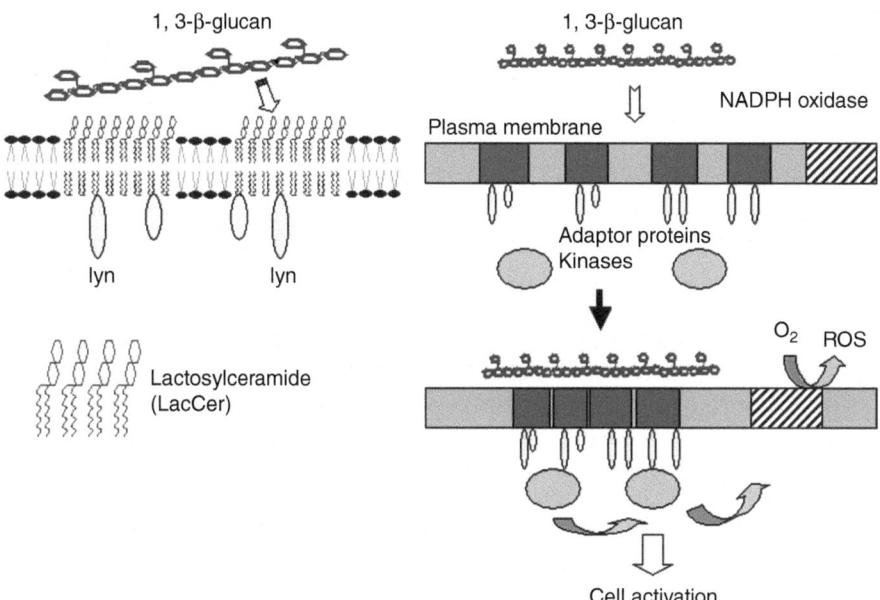

FIGURE 5.1 (See color insert following page 20.) LacCer can transmit signals for activation of neutrophils by 1,3-β-glucans.

molecule-1 expression on human umbilical vein endothelial cells (Bhunia et al., 1998) and to induce the proliferation of human aortic smooth muscle cells (Bhunia et al., 1997). Neutrophils also possess an extremely high amount of LacCer on their plasma membranes (Symington et al., 1985). LacCer is not expressed on the surface of myeloblasts or promyelocytes (Brackman et al., 1995). Moreover, it has been demonstrated that cross-linking of anti-LacCer antibodies induces superoxide production by neutrophils, and that LacCer itself up-regulates CD11/CD18 on neutrophils (Iwabuchi and Nagaoka, 2002). In consideration of this, the cross-linking of LacCer by 1,3-β-glucan with higher molecular weight may have a considerable physiological role in the activation of leukocytes (Figure 5.1).

5.4.2 Contribution of the Toll-Like Receptor to Zymosan-Mediated Inflammatory Cytokine Production

Zymosan is a yeast cell wall component of *Saccharomyces cerevisiae*, and contains α-mannan/mannoproteins and β-glucans (Di Carlo and Fiore, 1958; Lipke and Ovalle, 1998). It has long been used as a stimulus, influencing the β-glucan receptor on macrophages. The phagocytosis of zymosan particles can be inhibited by pretreatment with soluble 1,3-β-glucans, such as laminarin, implying that phagocyte activation is caused by a specific receptor for 1,3-β-glucosyl linkage. The stimulation of macrophages with zymosan also results in the production of inflammatory cytokines. However, zymosan derivatives obtained by alkaline-boiling (Gantner et al., 2003), hypochlorous acid oxidation (OX-ZYM) or chloroform/methanol-treatment

(CMIS) (Adachi et al., unpublished data) lose their ability to activate NF-κB and increase TNF-α production, despite the higher content of 1,3-β-glucan in the treated preparations compared to the original zymosan. It was concluded that highly purified 1,3-β-glucan was less active in inducing TNF-α production, and that other molecules in the zymosan may participate in the strong activation of inflammatory response.

Toll-like receptors (TLRs) mediate the recognition of a wide range of microbial products including lipopolysaccharides, lipoproteins, flagellin, and bacterial DNA. Signaling through TLRs leads to the production of inflammatory mediators (Takeda et al., 2003). The co-expression of TLR2 and the 1,3-β-glucan receptor, Dectin-1, on a HEK293 transfectant resulted in significant activation of NF-κB in response to zymosan compared to single TLR2-expressing transfectants (Gantner et al., 2003). However, transfectants expressing Dectin-1 alone showed no enhanced NF-κB activation in response to stimuli (Gantner et al., 2003). These findings demonstrate that certain substances in addition to 1,3-β-glucan in zymosan have a stimulatory effect on TLR2-mediated cell activation (Figure 5.2). The key molecules responsible for TLR2 ligation on zymosan are still unknown. However, the ligand may be heat-, alkaline-, hypochlorite-, or chloroform/methanol-labile (Adachi et al., unpublished data).

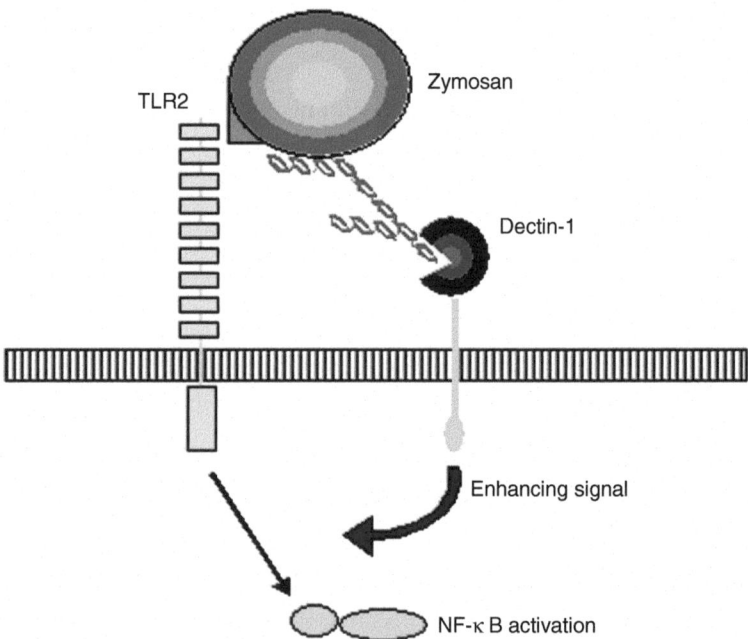

FIGURE 5.2 (See color insert following page 20.) Collaborative effect of Dectin-1 on TLR2-mediated NF-κB activation by zymosan.

5.4.3 DECTIN-1 MAY EXPLAIN THE BIOLOGICAL ACTIVITIES OF FUNGAL 1,3-β-GLUCAN

Dectin-1 is a type II transmembrane protein possessing a short cytoplasmic tail with a putative ITAM-like amino acid sequence (Ariizumi et al., 2000). The cytoplasmic region also has a series of triple acidic amino acid residues, which may be involved in the transportation of captured ligands to the lysosomal cell compartment (Cambi and Figdor, 2003). An extracellular CRD motif is responsible for specific binding to 1,3-β-glucan (Willment et al., 2001). Engagement of the CRD domain with particulate 1,3-β-glucan results in phosphorylation on the tyrosine residue and superoxide generation (Gantner et al., 2003). The production of a reactive oxygen species (ROS) by zymosan was observed in bone marrow-derived macrophages prepared from MyD88-, and TLR2-deficient mice (Gantner et al., 2003). This activity was diminished by overexpressing a Dectin-1 mutant lacking an ITAM-like domain and a lysosomal target sequence (Gantner et al., 2003). This suggests that Dectin-1 is a major functional 1,3-β-glucan receptor triggering signals for the production of ROS by macrophages.

A defensive role of Dectin-1 in fungal infection was also reported in several papers. The murine alveolar macrophage (AM), which constitutively expresses Dectin-1 molecules, has killing activity against *Pneumocystis carinii* (Steele et al., 2003). The anti-Dectin-1 monoclonal antibody treatment of AM reduced this fungicidal activity. Recognition of *Candia albicans* by macrophages is also mediated by Dectin-1 (Brown et al., 2003). The interaction of live *Candida* cells with Dectin-1-overexpressing macrophages, which is transduced by a retrovirus vector with Dectin-1 cDNA, caused a significant increase of TNF-α secretion (Brown et al., 2003). TNF-α production was not induced by transducing a cytoplasmic tail minus or ITAM-mutated Dectin-1, suggesting that Dectin-1 provides an activation signal to induce a host-defense reaction, including the production of inflammatory cytokines and ROS, by using a cytoplasmic region (Brown et al., 2003; Gantner et al., 2003). The mechanism of signal transduction through this cytoplasmic domain is unclear. Some C-type lectins belonging to the NK-receptor group can associate with a signaling adaptor protein, DAP12, through their intrinsic amino acid residues in the transmembrane (TM) domain (Gilfillan et al., 2002). Dectin-1 does not have this amino acid sequence in the TM, but, exhibits another characteristic cationic amino acid residue, which may be involved in the association with an Fc R gamma chain existing in the TM of Dectin-1 (Ariizumi et al., 2000; Kanazawa et al., 2003). Therefore, some *src*-family tyrosine kinases maybe candidates as the Dectin-1-mediated signal transmitter. The characterization of signaling mechanisms followed by Dectin-1 engagement remains an important data gap.

5.5 CONCLUDING REMARKS

The receptors or binding molecules specific for 1,3-β-glucan have been extensively examined in recent studies. Since these molecules seem to be expressed or coexist in similar cell populations, such as the myeloid cell lineage, they may be orchestrated

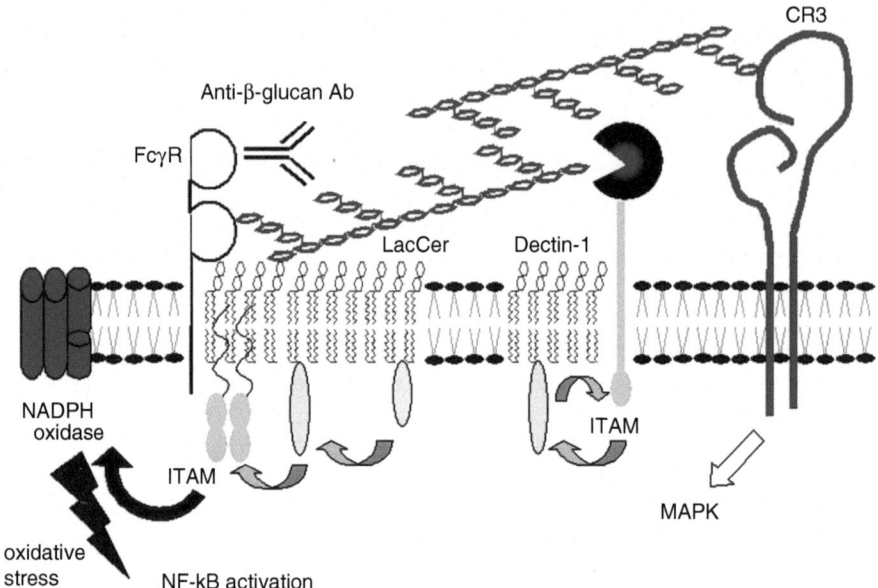

FIGURE 5.3 (See color insert following page 20.) Possible mechanisms in leukocyte activation signaling through multiple receptors.

by the cross-talk of each receptor on the cell surface. β-glucans generally have higher molecular weights with structure of repeating units. Thus, leukocytes exhibit the diverse biological activities in response to 1,3-β-glucans, depending on their molecular weight, branching structure, conformation, and solubility. In addition to multiple recognition of the β-glucan polymer, other immunological responses, such as the activation of a complement system resulting in the generation of anaphylatoxins, can influence the biological activities of β-glucans. It has also been demonstrated that some fungal 1,3-β-glucans can be epitopes of specific antibodies in the human immunological system. This expands the possibility that the activation signal via Fc receptors may participate in the activity of 1,3-β-glucan (Figure 5.3). Therefore, we should, consider that the β-glucan-mediated immunological function results from integration of the response to each functionally different receptor signal. It is necessary to elucidate the complete activation mechanism induced by 1,3-β-glucan and the synergistic or supplemental effect of other receptors, such as Fcγ receptors, cell adhesion or chemokine receptors, on signaling in response to β-glucan exposure.

REFERENCES

Adachi, Y., Ishii, T., Ikeda, Y., Hoshino, A., Tamura, H., Aketagawa, J., Tanaka, S., and Ohno, N. (2004). Characterization of β-glucan recognition site on C-type lectin, Dectin-1. *Infect. Immunity* 72(7), 4159–4171.

Adachi, Y., Ohno, N., and Yadomae, T. (1994). Preparation and antigen specificity of an anti-(1→3)-β-D-glucan antibody. *Biol Pharm. Bull.*, 17, 1508–1512.

β-Glucan Receptor(s) and Their Signal Transduction

Allen, M.J., Laederach, A., Reilly, P.J., and Mason, R.J. (2001). Polysaccharide recognition by surfactant protein D: novel interactions of a C-type lectin with nonterminal glucosyl residues. *Biochem.* 40, 7789–7798.

Anaissie, E. (1992). Opportunistic mycoses in the immunocompromised host: experience at a cancer center and review. *Clin. Infect. Dis.*, 14, S43–S53.

Ariizumi, K., Shen, G.L., Shikano, S., Xu, S., Ritter, R., III, Kumamoto, T., Edelbaum, D., Morita, A., Bergstresser, P.R., and Takashima, A. (2000). Identification of a novel, dendritic cell-associated molecule, dectin-1, by subtractive cDNA cloning. *J. Biol. Chem.* 275, 20157–20167.

Beschin, A., Bilej, M., Hanssens, F., Raymakers, J., Van Dyck, E., Revets, H., Brys, L., Gomez, J., De Baetselier, P., and Timmermans, M. (1998). Identification and cloning of a glucan- and lipopolysaccharide-binding protein from *Eisenia foetida* earthworm involved in the activation of prophenoloxidase cascade. *J. Biol. Chem.* 273, 24948–24954.

Bhunia, A.K., Arai, T., Bulkley, G., and Chatterjee, S. (1998). Lactosylceramide mediates tumor necrosis factor-alpha-induced intercellular adhesion molecule-1 (ICAM-1) expression and the adhesion of neutrophil in human umbilical vein endothelial cells. *J. Biol. Chem.* 273, 34349–34357.

Bhunia, A.K., Han, H., Snowden, A., and Chatterjee, S. (1997). Redox-regulated signaling by lactosylceramide in the proliferation of human aortic smooth muscle cells. *J. Biol. Chem.* 272, 15642–15669.

Brackman, D., Lund-Johansen, F., and Aarskog, D. (1995). Expression of leukocyte differentiation antigens during the differentiation of HL-60 cells induced by 1,25-dihydroxyvitamin D3: comparison with the maturation of normal monocytic and granulocytic bone marrow cells. *J. Leuko. Biol.* 58, 547–555.

Brown, G.D. and Gordon, S. (2001). Immune recognition: a new receptor for β-glucans. *Nature* 413, 36–37.

Brown, G.D., Herre, J., Williams, D.L., Willment, J.A., Marshall, A.S. and Gordon, S. (2003). Dectin-1 mediates the biological effects of β-glucans. *J. Exp. Med.* 197, 1119–1124.

Brown, M.S. and Goldstein, J.L. (1983). Lipoprotein metabolism in the macrophage: implications for cholesterol deposition in atherosclerosis. *Annu. Rev. Biochem.* 52, 223–261.

Cambi, A. and Figdor, C.G. (2003). Dual function of C-type lectin-like receptors in the immune system. *Curr. Opin. Cell Biol.* 15, 539–546.

Chihara, G., Hamuro, J., Maeda, Y., and Arai, Y. (1970). Antitumor polysaccharides, lentinan and pachymaran. *Saishin Igaku* 25, 1043–1048.

Czop, J.K. and Austen, K.F. (1985). A β-glucan inhibitable receptor on human monocytes: its identity with the phagocytic receptor for particulate activators of the alternative complement pathway. *J. Immunol.* 134, 2588–2593.

Czop, J.K., Gurish, M.F., and Kadish, J.L. (1990). Production and isolation of rabbit anti-idiotypic antibodies directed against the human monocyte receptor for yeast β-glucans. *J. Immunol.* 145, 995–1001.

Dejagar, S., Mietus-Snyder, M., and Pitas, R.E. (1993). Oxidized low density lipoproteins bind to the scavenger receptor expressed by rabbit smooth muscle cells and macrophages. *Aterioscler. Thromb. Vasc. Biol.* 13, 371–378.

Di Carlo, F.J. and Fiore, J.V. (1958). On the composition of zymosan. *Science* 127, 756–757.

Drickamer, K. (2002). C-type lectin-like domain databases. <http://ctld.glycob.ox.ac.uk/ctld/mammals/humandata.html> (Updated June 28, 2002).

Fliegmann, J., Mithofer, A., Wanner, G., and Ebel, J. (2004). An ancient enzyme domain hidden in the putative β-glucan elicitor receptor of soybean may play an active part in the perception of pathogen-associated molecular patterns during broad host resistance. *J. Biol. Chem.* 279, 1132–1140.

Gantner, B.N., Simmons, R.M., Canavera, S.J., Akira, S., and Underhill, D.M. (2003). Collaborative induction of inflammatory responses by dectin-1 and Toll-like receptor 2. *J. Exp. Med.* 197, 1107–1117.

Gilfillan, S., Ho, E.L., Cella, M., Yokoyama, W.M., and Colonna, M. (2002). NKG2D recruits two distinct adapters to trigger NK cell activation and costimulation. *Nat. Immunol.* 3, 1150–1155.

Grunebach, F., Weck, M.M., Reichert, J., and Brossart, P. (2002). Molecular and functional characterization of human Dectin-1. *Exp. Hematol.* 30, 1309–1315.

Hahn, P.Y. and Limper, A.H. (2001). Pneumocystis carinii β-glucan induces release of macrophage inflammatory protein-2 from primary rat alveolar epithelial cells via a receptor distinct from CD11b/CD18. *J. Eukaryot. Microbiol.* (suppl, 157S).

Harder, T. and Simons, K. (1997). Caveolae, DIGs, and the dynamics of sphingolipid-cholesterol microdomains. *Curr. Opin. Cell Biol.* 9, 534–542.

Herre, J., Gordon, S., and Brown, G.D. (2004). Dectin-1 and its role in the recognition of β-glucans by macrophages. *Mol. Immunol.* 40, 869–876.

Iwabuchi, K. and Nagaoka, I. (2002). Lactosylceramide-enriched glycosphingolipid signaling domain mediates superoxide generation from human neutrophils. *Blood* 100, 1454–1464.

Kanazawa, N., Tashiro, K., Inaba, K., and Miyachi, Y. (2003). Dendritic cell immunoactivating receptor, a novel C-type lectin immunoreceptor, acts as an activating receptor through association with Fc receptor gamma chain. *J. Biol. Chem.* 278, 32645–32652.

Kim, Y.S., Ryu, J.H., Han, S.J., Choi, K.H., Nam, K.B., Jang, I.H., Lemaitre, B., Brey, P.T., and Lee, W.J. (2000). Gram-negative bacteria-binding protein, a pattern recognition receptor for lipopolysaccharide and β-1,3-glucan that mediates the signaling for the induction of innate immune genes in *Drosophila melanogaster* cells. *J. Biol. Chem.* 275, 32721–32727.

Kishor, U., Madan, T., Sarma, P.U., Singh, M., Urban, B.C., and Reid, K.B. (2002). Protective roles of pulmonary surfactant proteins, SP-A and SP-D, against lung allergy and infection caused by *Aspergillus fumigatus*. *Immunobiol.* 205, 610–618.

Kuroki, Y. and Voelker, D.R. (1994). Pulmonary surfactant proteins. *J. Biol. Chem.* 269, 25943–25946.

Lipke, P.N. and Ovalle, R. (1998). Cell wall architecture in yeast: new structure and new challenges. *J. Bacteriol.* 180, 3735–3740.

Masuzawa, S., Yoshida, M., Ishibashi, K., Saito, N., Akashi, M., Yoshikawa, N., Suzuki, T., Nameda, S., Miura, N.N., Adachi, Y., and Ohno, N. (2003). Solubilized Candida cell wall β-glucan, CSBG, is an epitope of natural human antibody. *Drug Devel. Res.* 58, 179–189.

Miyazaki K., Mizutani, H., Katabuchi, H., Fukuma, K., Fujisaki, S., and Okamura, H. (1995). Activated (HLA-DR⁺) T-lymphocyte subsets in cervical carcinoma and effects of radiotherapy and immunotherapy with sizofiran on cell-mediated immunity and survival. *Gynecol. Oncol.* 56, 412–420.

Myskowski, P.L., White, M.H., and Ahkami, R. (1997). Fungal disease in the immunocompromised host. *Dermatol. Clin.* 15, 295–305.

Natarajan, K., Dimasi, N., Wang, J., Mariuzza, R.A., and Margulies, D.H. (2002). Structure and function of natural killer cell receptors: multiple molecular solutions to self, nonself discrimination. *Annu. Rev. Immunol.* 20, 853–885.

Ochiai, M. and Ashida, M. (1988). Purification of a β-1,3-glucan recognition protein in the prophenoloxidase activating system from hemolymph of the silkworm, *Bombyx mori*. *J. Biol. Chem.* 263, 12056–12062.

Ochiai, M., Niki, T., and Ashida, M. (1992). Immunocytochemical localization of β-1,3-glucan recognition protein in the silkworm, *Bombyx mori*. *Cell Tissue Res.* 268, 431–437.

Pearson, A., Lux, A., and Krieger, M. (1995). Expression cloning of dSR-CI, a class C macrophage-specific scavenger receptor from *Drosophila melanogaster*. *Proc. Nat. Acad. Sci. U.S.A.* 92, 4056–4060.

Rezzonico, R., Chicheportiche, R., Imbert, V., and Dayer, J.M. (2000). Engagement of CD11b and CD11c β2 integrin by antibodies or soluble CD23 induces IL-1β production on primary human monocytes through mitogen-activated protein kinase-dependent pathways. *Blood* 95, 3868–3877.

Rice, P.J., Kelley, J.L., Kogan, G., Ensley, H.E., Kalbfleisch, J.H., Browder, I.W., and Williams, D.L. (2002). Human monocyte scavenger receptors are pattern recognition receptors for (1→3)-β-D-glucans. *J. Leuko. Biol.* 72, 140–146.

Ross, G.D. (2000). Regulation of the adhesion versus cytotoxic functions of the Mac-1/CR3/alphaMβ2-integrin glycoprotein. *Crit. Rev. Immunol.* 20, 197–222.

Ross, G.D., Cain, J.A., Myones, B.L., Newman, S.L., and Lachmann, P.J. (1987). Specificity of membrane complement receptor type three (CR3) for β-glucans. *Complement* 4, 61–74.

Roubey, R.A., Ross, G.D., Merrill, J.T., Walton, F., Reed, W., Winchester, R.J., and Buyon, J.P. (1991). Staurosporine inhibits neutrophil phagocytosis but not iC3b binding mediated by CR3 (CD11b/CD18). *J. Immunol.* 146, 3557–3562.

Roux, M.M., Pain, A., Klimpel, K.R., and Dhar, A.K. (2002). The lipopolysaccharide and β-1,3-glucan binding protein gene is upregulated in white spot virus-infected shrimp (*Penaeus stylirostris*). *J. Virol.* 76, 7140–7149.

Sharp, J.K., Valent, B., and Albersheim, P. (1984). Purification and partial characterization of a β-glucan fragment that elicits phytoalexin accumulation in soybean. *J. Biol. Chem.* 259, 11312–11320.

Sritunyalucksana, K., Lee, S.Y., and Soderhall, K. (2002). A β-1,3-glucan binding protein from the black tiger shrimp, *Penaeus monodon*. *Dev. Comp. Immunol.* 26, 237–245.

Steele, C., Marrero, L., Swain, S., Harmsen, A.G., Zheng, M., Brown, G.D., Gordon, S., Shellito, J.E., and Kolls, J.K. (2003). Alveolar macrophage-mediated killing of *Pneumocystis carinii* f. sp. muris involves molecular recognition by the Dectin-1 β-glucan receptor. *J. Exp. Med.* 198, 1677–1688.

Steinberg, D. (1995). Role of oxidized LDL and antioxidants in atherosclerosis. *Adv. Exp. Med. Biol.* 369, 39–48.

Symington, F.W., Hedges, D.L., and Hakomori, S. (1985). Glycolipid antigens of human polymorphonuclear neutrophils and the inducible HL-60 myeloid leukemia line. *J. Immunol.* 134, 2498–2506.

Szabo, T., Kadish, J.L., and Czop, J.K. (1995). Biochemical properties of the ligand-binding 20-kDa subunit of the β-glucan receptors on human mononuclear phagocytes. *J. Biol. Chem.* 270, 2145–2151.

Takaki, Y., Seki, N., Kawabata, Si, S., Iwanaga, S., and Muta, T. (2002). Duplicated binding sites for (1→3)-β-D-glucan in the horseshoe crab coagulation factor G: implications for a molecular basis of the pattern recognition in innate immunity. *J. Biol. Chem.* 277, 14281–14287.

Takeda, K., Kaisho, T., and Akira, S. (2003). Toll-like receptors. *Annu. Rev. Immunol.* 21, 335–376.

Tanaka, S., Aketagawa, J., Takahashi, S., Shibata, Y., Tsumuraya, Y., and Hashimoto, Y. (1993). Inhibition of high-molecular-weight-(1→3)-β-D-glucan-dependent activation of a limulus coagulation factor G by laminaran oligosaccharides and curdlan degradation products. *Carbohydr. Res.* 244, 115–127.

Taylor, P.R., Brown, G.D., Herre, J., Williams, D.L., Willment, J.A., and Gordon, S. (2004). The role of SIGNR1 and the β-glucan receptor (Dectin-1) in the nonopsonic recognition of yeast by specific macrophages. *J. Immunol.* 172, 1157–1162.

Umemoto, N., Kakitani, M., Iwamatsu, A., Yoshikawa, M., Yamaoka, N. and Ishida, I. (1997). The structure and function of a soybean β-glucan-elicitor-binding protein. *Proc. Nat. Acad. Sci. U.S.A.* 94, 1029–1034.

Vetvicka, V., Thornton, B.P., Wieman, T.J. and Ross, G.D. (1997). Targeting of natural killer cells to mammary carcinoma via naturally occurring tumor cell-bound iC3b and β-glucan-primed CR3 (CD11b/CD18). *J. Immunol.* 159, 599–605.

Willment, J.A., Gordon, S., and Brown, G.D. (2001). Characterization of the human β-glucan receptor and its alternatively spliced isoforms. *J. Biol. Chem.* 276, 43818–43823.

Xia, Y., Borland, G., Huang, J., Mizukami, I.F., Petty, H.R., Todd, R.F., III, and Ross, G.D. (2002). Function of the lectin domain of Mac-1/complement receptor type 3 (CD11b/CD18) in regulating neutrophil adhesion. *J. Immunol.* 169, 6417–6426.

Xia, Y. and Ross, G.D. (1999). Generation of recombinant fragments of CD11b expressing the functional β-glucan-binding lectin site of CR3 (CD11b/CD18). *J. Immunol.* 162, 7285–7293.

Yadomae, T. and Ohno, N. (1996). Structure-activity relationship of immunomodulating (1→3)-β-D-glucans. *Recent Res. Devel. Chem. Pharmacal. Sci.* 1, 23–33.

Yokota, K., Takashima, A., Bergstresser, P.R. and Ariizumi, K. (2001). Identification of a human homologue of the dendritic cell-associated C-type lectin-1, dectin-1. *Gene* 272, 51–60.

Young, R.M., Holowka, D., and Baird, B. (2003). A lipid raft environment enhances Lyn kinase activity by protecting the active site tyrosine from dephosphorylation. *J. Biol. Chem.* 278, 20746–20752.

Zaffran, Y., Escallier, J.C., Ruta, S., Capo, C., and Mege, J.L. (1995). Zymosan-triggered association of tyrosine phosphoproteins and lyn kinase with cytoskeleton in human monocytes. *J. Immunol.* 154, 3488–3497.

Zimmerman, J.W., Lindermuth, J., Fish, P.A., Palace, G.P., Stevenson, T.T., and DeMong, D.E. (1998). A novel carbohydrate-glycosphingolipid interaction between a β-(1→3)-glucan immunomodulator, PGG-glucan, and lactosylceramide of human leukocytes. *J. Biol. Chem.* 273, 22014–22020.

6 Fate of β-Glucans *In Vivo*: Organ Distribution and Degradation Mechanisms of Fungal β-Glucans in the Body

Noriko N. Miura

CONTENTS

6.1 Introduction 109
6.2 Study of Organ Distribution Using a Metabolically Labeled
 Form of SSG from *Sclerotinia sclerotiorum* 110
6.3 Study of Blood β-Glucan Concentrations Using the *Limulus* Test 114
6.4 Analysis of Internal Accumulation Using a Metabolically Labeled
 Form of *Candida* 116
6.5 Measurement of Amounts of *Candida* Cells Accumulating
 in Organs 117
6.6 Relationship between β-Glucan Dosage and Duration of
 Antitumor Activity 118
6.7 Solubilization of β-Glucans from *Candida* Cells 119
6.8 Conclusion 121
References 124

6.1 INTRODUCTION

The average life span in developed countries has increased considerably with the progress in the field of health care. However, the number of immunocompromised individuals has increased as well. There is a high incidence of deep mycoses among immunocompromised hosts, including patients with malignancies of the hematopoietic organs (such as leukemia and malignant lymphoma) as well as AIDS patients, where fungal infection often has a considerable effect on the patient's prognosis. The treatment of deep mycoses began with the clinical use of such drugs as fluconazole, itraconazole, and terbinafine starting in the end of the 1980s through the early 1990s, and those drugs have afforded effective and safe treatment (Vincent, 1999). Moreover, with the recent approval of micafungin, even quite serious cases of deep mycoses can now be treated without the use of amphotericin B, which has a high level of side effects (Ikeda, 2003; Walsh et al., 2000; Stone et al., 2002). However, there is still a significant number of patients who die of deep mycoses.

Early diagnosis is vital for the effective treatment of deep mycoses. One method for diagnosing deep mycoses is the Limulus test (Obayashi et al., 1995 and 1992; Tamura et al., 1997; Yoshida et al., 1997 and 1999), which consists of detecting (1→3)-β-D-glucan in blood samples, as β-glucans are present in the blood at the time of fungal infection (Kato et al., 2001). The major cell wall components of fungi are mannoprotein (20–30%), β-(1→3)-glucans (25–35%), β-(1→6)-glucans (35–45%), chitin (0.6–2.7%), protein (5–15%), and lipid (2–5%) (Bishop et al., 1960; Yu et al., 1967; Shepherd, 1991; Osumi, 1998; Chaffin et al., 1998). Although it is thought that a portion of the cell wall is released into the blood at the time of fungal infection (Nagi et al., 1992), thereby enabling the detection of β-glucans in the blood, the details behind this phenomenon are not clear.

On the other hand, β-glucans are also used as biological response modifiers (BRMs), the examples of which are krestin, schizophyllan, and lentinan, and those BRMs have been administered to humans as immunomodulators for cancer treatment (Fisher and Yang, 2002; Hayakawa et al., 1997; Borchers et al., 1999; Kidd, 2000; Wasser, 2002). Although the pharmacokinetics of these β-glucans in the body is thought to resemble that at the time of fungal infection, humans, mice, and other mammals do not possess enzymes that degrade β-glucans. It is for that reason that the β-glucans administered to the body are thought to be processed by oxidative degradation (Nono et al., 1991), although the details of this mechanism remain unclear. In addition, as the preferred sites of deep mycoses are the liver and kidneys, a large number of fungal cells accumulate in these organs. In the diagnosis of mycoses, it is extremely important to analyze the elution mechanism of β-glucans from the fungal cells, and the mechanism of degradation of the fungal cells that have accumulated in these organs.

Therefore, it is vital to elucidate the distribution and degradation mechanisms of β-glucans in the body in both cases of deep mycoses and after using β-glucans as BRMs. We describe herein the results of a comparative study of the metabolism, distribution, and degradation of β-glucans using the Limulus test, tritium-labeled SSG (a β-glucan obtained from *Sclerotinia sclerotiorum* IFO 9395), and tritium-labeled *Candida* cells.

6.2 STUDY OF ORGAN DISTRIBUTION USING A METABOLICALLY LABELED FORM OF SSG FROM *SCLEROTINIA SCLEROTIORUM*

SSG is a soluble β-glucan obtained from the culture supernatant of *Sclerotinia sclerotiorum*. It contains β-(1→6)-D-glucosyl side chains at the ratio of one residue per two main chain residues consisting of β-(1→3) bonds. SSG differs from other glucans in that it possesses antitumor effects even when administered orally (Ohno et al., 1986; Suzuki et al., 1991). The distribution of SSG was examined by preparing a tritium labeled form of SSG (^{3}H-SSG) (Suda et al., 1994), administering it i.p.ly to mice, and excising the organs over time (Suda et al., 1992). As a result, SSG was found to be distributed primarily in the liver and spleen following administration. Radioactivity slowly disappeared from peritoneal exudate (supernatant) with roughly 10% of the dose remaining in the peritoneal cavity 48 hours after administration. The half-life of SSG in the peritoneal cavity was estimated to be 8–10 hours. Although the activation of alveolar macrophage has been reported in the case of i.p. administration (Sakurai et al., 1991), there was little deposition in the lungs, and excretion into the urine and feces was low. In addition, there was also very little radioactivity present in peritoneal exudated cells (PECs) that were thought to be the first to interact with the administered SSG (Figure 6.1A). In contrast, high levels of radioactivity were observed in the liver and spleen even at 4 weeks after administration during the course of long-term observation, with the levels being roughly the same as those observed at 1 week after administration. The administered SSG disappeared from the circulation after 1 week (Figure 6.1B). The results suggest that the metabolic degradation and excretion of SSG are extremely slow. Normally, BRM activity does not persist even for 1 month in the case of i.p. administration of SSG to mice at a dose of 250 µg per mouse. For example, changes in the number of spleen cells constitute a typical indicator of BRM activity; those changes reach a maximum 1 week after SSG administration and return to the control level after 1 month (Suda et al., 1992). These results suggest that there is no direct correlation between the biological activity of SSG and its residual levels.

An additional study was conducted to determine the form in which the SSG trapped in organs exists (Suda et al., 1996). The liver is the largest organ in the animal body and plays a central role in the processing of foreign substances. It is also the organ with the largest distribution of SSG. Hepatocytes are broadly divided into parenchymal cells and nonparenchymal cells (hepatic sinusoidal cells), and Kupffer cells are nonparenchymal cells that serve to fix hepatic macrophage. Macrophages have receptors that specifically recognize and bind with β-glucans. Therefore, the distribution of radiolabeled SSG can be studied by perfusing the liver with collagenase solution to obtain single cells, fractionating the cells into parenchymal and nonparenchymal cells by density gradient centrifugation, and examining the radioactivity distribution. Although the radioactivity distribution in the entire liver was approximately 35%, more than half of the radioactivity disappeared due to perfusion with collagenase solution. As the disappearance of radioactivity was not observed upon perfusion with solution that contained no collagenase, an interaction with some type of protein that is solubilized by collagenase is suggested to be important for the distribution of SSG.

On the other hand, SSG was also distributed mainly in the liver and spleen in the case of intravenous administration, and remained in these organs for at least 1 month (Figure 6.2). A comparison with the organ distribution during i.p. administration revealed no remarkable differences with the exception of the distribution in the kidneys being approximately 1.5–2.0 times higher for intravenous administration compared to i.p. treatment.

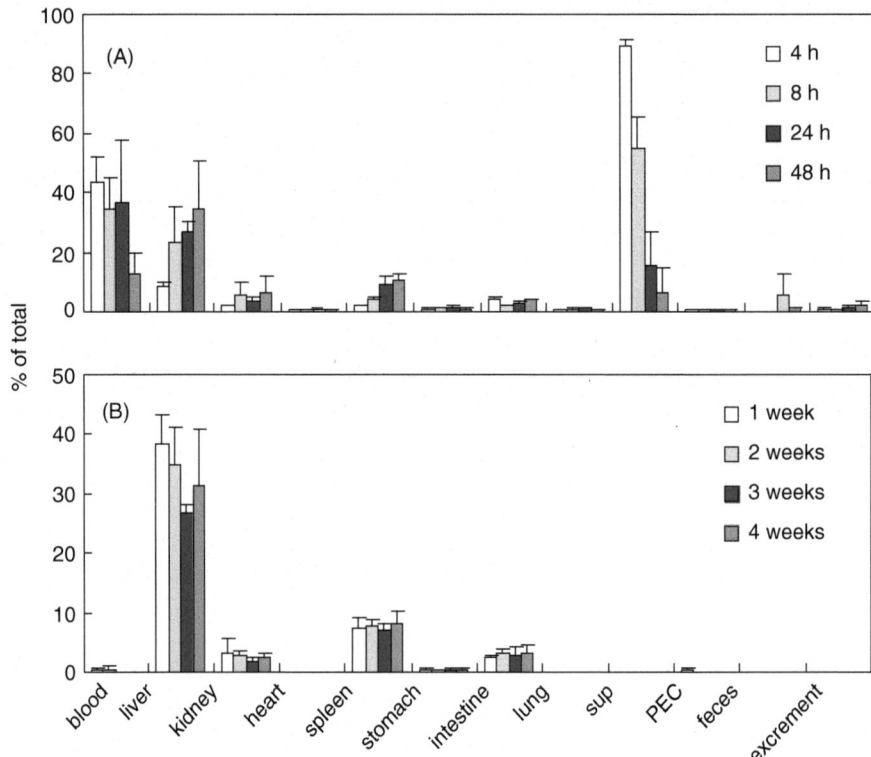

FIGURE 6.1 Distribution of ^{3}H-SSG in various tissues. A 250 μg dose of ^{3}H-SSG (about 87000 dpm [total ^{3}H-SSG]) was administered i.p. to ICR mice. Four, 8, 24, or 48 hours later (A), or 7, 14, 21, or 28 days later (B), tissues from mice were removed and oxidized. The data were expressed as follows: percent of total = [dpm (dried tissue)/dpm (total ^{3}H-SSG)] × 100. "Sup" is the supernatant fraction after centrifugation of the peritoneal exudate cell suspension recovered by washing peritoneal cavities of mice with Hanks solution. PEC is the peritoneal exudate cell pellet obtained after centrifugation.

In contrast, 98% of orally administered SSG was excreted into the feces within 24 hours after administration, suggesting that hardly any SSG is absorbed from the gastrointestinal system by the body (Table 6.1).

SSG exhibits potent antitumor activity using all (i.p., intravenous, or oral) routes of administration (Suzuki et al., 1988; Ohno et al., 1987; Sakurai et al., 1995). However, even in the cases of i.p. administration and intravenous administration for which there were no remarkable differences observed in organ distribution, different expression patterns of immunopharmacological activity have been reported (Sakurai et al., 1992). Therefore, the correlation between SSG activity and its organ distribution raises some extremely interesting questions.

Fate of β-Glucans *In Vivo*

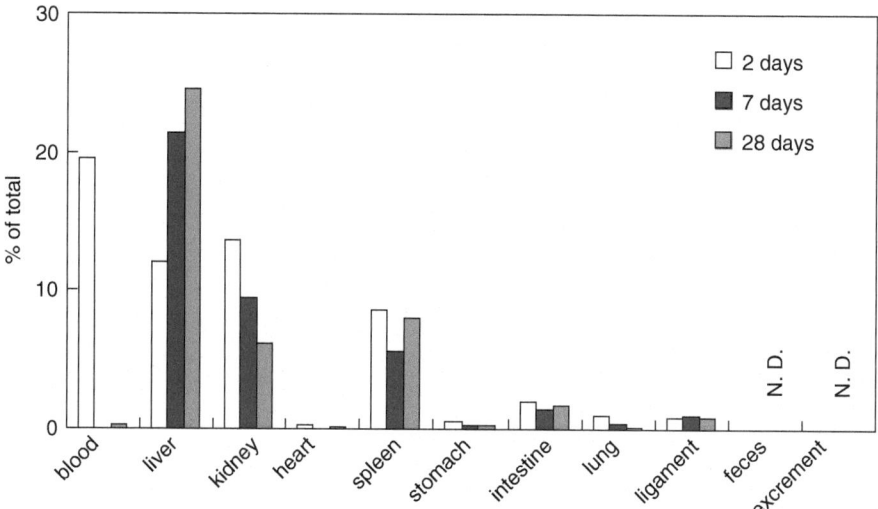

FIGURE 6.2 Tissue distribution of intravenously administered ^{3}H-SSG. A 250 μg dose of ^{3}H-SSG was administered intravenously to ICR mice. The data are expressed as follows: % of total = [dpm (dried tissue)/dpm (total ^{3}H-SSG)] × 100.

TABLE 6.1
Tissue Distribution of Orally Administered ^{3}H-SSG

	% Recovery
Blood	0.04
Liver	0.25
Kidney	0.07
Heart	n.d.
Spleen	n.d.
Stomach	0.94
Intestine	0.25
Lung	n.d.
Feces	98.41
Excrement	n.d.

A 2 mg dose of ^{3}H-SSG was orally administered to ICR mice, and 24 h later each tissue was removed.

6.3 STUDY OF BLOOD β-GLUCAN CONCENTRATIONS USING THE *LIMULUS* TEST

SSG administered i.p.ly is mainly distributed in the liver after being transported by the blood. A study of the rate of disappearance of β-glucans from the blood was conducted using the *Limulus* test by administering sonifilan (schizophyllan, SPG), which is also used clinically, to ICR mice (Miura et al., 1995). When SPG was administered intravenously to the mice at a dose of 1, 10, or 100 μg per mouse, blood concentrations could be measured even at 24 hours after administration at 10 and 100 μg, whereas blood concentrations were below the detection limit (10 ng/mL) following administration at 1 μg (Figure 6.3A). If the disappearance of SPG from the blood were assumed to follow a one-compartment model, the half-life for administration at 100 μg would be approximately 5.0 hours. On the other hand, when SPG was administered i.p.ly at doses of 10, 100, and 1000 μg per mouse, maximum blood concentrations were demonstrated at approximately 4 hours, after which the SPG gradually disappeared from the blood (Figure 6.3B).

β-Glucan preparations are repeatedly administered to cancer patients in the clinical setting. Therefore, grifolan (GRN), a β-glucan obtained from *Grifola frondosa*, or SSG was repeatedly administered to an immunocompromised model of MRL *lpr/lpr* mice presenting with a systemic lupus erythematosus (SLE)-like immune disease to study its disappearance from the blood. Blood was sampled over the course of 2 weeks from mice that were intraperitoneally administered 250 μg of GRN or SSG once a week for 37 weeks, and blood concentration was measured using the *Limulus* test (Miura et al., 1996a). The mice were unable to eliminate the β-glucans from the blood over the course of seven days, and blood β-glucan

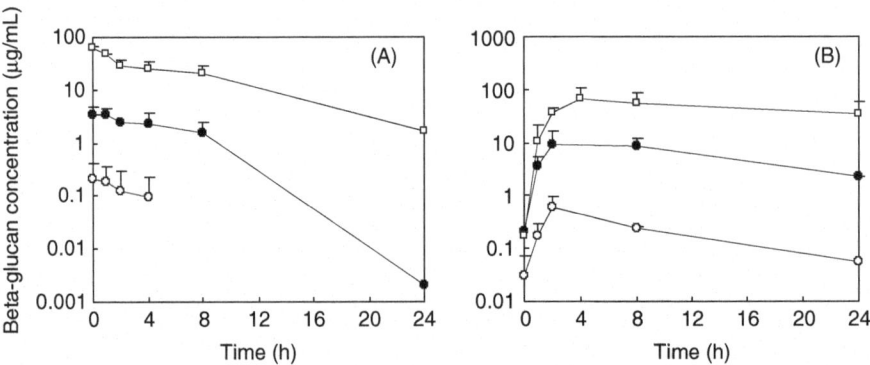

FIGURE 6.3 Dose response of β-glucan clearance from plasma following i.p. or intravenous administration of SPG. Several doses of SPG (1–100 μg per mouse for intravenous and 10–1000 μg per mouse for i.p.) were administered intravenously (i.v.) (A) or i.p.ly (i.p.) (B) to mice (n = 2), and blood samples were collected at various times. ○, i.v. 1 μg, i.p. 10 μg; ●, i.v. 10 μg, i.p. 100 μg; □, i.v. 100 μg, i.p. 1000 μg. Plasma was obtained and the β-glucan concentration measured by the *Limulus* test. Before carrying out the test, each sample was treated with 0.5 N NaOH.

Fate of β-Glucans *In Vivo*

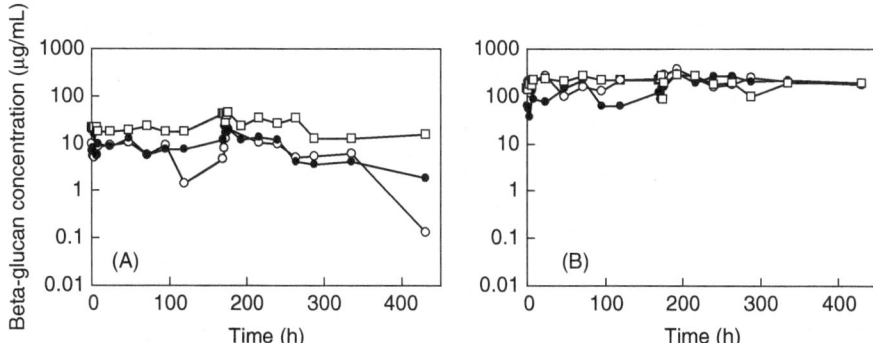

FIGURE 6.4 Changes in β-glucan concentration in plasma after multiple administrations of GRN or SSG in MRL *lpr/lpr* mice. 250 μg of GRN (A) or SSG (B) was administered to each MRL *lpr/lpr* mouse (3 mice per glucan; GRN-1[○], GRN-2[●], GRN-3[□], SSG-1[○], SSG-2[●], SSG-3[□]) once a week for about 37 weeks i.p.ly. For the final 2 weeks, we measured the β-glucan concentration in these mice by the Limulus test.

concentrations remained high (Figure 6.4). On the other hand, when β-glucan concentration in the mouse organs was measured, roughly 56% of the repeatedly administered GRN was recovered from the liver, whereas roughly 7% was recovered from the spleen. In comparison, 1 week after a single administration of GRN, roughly 21% was recovered from the liver, whereas roughly 8% was recovered from the spleen (Table 6.2). These results indicate that a substantial amount of the β-glucans administered was accumulated in those organs. In addition, β-glucan levels that exceeded the administered dose were recovered from the liver and spleen in the case of SSG. This is surmised to be a result of the SSG having undergone some form of structural change in the liver (this will be described in Section 6.7).

TABLE 6.2
Accumulation of β-Glucan in Liver and Spleen in MRL *lpr/lpr* Mice[a]

β-Glucan (Dose)	Liver (mg)	Spleen (mg)
Multiple administration		
GRN (10 mg)	5.6 ± 0.9	0.7 ± 0.1
SSG (10 mg)	27.8 ± 11.1	5.4 ± 0.4
Single administration		
GRN (250 mg)	0.053 ± 0.023	0.019 ± 0.006

[a] Each MRL *lpr/lpr* mice was administered β-glucan and these organs were collected. Samples were stored at –4°C and treated with 0.5 N NaOH before measurement of β-glucan concentration by the *Limulus* test.

6.4 ANALYSIS OF INTERNAL ACCUMULATION USING A METABOLICALLY LABELED FORM OF *CANDIDA*

In order to investigate the distribution of *Candida* cells in the body during fungal mycoses, ³H-glucose was added to a liquid culture to obtain ³H-labeled *Candida* cells, followed by i.p. administration of the whole cells to mice, and excision of the organs 2 days later to compare the accumulated amounts (Miura et al., 1998). When the tritiated SSG, a soluble β-glucan, was administered to mice, the majority of the administered amount was distributed in the liver and spleen. However, as shown in Table 6.3, in the case of i.p. administration of whole *Candida* cells, the distribution differed considerably, with a large amount (20.2%) being present in ligament such as the omentum having milky spots. On the other hand, in the case of intravenous administration of whole *Candida* cells, nearly the entire amount accumulated in the liver, and accumulation in the omentum was very low. Therefore, the omentum may be intimately involved in the processing of foreign objects in the peritoneal cavity.

TABLE 6.3
Distribution of ³H-SSG or Labeled *Candida* Extracts in Various Tissues of Mice

	SSG i.p.	Whole Cells i.p.	Mannan i.p.	Soluble Glucan i.p.	Insoluble Glucan i.p.	Residue i.p.	α-Amylase Treated Residue i.p.	Whole Cells i.v.
Blood	37.3	0.5	0.1	0.4	0.2	0.3	3.5	0.8
Liver	25.4	15.9	10.2	4.9	44.8	8.6	14.0	49.5
Kidney	15.0	1.1	0.6	0.5	0.6	0.6	2.3	1.2
Heart	0.6	0.0	0.0	0.0	0.1	0.0	0.1	0.1
Spleen	12.0	2.9	3.9	0.7	4.0	2.4	12.4	3.4
Stomach	0.4	0.7	0.4	0.2	1.8	4.3	3.4	0.2
Small Intestine	1.8	1.3	1.0	0.8	1.5	1.2	3.4	1.3
Ligament	0.5	20.2	1.4	1.7	4.7	10.4	23.0	0.3
Large Intestine	1.0	1.5	0.4	0.4	0.4	0.4	0.4	0.5
Lung	1.5	0.6	0.1	0.1	0.7	0.4	0.1	2.2
Sup	3.6	5.8	2.0	3.6	1.9	4.0	0.6	4.9
PEC	1.3	7.9	0.1	0.7	1.8	2.7	2.2	0.0
Feces	0.1	0.2	1.1	0.9	1.6	1.5	0.1	0.8
Excrement	1.0	9.4	9.0	5.4	10.9	28.9	11.7	9.0
Total		68.1	30.5	20.3	74.9	65.6	77.1	74.3

³H-SSG or ³H labeled *Candida* cells and extracts were administered intraperitoneally or intravenously to normal ICR mice. Two days later, each tissue from mice was removed and oxidized. The data are expressed as follows: % of total administered dose = {dpm (dried tissue)/dpm (total administered dose)} × 100.

Fate of β-Glucans In Vivo

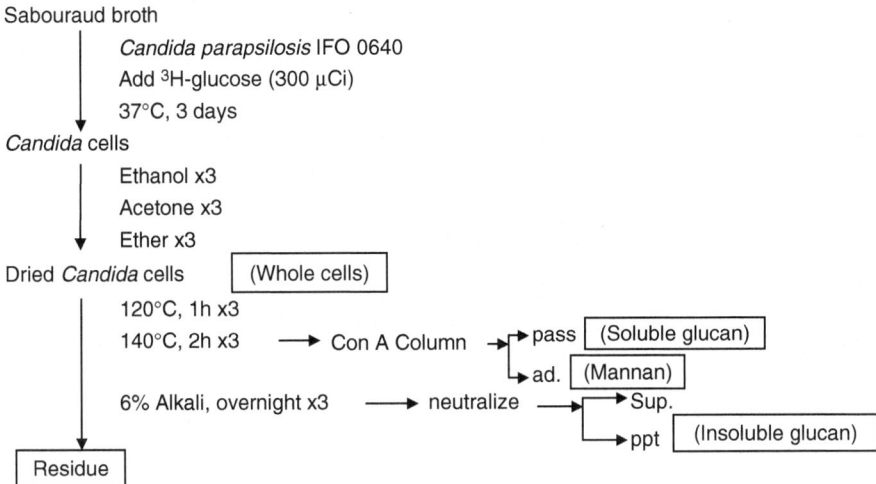

CHART 6.1

In order to compare the metabolic rates for each component, fungal components were fractioned as shown in Chart 6.1, followed by administration of each component to investigate its distribution. The insoluble fraction (residue) was treated with α-amylase to remove α-glucans. Although administration of the soluble glucan fraction resulted in a high distribution in the liver in the same manner as SSG, the amount excreted was greater than that of SSG. The residue was also distributed at high levels in the liver and excreted, and it was also distributed in the omentum (Table 6.3).

The β-glucan itself is labeled in the case of the soluble β-glucan, SSG. However, in the case of *Candida*, not only the β-glucan portion but also all other cell components are thought to be labeled. The pharmacokinetics following i.p. administration of whole cells are similar to that of particulate substances, such as carbon particles, with a large amount accumulating in the omentum, whereas the portion that is excreted is either the low-molecular-weight portion, that is immediately released by the cells, or the portion that is rapidly converted to a lower molecular weight.

6.5 MEASUREMENT OF AMOUNTS OF *CANDIDA* CELLS ACCUMULATING IN ORGANS

In order to study the behavior of *Candida* cell wall β-glucans after fungal infections have been cured, 1 mg of whole *Candida* cells was administered to mice either i.p. or intravenously, followed by the measurement of their disappearance from the blood and accumulation in the liver and spleen over the course of 6 months using the *Limulus* test (Figure 6.5) (Miura et al., 1998). As a result, the whole *Candida* cells were found to accumulate in the body for at least longer than 6 months regardless of the administration route. In the case of intravenous administration of whole *Candida* cells, it was found that nearly all of the cells accumulated in the liver and

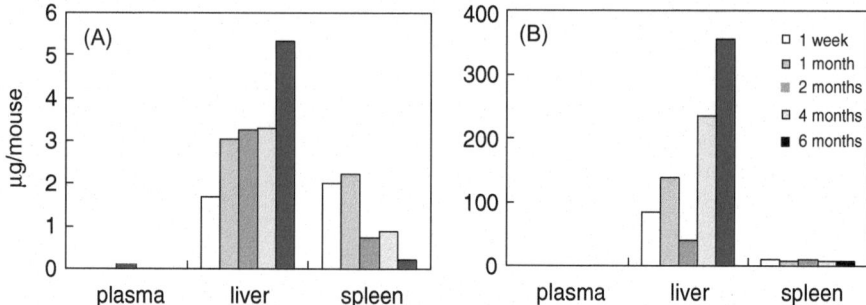

FIGURE 6.5 Accumulation of β-glucan in organs of *Candida* whole cell administered mice. *Candida* whole cells (1 mg) were administered i.p. (A) or i.v. (B) to mice. One week, 1, 2, 4, or 6 months later, tissues from mice were removed and β-glucan concentration was measured by *Limulus* test. Each sample was treated with 0.5 N NaOH twice and then diluted with 0.01 N NaOH before this assay.

gradually increased in amount. The *Limulus* reactivity of the liver was high and exceeded the dose of *Candida* cells prior to administration. This was thought to indicate an apparent increase in reactivity due to the oxidative degradation of *Candida* cells in the liver, as will be described later. Although a majority of *Candida* cells accumulated in the liver during i.p. administration, the accumulated amount was less than 1/50 of that during intravenous administration, indicating that a majority of the cells do not leave the peritoneal cavity but remain therein.

6.6 RELATIONSHIP BETWEEN β-GLUCAN DOSAGE AND DURATION OF ANTITUMOR ACTIVITY

The antitumor activity of β-glucans against the mouse transplanted tumor Sarcoma 180 is such that 3 administrations of β-glucans at a dose of 50 to 250 μg per mouse 1 week after tumor transplant are considered to yield potent antitumor activity. However, β-glucans differ from other typical drugs in that their antitumor activity does not increase with an increase in dose, and there is hardly any occurrence of side effects. SPG was administered at five to ten times the optimum dose (1–2 mg per mouse) and the resultant effects were compared (Chart 6.2) (Miura et al., 2000). Antitumor activity was decreased when excess SPG was administered 1 week after tumor transplant. However, antitumor activity was demonstrated when SPG was administered 60 or 90 days prior to tumor transplant (Table 6.4). On the basis of these results, it is suggested that some form of biological deactivating mechanism functions temporarily when SPG is administered in large doses. When blood concentrations of SPG during large-dose administration were measured using the *Limulus* test, the concentration in mice given three administrations at a dose of 2 mg per mouse was 731.5 μg/mL after 1 day. This fell to 5.0 μg/mL after 1 month. The blood concentration in mice given three administrations of 50 μg per mouse was 9.9 μg/mL after 1 day. The blood concentration after 1 month approached the values measured after administration at the optimum dose.

Fate of β-Glucans *In Vivo* 119

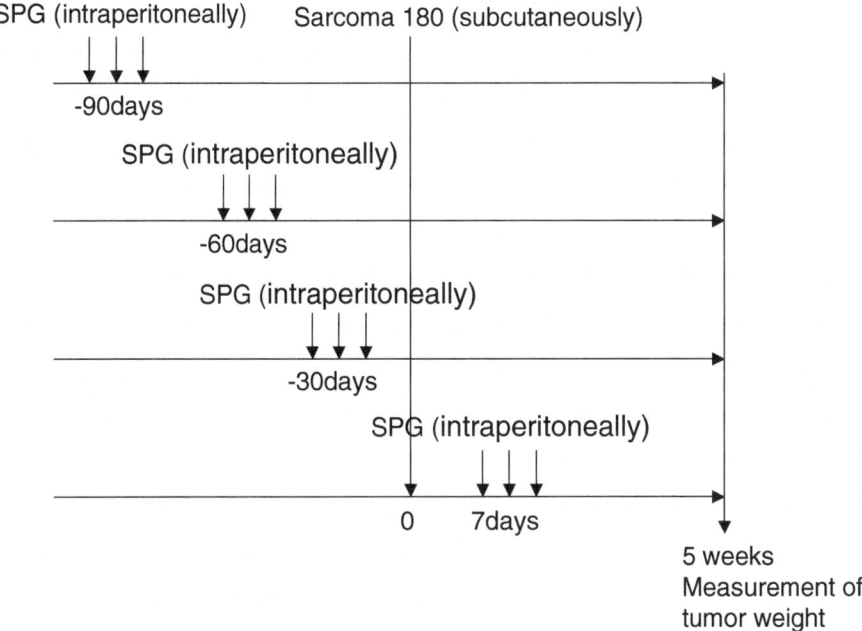

CHART 6.2

6.7 SOLUBILIZATION OF β-GLUCANS FROM *CANDIDA* CELLS

As the amount of Limulus-test-positive substances that accumulated in the mouse liver following administration of *Candida* cells increased over time (as previously described in Section 6.5), a study was conducted to determine whether or not this was a result of solubilization of *Candida* cells using a hypochlorous acid oxidation system (Miura et al., 1998). Hypochlorous acid oxides of *Candida* cells, the alkaline extract, or the hot alkaline extract of *Candida* cells were prepared and examined for their reaction in the Limulus test (Figure 6.6). As a result, the degree of Limulus reactivity was temporarily increased as the oxidation of *Candida* cells progressed, and then decreased as the oxidation progressed further. However, the Limulus reactivity of the alkaline or the hot alkaline extract of *Candida* cells did not increase as a result of oxidation, and decreased as oxidation progressed.

In order to explain this phenomenon, hypochlorous acid oxides were prepared from the linear β-glucan, curdlan (CRD), the branched β-glucan, SSG, and the particulate β-glucan preparation, zymosan (ZYM), followed by examination of their Limulus reactivities (Miura et al., 1996b). The Limulus reactivities of SSG and ZYM were increased considerably by hypochlorous acid oxidation (Figure 6.7). The reactivity of the oxidized SSG obtained by oxidizing for 1 day with 12.5 mL of hypochlorous acid was 100 times higher than that of untreated SSG, whereas ZYM oxidized under the same conditions exhibited reactivity that was roughly 50 times higher than that of untreated ZYM. However, the reactivity was decreased considerably when

TABLE 6.4
Antitum or Activity of Various Doses of SPG in ICR Mice

Day	Dose (μg/mouse ×3)	Tum or weight (g, mean ± S.D.)	Inhibition (%)	CR/Total
−90	50	2.2 ± 1.5	62.7	4/8
	250	0.7 ± 0.7	88.1	3/7
	1000	0.2 ± 0.4		5/9
	2000	0.7 ± 1.0	88.1	1/10
−60	50	2.0 ± 3.6	66.1	5/9
	250	0.4 ± 1.0	93.2	5/7
	1000	0.8 ± 1.1	86.4	1/7
	2000	0.9 ± 1.0	84.7	3/9
−30	50	2.2 ± 3.3	62.7	5/10
	250	0	100.0	10/10
	1000	1.5 ± 1.4	74.6	3/10
	2000	2.7 ± 3.7	54.2	1/10
7	50	0.6 ± 1.5	89.8	8/10
	250	0.2 ± 0.5	96.6	9/10
	1000	1.3 ± 2.1	78.0	2/10
	2000	2.6 ± 1.9	55.9	1/10
Nil	0	5.9 ± 3.1	0.0	0/10

SPG was administered three times every 90, 60, 30 days before or 7 days after Sarcoma 180 inoculation. Doses of SPG were 50, 250, 1000, 2000 μg/mouse (n = 7 – 10). Sarcoma 180 tumor cells (5×10^6 cells/mouse) were inoculated subcutaneously into the right groin of ICR mice on day 0. Five weeks after tumor inoculation, the mice were sacrificed and tumors were weighed. CR, complete regression.

oxidation was carried out under more severe conditions (12.5 mL; 6 days). On the other hand, the reactivity of the oxides of the linear β-glucan, CRD, did not increase. The results strongly suggest that the increase in *Limulus* reactivity of SSG is caused by a reduction in the number of side chains. In addition, the increase in reactivity is thought to be related to the solubilization of cell wall β-glucans caused by oxidation in the case of ZYM.

The *Limulus* reactivity of the products of alkaline or hot alkaline extraction of *Candida* cells was similar to that of linear β-glucans such as CRD. This clearly demonstrates that the increase in *Limulus* reactivity exhibited by *Candida* cells is due to solubilization.

Fate of β-Glucans In Vivo

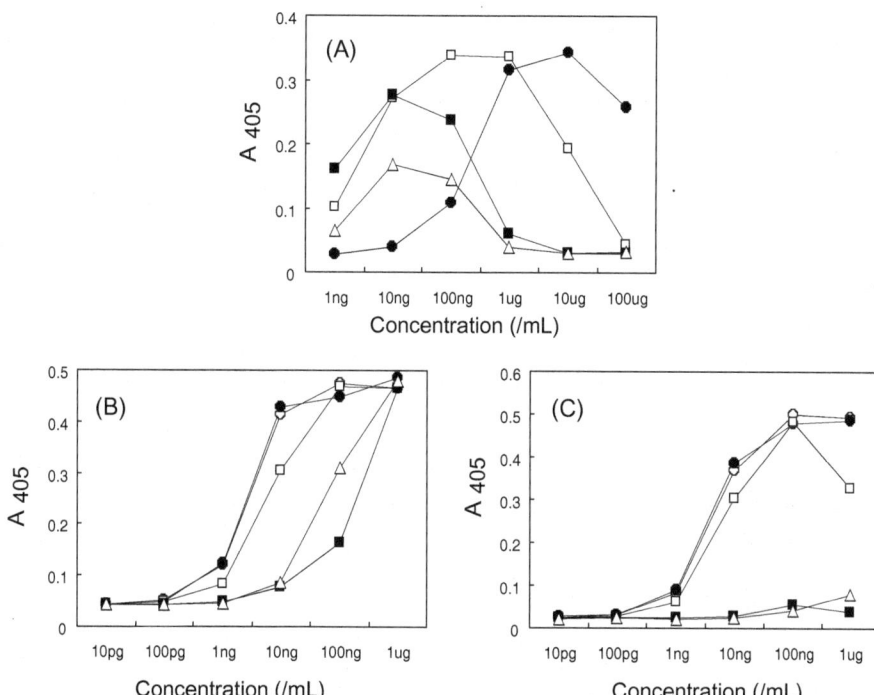

FIGURE 6.6 *Limulus* reactivity of oxidized *Candida* whole cells, alkali extract, and hot alkali extract. *Candida* whole cell, alkali extract, and hot alkali extract were oxidized with three concentrations (0[○], 1.25[●], 6.25[■], or 12.5[Δ] mL) of sodium hypochlorite solution (NaClO) added to 45 mL of glucans for 1 day at room temperature. After the reaction was completed, the reaction mixture was dialyzed extensively with distilled water, and then the non-dialyzable fraction (NDF) was lyophilized. The reactivity of each of these oxidized fractions was measured by the *Limulus* test. (A) *Candida* whole cells; (B) alkali extract of *Candida* whole cell (○); alkali extract dissolved in 0.5 N NaOH; and *Limulus* reactivity were measured. (C) Hot alkali extract of *Candida* whole cells, ○; hot alkali extract dissolved in 0.5N NaOH, and *Limulus* reactivity were measured.

6.8 CONCLUSION

Following administration into the peritoneal cavity, the soluble β-glucan SSG is distributed primarily in organs having a prominent reticuloendothelial system, such as the liver and spleen, and remains in those organs for at least 1 month. As enzymes capable of hydrolyzing glucans have not been found in mice, their metabolism and degradation is presumed to take place as a result of oxidative degradation by active oxygen or nitrite ions that are produced by macrophages or polymorphonuclear (PMN) leukocyte. However, considering their long retention time in the body, both metabolism and degradation are thought to occur extremely slowly.

Drugs that are administered into the body are typically able to migrate to both blood vessels and lymphatic vessels due to the low permeability of the capillary

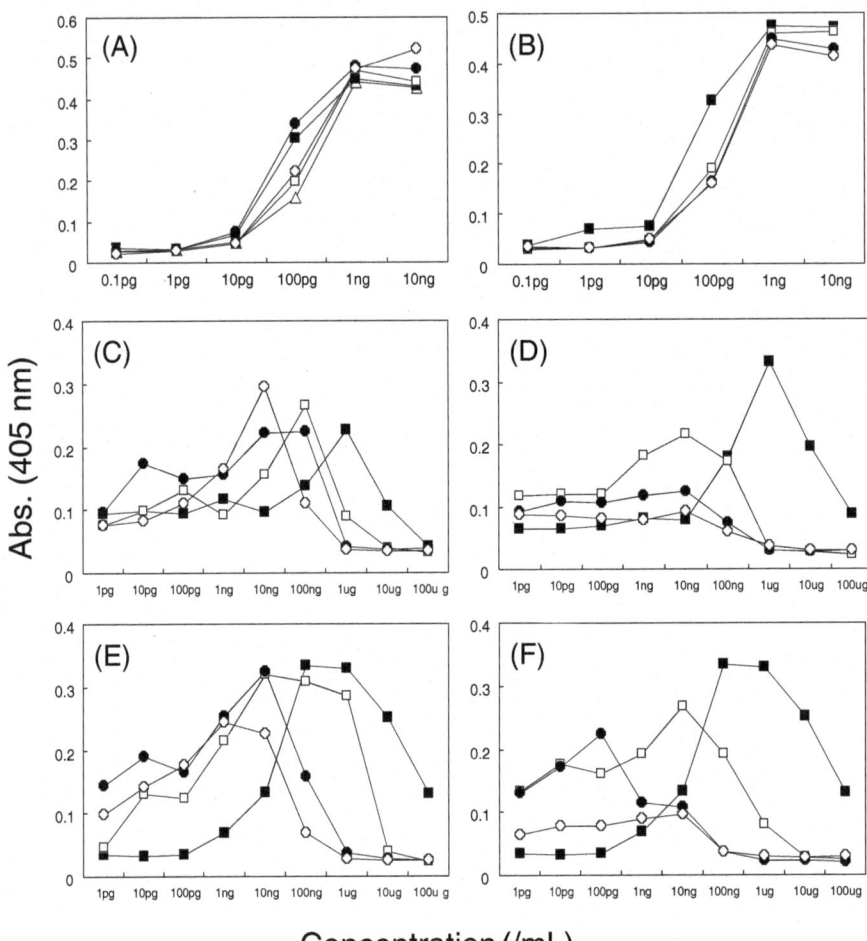

FIGURE 6.7 Limulus reactivity of oxidized curdlan, SSG, and zymosan. Curdlan, SSG, and zymosan were oxidized with three concentrations (added 12.5[○], 6.25[●], 1.25[□], 0[■] ml) of sodium hypochlorite solution (NaClO) added to 45 mL of glucans for 1 day (A, C, E) or 6 days (B, D, F) at room temperature. After the reaction was completed, the reaction mixture was dialyzed extensively with distilled water, and the non-dialyzable fraction (NDF) was lyophilized. The reactivity of each of these oxidized fractions was measured by the Limulus test. (A) Curdlan, (B) Curdlan (□), Curdlan dissolved in 0.5 N NaOH, and Limulus reactivity were measured. (C) SSG, (D) SSG, (E) zymosan, and (F) zymosan.

walls, and high molecular weight substances are known to preferentially migrate through lymphatic vessels. As SSG is also a high molecular weight substance having a molecular weight of 5 million or more, it is thought to be absorbed through the lymphatic system. Glucans that are administered i.p. are surmised to migrate into the bloodstream from the left subclavian vein after passing through the thoracic duct, after which they circulate throughout the body and are distributed in the respective organs. The process of distribution in each organ from the blood in the case of intravenous administration is thought to be the same as that for i.p. administration except for the absence of the step in which they migrate into the bloodstream from the peritoneal cavity.

Moreover, an additional study was conducted on the metabolism of β-glucans that either appeared in the blood accompanying fungal infection or originated in the grown cells themselves. The milky spots in the omentum have been reported to play an important role in the removal of bacteria from the peritoneal cavity, and insoluble β-glucans are also thought to be distributed in the omentum in the case of i.p. administration (Cranshaw and Leak et al., 1990; Wijffels et al., 1992; Doherty et al., 1995; Van Vugt et al., 1996). In actuality, *Candida* cells or the insoluble fraction from *Candida* cells is distributed in large amounts in the omentum and the intestinal membrane following i.p. administration, exhibiting a distribution pattern that differs from that of the soluble β-glucan, SSG. In addition, some accumulation in the liver was observed following i.p. administration of an insoluble fraction in experiments using tritium-labeled *Candida* cells. Therefore, distribution accompanying solubilization of the cells is possible. The *Limulus* test revealed that the amount that distributed into the liver during i.p. administration was markedly smaller than the amount that distributed during intravenous administration, thereby clearly demonstrating that little insoluble β-glucans leave the peritoneal cavity.

It was clearly shown in the case of intravenous administration that the *Candida* cells accumulate in the liver and that *Limulus* reactivity is increased gradually over the course of 6 months. Oxidation degradation was conducted using hypochlorous acid to determine whether this increase in *Limulus* reactivity is due to the oxidative degradation of side chains as is observed in soluble, branched β-glucans, or to the solubilization of the cells themselves. As a result, although an increase in *Limulus* reactivity was not observed following degradation by hypochlorous acid for soluble fractions obtained by alkaline extraction or hot alkaline extraction from *Candida* cells, the hypochlorous acid degradation products of *Candida* cells exhibited an increase in *Limulus* reactivity. In other words, it was clearly demonstrated that the β-glucans in *Candida* cells are mostly linear and do not have a structure in which the degradation of the side chains has an effect on *Limulus* reactivity. It is also strongly suggested from these findings that the increase in *Limulus* reactivity is a result of solubilization of the cells.

Candida cells are thought to accumulate in the liver and other organs even after candidiasis has been cured. However, as β-glucan concentrations in the blood are decreased, it is unlikely that they are released into the blood even after having been solubilized in the organs. This study demonstrates that there was little release of β-glucans into the blood even though organ accumulation and an increase in *Limulus* reactivity were observed. Considering the results collectively, the increase in blood β-glucan concentration during deep mycoses is thought to occur when the fungus has infected tissue and grown therein, and thus the results are important in terms of the clinical application of the *Limulus* test.

Although the biological activity of SSG in the host disappeared within 1 week following administration, the administered SSG maintained its structure for more than 1 month in the body. In addition, in a system that was used to examine the duration of antitumor activity, antitumor activity was detected when β-glucan was administered at the optimum dose 1 week prior to tumor transplant but not when administered at more than the optimum dose 1 week prior to tumor transplant. The activity was also detected when β-glucan was administered 5 weeks prior to tumor transplant. Because the degradation and excretion of β-glucans in the body are quite slow, the effects of accumulated β-glucans on the host are a subject of great interest and warrant additional studies in the future.

REFERENCES

Bishop, C.T., Blank, F., and Gardner, P.E. (1960). The cell wall polysaccharides of *Candida albicans*: glucan, mannan and chitin, *Can. J. Chem.* 38, 869–881.

Borchers, A.T., Stern, J.S., Hackman, R.M., Keen, C.L., and Gershwin, M.E. (1999). Mushrooms, tumors, and immunity, *Proc. Soc. Exp. Biol. Med.* 221, 281–293.

Chaffin, W.L., Lopez Ribot, J.L., Casanova, M., Gozalbo, D., and Martinez, J.P. (1998). Cell wall and secreted proteins of *Candida albicans*: identification, function, and expression, *Microbiol. Mol. Biol. Rev.* 62, 130–180.

Cranshaw, M. and Leak, L.V. (1990). Milky spots of the omentum: a source of peritoneal cells in the normal and stimulated animal, *Arch. Histol. Cytol.* 53, 165–177.

Doherty, N.S., Griffiths, R.J., Hakkinen, J.P., Scampoli, D.N., and Milici, A.J. (1995). Postcapillary venules in the "milky spot" of the greater omentum are the major site of plasma protein and leukocyte extravasation in rodent models of peritonitis. *Inflamm. Res.* 44, 169–177.

Fisher, M. and Yang, L.X. (2002). Anticancer effects and mechanisms of polysaccharide-K (PSK): implications of cancer immunotherapy, *Anticancer Res.* 22, 1737–1754.

Hayakawa, K., Mitsuhashi, N., Saito, Y., Nakayama, Y., Furuta, M., Nakamoto, S., Kawashima, M., and Niibe, H. (1997). Effect of Krestin as adjuvant treatment following radical radiotherapy in non-small cell lung cancer patients, *Cancer Detect Prev.* 21, 71–77.

Ikeda, F. (2003). Antifungal activity and clinical efficacy of micafungin sodium (Funguard), *Jpn. J. Pharmacol.* 122, 339–344.

Kato, A., Takita, T., Furuhashi, M., Takahashi, T., Maruyama, Y., and Hishida, A. (2001). Evaluation of blood (13)-beta-D-glucan concentration in hemodialysis patients. *Nephron*, 89, 15–19.

Kidd, P.M. (2000). The use of mushroom glucans and proteoglycans in cancer treatment, *Altern. Med. Rev.* 5, 4–27.

Miura, N.N., Ohno, N., Adachi, Y., Aketagawa, J., Tamura, H., Tanaka, S., and Yadomae, T. (1995). Comparison of clearance rate from blood stream of triple- and single-helical schizophyllan in mice, *Biol. Pharm. Bull.* 18, 185–189.

Miura, N.N., Ohno, N., Aketagawa, J., Tamura, H., Tanaka, S., and Yadomae, T. (1996a). Blood clearance of (1→3)-beta-D-glucan in MRL *lpr/lpr* mice, *FEMS Immunol. Med. Microbiol.* 13, 51–57.

Miura, N.N., Ohno, N., Adachi, Y., and Yadomae, T. (1996b). Characterization of sodium hypochlorite degradation of beta-glucan in relation to its metabolism *in vivo*. *Chem. Pharm. Bull.* 44, 2137–2141.

Miura, N.N., Miura, T., Ohno, N., Adachi, Y., Watanabe, H., Tamura, H., Tanaka, S., and Yadomae, T. (1998). Gradual solubilization of *Candida* cell wall beta-glucan by oxidative degradation in mice. *FEMS Immunol. Med. Microbiol.* 21, 123–129.

Miura, T., Miura, N.N., Ohno, N., Adachi, Y., Shimada, S., and Yadomae, T. (2000). Failure in antitumor activity by overdose of an immunomodulating beta-glucan preparation, sonifilan. *Biol. Pharm. Bull.* 23, 249–253.

Nagi, N., Ohno, N., Tanaka, S., Aketagawa, J., Shibata, Y., and Yadomae, T. (1992). Solubilization of limulus test reactive material(s) from *Candida* cells by murine phagocytes. *Chem. Pharm. Bull.* 40, 1532–1536.

Nagi, N., Ohno, N., Adachi, Y., Aketagawa, J., Tamura, H., Shibata, Y., Tanaka, S., and Yadomae, T. (1993). Application of limulus test (G pathway) for the detection of different conformers of (1→3)-beta-D-glucans, *Biol. Pharm. Bull.* 16, 822–828.

Nono, I., Ohno, N., Masuda, A., Oikawa, S., and Yadomae, T. (1991). Oxidative degradation of an antitumor (1→3)-beta-D-glucan, grifolan. *J. Pharmacobio-Dyn.* 14, 9–19.

Obayashi, T., Yoshida, M., Tamura, H., Aketagawa, J., Tanaka, S., and Kawai, T. (1992). Determination of plasma (1→3)-beta-D-glucan: a new diagnostic aid to deep mycosis. *J. Med. Vet. Mycol.* 30, 275–280.

Obayashi, T., Yoshida, M., Mori, T., Goto, H., Yasuoka, A., Iwasaki, H., Teshima, H., Kohno, S., Horiuchi, A., Ito, A., Yamaguchi, H., Shimada, K., and Kawai, T. (1995). Plasma (1→3)-beta-D-glucan measurement in diagnosis of invasive deep mycosis and fungal febrile episodes. *Lancet* 345, 17–20.

Ohno, N., Suzuki, I., and Yadomae, T. (1986). Structure and antitumor activity of a beta-1,3-glucan isolated from the culture filtrate of *Sclerotinia sclerotiorum* IFO 9395. *Chem. Pharm. Bull.* 34, 1362–1365.

Ohno, N., Kurachi, K., and Yadomae, T. (1987). Antitumor activity of a highly branched (1→3)-beta-D-glucan, SSG, obtained from *Sclerotinia sclerotiorum* IFO 9395. *J. Pharmacobio-Dyn.* 10, 478–486.

Osumi, M. (1998). The ultrastructure of yeast: cell wall structure and formation. *Micron.* 29, 207–233.

Sakurai, T., Suzuki, I., Kinoshita, A., Oikawa, S., Masuda, A., Ohsawa, M., and Yadomae, T. (1991). Effect of intraperitoneally administered beta-1,3-glucan, SSG, obtained from *Sclerotinia sclerotiorum* IFO 9395 on the functions of murine alveolar macrophages. *Chem. Pharm. Bull.* 39, 214–217.

Sakurai, T., Ohno, N., and Yadomae, T. (1992). Intravenously administered (1→3)-beta-D-glucan, SSG, obtained from *Sclerotinia sclerotiorum* IFO 9395 augments murine peritoneal macrophages function *in vivo*. *Chem. Pharm. Bull.* 40, 2120–2124.

Sakurai, T., Ohno, N., Suzuki, I., and Yadomae, T. (1995). Effect of solble fungal (1→3)-beta-D-glucan, obtained from *Sclerotinia sclerotiorum* on alveolar macrophage activation. *Immunopharmacol.* 30, 157–166.

Shepherd, M.G. (1991). The structure and function of *Candida albicans* cell wall. *Jpn. J. Med. Mycol.* 32, 63–73.

Stone, E.A., Fung, H.B., and Kirschenbaum, H.L (2002). Caspofungin: an echinocandin antifungal agent. *Clin. Ther.* 24, 351–377.

Suda, M., Ohno, N., Adachi, Y., and Yadomae, T. (1992). Tissue distribution of intraperitoneally administered (1→3)-beta-D-glucan (SSG), a highly branched antitumor glucan, in mice. *J. Pharmacobio-Dyn.* 15, 417–426.

Suda, M., Ohno, N., Adachi, Y., and Yadomae, T. (1994). Preparation and properties of metabolically ^{3}H- or ^{13}C-labeled (1→3)-beta-D-glucan, SSG, from *Sclerotinia sclerotiorum* IFO 9395. *Carbohydr. Res.* 254, 213–219.

Suda, M., Ohno, N., Hashimoto, T., Adachi, Y., and Yadomae, T. (1996). Kupffer cells play important roles in the metabolic degradation of a soluble anti-tumor (1→3)-beta-D-glucan, SSG, in mice. *FEMS Immunol. Med. Microbiol.* 15, 93–100.

Suzuki, I., Hashimoto, K., and Yadomae, T. (1988). The effect of a highly branced beta-1,3-glucan, SSG, obtained from *Sclerotinia sclerotiorum* IFO 9395 on the growth of syngenic tumors in mice. *J. Pharmacobio-Dyn.* 11, 527–532.

Suzuki, I., Sakurai, T., Hashimoto, K., Oikawa, S., Masuda, A., Ohsawa, M., and Yadomae, T. (1991). Inhibition of experimental pulmonary metastasis of Lewis lung carcinoma by orally administered beta-glucan in mice, *Chem. Pharm. Bull.* 39, 1606–1608.

Tamura, H., Tanaka, S., Ikeda, T., Obayashi, T., and Hashimoto, Y. (1997). Plasma (1→3)-beta-D-glucan assay and immunohistochemical staining of (1→3)-beta-D-glucan in the fungal cell walls using a novel horseshoe crab protein (T-GBP) that specifically binds to (1→3)-beta-D-glucan. *J. Clin. Lab. Anal.* 11, 104–109.

Van Vugt, E., Van Rijthoven, E.A., Kamperdijk, E.W., and Beelen, R.H. (1996). Omental milky spot in the local immune response in the peritoneal cavity of rats. *Anat. Rec.* 244, 235–245.

Vincent, T.A. (1999). Current and future antifungal therapy: new targets for antifungal agents, *J. Antimicrob. Chemother.* 44, 151–162.

Walsh, T.J., Viviani, M.A., Arathoon, E., Chiou, C., Ghannoum, M., Groll, A.H., and Odds, F.C. (2000). New targets and delivery system for antifungal therapy, *Med. Mycol.* 38, 335–347.

Wasser, S.P. (2002). Medical mushrooms as a source of antitumor and immunomodulating polysaccharides. *Appl. Microbiol. Biotechnol.* 60, 258–274.

Wijffels, J.F., Hendrickx, R.J., Steenbergen, J.J., Eestermans, I.L., and Beelen, R.H. (1992). Milky spots in the mouse omentum may play an important role in the origin of peritoneal macrophages, *Res. Immunol.* 143, 401–409.

Yoshida, M., Obayashi, T., Iwama, A., Ito, M., Tsunoda, S., Suzuki, T., Muroi, K., Ohta, M., Sakamoto, S., and Miura, Y. (1997). Detection of plasma (1→3)-beta-D-glucan in patients with *Fusarium, Trichosporon, Saccharomyces,* and *Acremonium fungaemias. J. Med. Vet. Mycol.* 35, 371–374.

Yoshida, M., Tsubaki, K., Kobayashi, T., Tanimoto, M., Kuriyama, K., Murakami, H., Minami, S., Hiraoka, A., Takahashi, I., Naoe, T., Asou, N., Kageyama, S., Tomonaga, M., Saito, H., and Ohno, R. (1999). Infectious complications during remission induction therapy in 577 patients with acute myeloid leukemia in the Japan Adult Leukemia Study Group studies between 1987 and 1991, *Int. J. Hematol.* 70, 261–267.

Yu, R.J., Bishop, C.T., Cooper, F.P., Hasenclever, H.F., and Blank, F. (1967). Structural studies of mannan from *Candida albicans* (serotypes A and B), *Candida parapsilosis, Candida stellatoidea,* and *Candida tropicalis, Can. J. Chem.* 45, 2205–2211.

7 Adjuvant Effects of β-Glucans in a Mouse Model for Allergy

Heidi Ormstad and Geir Hetland

CONTENTS

7.1 Abstract 127
7.2 Introduction 128
7.3 Materials and Methods 130
    7.3.1 Animals 130
    7.3.2 Preparations for Immunization 131
    7.3.3 Experimental Design 131
    7.3.4 PLN Assay 131
    7.3.5 Assay for Serum OA-Specific IgE 132
    7.3.6 Assay for Serum OA-Specific IgG1 and IgG2a 132
    7.3.7 Statistical Analysis 133
7.4 Results 133
    7.4.1 Effect of β-Glucan on the PLN Weight, Cell Numbers, and Proliferation 133
    7.4.2 Effect of β-Glucan on the Antibody Response to OA 133
7.5 Discussion 134
Acknowledgments 138
References 139

7.1 ABSTRACT

The polyglucose 1→3-β-D-glucan is a major structural component of the cell wall of yeast and fungi. We have studied the adjuvant activity of different β-glucans on the response to the model allergen ovalbumin (OA), using the popliteal lymph node (PLN) assay in Balb/c mice. The adjuvant activity on local cellular response was determined by measuring the weight, cell number, and proliferation of the extracted PLNs. The levels of OA-specific IgE, IgG1 (Th2 response) and IgG2a (Th1 response) in serum were measured by enzyme-linked immunoabsorbent assay (ELISA). Taken together, our results show that whereas β-glucan had little effect on the local inflammatory response, it revealed a strong adjuvant activity on the systemic allergic immune response in a mouse model. This supports the reported observations that inhaled β-1,3-glucan is an enhancer and inducer of allergic disease in man.

7.2 INTRODUCTION

Although the reason remains unclear, most investigators today agree that the prevalence of asthma and allergic diseases has increased (Mushinski, 1997; Jarvis and Burney, 1998; Shy, 1999). The increase is most likely attributable to environmental factors and not to changes in the genetic constitution of Western populations during the last decades (Leikauf et al., 1995; Sears, 1997). Some relevant environmental factors are changes in building insulation, reduction in the turnover of indoor air, increased exposure to indoor pollutants (especially environmental cigarette smoke), and exposure to potent allergens (Boushey and Fahy, 1995; Newman-Taylor, 1995). It is likely that several factors acting together are responsible. Since Western populations today spend 95% of their time indoors (Platts-Mills, 1995), indoor air pollution should be of special interest.

It is not fully known in what way and to what extent airborne dust particles may contribute to asthma and respiratory symptoms. One possibility is that suspended particulate matter (SPM) may modify the immune response by promoting IgE production. Previously, we have found that SPM in indoor air has an adjuvant activity on the IgE production in response to a model allergen (OA) (Ormstad et al., 1998a). This means that SPM, either in its entirety or only individual particle types, may increase the production of specific IgE in response to environmental allergens. Since SPM contains many different types of particles and components (Ormstad et al., 1997), it is not yet clear what components provide this effect. In addition to different types of inorganic and organic particles, SPM also contains various biological substances. For example, we have shown earlier that different allergens are carried by soot particles in SPM samples (Ormstad et al., 1998b). These are likely to cause an allergic reaction when inhaled by susceptible individuals. However, it is also an interesting question whether other airborne biological components may affect an allergic immune response without acting as allergens themselves, but rather as adjuvants of the allergic reaction to other allergens.

Material from bacteria and molds (fungi) are biological components present in airborne house dust. Two agents that are ubiquitous in the air of work, school, and home environments are endotoxin (also known as lipopolysaccharide [LPS]) and glucan (Young et al., 1998). LPS is present in the outer cell membrane of all Gram-negative bacteria and also in some blue-green algae (Young et al., 1998). It is a well-known stimulator of macrophages (Nathan, 1987), promoter of differentiation of promyelocytic cells (Breitman et al., 1986) and activator of complement (Wilson and Morrison, 1982).

D-Glucans are D-glucopyranose (glucose) subunits joined by glucosidic α or β linkages (Stone and Clarke, 1992). 1→3-β-D-glucans are natural products widely distributed in microorganisms and plants, and are particularly a major structural constituent of the cell wall of yeast and fungi. The compounds have a (1→3)-β-D-glucan backbone with (1→6)-β-linked side-chains of single (e.g., lentinan from

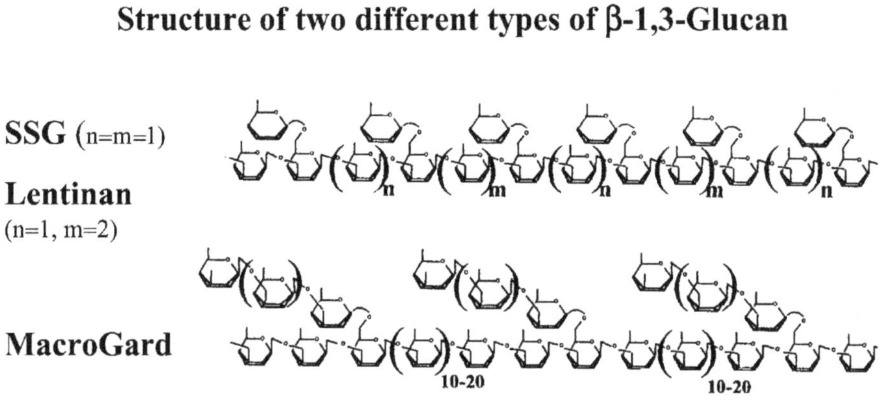

Structure of two different types of β-1,3-Glucan

SSG (n=m=1)
Lentinan
(n=1, m=2)

MacroGard

FIGURE 7.1 Structural composition of SSG (scleroglucan), lentinan, and MacroGard® showing the β-1,3-linked backbone and side-chains (β-1,6-linked to backbone), which are responsible for the binding to the glucan receptor(s) and the immunostimulatory effect. The figure is a modification of an illustration in the publication "MacroGard®: structural aspects and basic mode of action on phagocytes," from BioTec ASA, Tromsø, Norway, by R.E. Engstad.

mushrooms) or multiple (1→3)-β-linked glucose molecules (e.g., MacroGard® from baker's yeast) (Figure 7.1). 1→3-β-D-glucan appears in three structurally different forms: random coil, single helix, or triple helix, the latter being the most frequent in nature and water-insoluble (Rylander and Lin, 2000). It is a known immunomodulator and potent nonspecific stimulator of the innate immune system (Riggi and Di Luzio, 1961). It stimulates phagocytic defense mechanisms in macrophages (Bøgwald et al., 1984) and activates complement (Czop and Austen, 1985). β-Glucans induce protective activities against bacterial (Kokoshis et al., 1978), viral, protozoal (Cook et al., 1982), and fungal (Williams et al., 1978) infections. Recently, we have shown that β-glucans protect against encapsulated bacteria like streptococci or intracellular mycobacteria as well (Hetland et al., 1998 and 2000; Hetland and Sandven, 2002). β-Glucans also stimulate tumoricidal activity in polymorphonuclear leukocytes (PMN), macrophages, and natural killer (NK) cells (Di Luzio et al., 1976; Amino et al., 1983; Morikawa et al., 1995), and exhibit antitumor (Ohno et al., 1987) effects in rodent models. These effects are mediated via the lectin binding site for β-glucan in complement receptor 3 (CR3) (CD11b/CD18) on mononuclear phagocytes (Czop and Kay, 1991), PMN, and NK cells (Vetvicka et al., 1996). Recently, a novel receptor for β-glucan, the dectin-1 receptor (Brown and Gordon, 2001), has been discovered on macrophages. The receptor recognizes and binds the nonreducing terminal end of β-1,3-glucan chains containing two or more glucose molecules (Engstad and Robertsen, 1995; Suzuki et al., 1989). Since vertebrates do not possess specific β-glucan hydrolases, β-glucans are slowly broken down by oxidative degradation within the cells (Ohno et al., 1999).

However, in relation to respiratory health, 1→3-β-D-glucan seems to have rather destructive effects. Acute exposure to the substance has previously been shown to induce symptoms of airways inflammation in normal individuals (Rylander, 1996).

An association has also been found between exposure of the airways to β-1,3-glucan and increased prevalence of atopy as well as decreased forced expiratory capacity (FEV_1) (Thorne and Rylander, 1998). In addition, it has recently been found that 1→3-β-D-glucan can abrogate inhalation induced IgE isotype-specific down-regulation and promote airway eosinophil infiltration to inhaled allergen in mice (Wan et al., 1999). Despite these findings of 1→3-β-D-glucan being an important factor in relation to atopy and airway symptoms, only a few investigators have monitored 1→3-β-D-glucan levels in indoor air (Wan and Li, 1999). However, two studies from Sweden reported levels of 1→3-β-D-glucan ranging from 0.1 ng/m^3 in office buildings to 106 ng/m^3 in moldy homes (Rylander et al., 1992 and 1994a). In a recent study, they measured an average level of 2.9 ng/m^3 in control schools and 15.3 ng/m^3 in so-called problem schools (Rylander et al., 1998). Most likely, 1→3-β-D-glucan is widespread in indoor air in other countries as well.

The scleroglucan sclerotinia sclerotiorum glucan (SSG) is a gel-forming but soluble β-1,3/1,6-*D*-glucan obtained from the culture broth of the fungus *Sclerotinia sclerotiorum* IFO 9395 (Ohno et al., 1987). It is highly branched with branches at every other main chain glucosyl (Ohno et al., 1986) (Figure 7.1), has a high molecular weight (> 5×10^6) and is more viscous than other β-1,3-*D*-glucans (Ohno and Yadomae, 1987). SSG has proven potent antitumor effects when administered systemically in mice (Ohno et al., 1986, 1987; Suda et al., 1995). In Japan, 1→3-β-D-glucans from mushrooms, such as lentinan, have been used to treat patients with cancer (Taguchi et al., 1983). MacroGard® (MG) is a β-1,3/1,6-*D*-glucan extracted from baker's yeast (U.S. Patent No. 5,401,727; E.P. Patent No. 0466037) with frequent extending terminal ends, containing more than two glucose molecules. It can be produced in a soluble and a particulate form. The frequency and nature of the side-chains is of great importance for the biological activity of β-glucans, rendering MG a very potent immunostimulant (Engstad and Robertson, 1995).

The purpose of the present investigation was to study whether β-glucans have an adjuvant effect on the allergic immune response to the model allergen, ovalbumin, in a mouse model for allergy. This model is based on the PLN assay, in which both adjuvant and allergen were injected *s.c.* into one footpad. After allergen boosting, local inflammatory effects and systemic effects were examined in the excised PLN and serum, respectively. Three different β-glucans were examined. One was a β-glucan from barley (BG), which is a mixture of approximately 70% β-1,4-glucan and 30% β-1,3-glucan. The two others were the soluble but viscous SSG and soluble MG. The present review is mainly based on reports on adjuvant activities of BG and SSG (Ormstad et al., 2000) and MG (Instanes et al., 2004).

7.3 MATERIALS AND METHODS

7.3.1 ANIMALS

Female Balb/c mice (6–8 weeks old) were obtained from Gl. Bomholgård Ltd., Ry, Denmark. They were housed eight per cage and given food and water *ad libitum*. The experiments were approved by the local officer of the Animal Board under the

Ministry of Agriculture in Norway and performed in conformation with the laws and rules regulating animal experiments in Norway.

7.3.2 PREPARATIONS FOR IMMUNIZATION

Medium — Hank´s Balanced Salt Solution (HBSS) (Gibco BRL, Paisley, Scotland) with 10% Balb/c mouse serum was used in these studies.

β-Glucans — BG (catalogue number G-6513, Sigma Chemical Co., St. Louis, MO, U.S.) was dissolved in medium (20 mg/ml). SSG from the culture broth of *Sclerotinia sclerotiorum* (FO 9395, a highly appreciated gift from Dr. N. Ohno, Tokyo University of Pharmacy and Life Science, Tokyo, Japan) was dissolved in PBS by sonication for 1 hour. Soluble MG was a highly appreciated gift from professor Jan Raa of BioTec ASA, Tromsø, Norway.

Ovalbumin Solution — OA (catalogue number A-7641, Sigma Chemical Co., St. Louis, MO, U.S.) was dissolved in medium (5 mg/mL). These batch preparations were diluted 1:1 either with medium (to make preparations of β-glucan alone or OA alone) or with each other (to make β-glucan + OA). The injected dose of BG was 200 μg in 20 μL, and that of SSG and MG was 100 μg in 20 μL. The dose of OA was 50 μg in 20 μL.

7.3.3 EXPERIMENTAL DESIGN

Groups of mice were injected on day 0 with 20 μL of one of the following preparations into the right hind footpad: SSG + OA, BG + OA, MG + OA, SSG alone, BG alone, MG alone, or OA alone. After 20 days, 8 mice from each group were exsanguinated, and the PLNs were removed. The PLN weight, cell numbers, and cell proliferation were determined, as well as serum OA-specific IgE, IgG1, and IgG2a. On day 21, all remaining mice were reinjected (boosted) with OA alone. On day 26 and day 33, another 8 mice from each group were exsanguinated, and the popliteal lymph nodes were removed (Figure 7.2). The same parameters were measured as on day 20.

7.3.4 PLN ASSAY

The mice were injected with 20 μL of the different immunization suspensions in the right hind footpad (heel-toe direction), using a 100 μL Hamilton syringe (Hamilton Bonaduz AG, Switzerland) with a 30 G needle (Becton, Dickinson and Co., Dublin, Ireland). The mice were unanesthetized during the injection, but exsanguinated under CO_2 anesthesia. Serum from each mouse was prepared and frozen for later measurement of OA specific Ig. The PLNs were removed from both hind legs and put into HBSS.

After removal of excessive fat, the lymph nodes were weighed using a Sartorius Research Series Electronic Balance MC1. Cell suspensions (1 x 10^5 cells) were prepared from each lymph node for ^{3}H-thymidine incorporation to study cell proliferation. This methodology has been described in more detail elsewhere (Ormstad et al., 1998b). The cells were harvested in a Skatron cell harvester onto Skatron

PLN scheme

☞ Mice were injected with test solution in footpad on day 0, re-injected with OA on day 21 and exsanguinated on day 26:

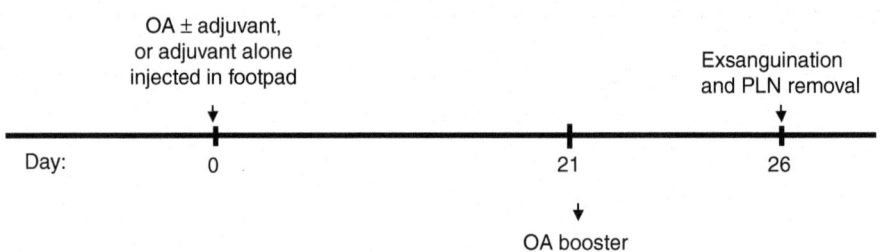

☞ Parameters:

☞ Serum IgE, IgG1, and IgG2a to OA

☞ Weight and cell numbers of popliteal lymph nodes

FIGURE 7.2 A schematic presentation of the study design.

Filtermats (Skatron, Lier, Norway), which were put into 3 ml of scintillation fluid (Opti-Fluor, Packard Instrument B.V., Groningen, The Netherlands) and counted in a scintillation counter.

End point measurements, as a measure for the local lymph node response, were the indices of the PLN weight, cell numbers, and cell proliferation. The index is the value of the right PLN (inoculated), divided by the value of the left PLN (untreated). Serum levels of IgE, IgG1, and IgG2a anti-OA antibodies were also measured (using ELISA).

7.3.5 ASSAY FOR SERUM OA-SPECIFIC IGE

The ELISA assay used for measurement of OA-specific IgE is described in more detail elsewhere (Ormstad et al., 1998a).

7.3.6 ASSAY FOR SERUM OA-SPECIFIC IGG1 AND IGG2A

Polystyrene microtiter plates (Immulon II M 129B, Dynatech Laboratories Ltd., Chantilly, VA, U.S.) were coated with 100 µL of 2 µg/mL rat anti-mouse IgG1 (clone LO-MG1-13, Serotec Ltd., Oxford, U.K.) or IgG2a antibody (clone LO-MG2a-7, Serotec Ltd., Oxford, U.K.) antibody in 0.05 M carbonate/bicarbonate buffer, pH 9.6. The coated plates were incubated for 1 hour at room temperature and thereafter at +4° C overnight. After washing with 0.1 M Tris/HCl buffer (pH 7.4) containing 0.05% Tween 20 (Tris/Tw), the plates were incubated with 280 µL blocking buffer (0.1 M Tris/HCL with 1% BSA) for 1 hour. After five washes with Tris/Tw, the plates were incubated with an OA-specific standard IgG1 or IgG2a serum and the unknown sera diluted in BSA/Tris/Tw, 100 µL to each well. Triplicates of a 1:20 dilution of the unknown sera were used. The plates were then incubated for 2 hours at room temperature followed by overnight incubation at 4° C. After five new washes with Tris/Tw, the plates were incubated for 2 hours at room temperature with biotin-labeled OA diluted to 0.5 µg/mL in Tris/Tw, 100 µL per well. After another washing procedure, the plates were incubated for 1 hour at room temperature with 1:50 dilutions in 0.05 M Tris/HCl of complexes of avidin and biotinylated alakaline phosphatase (ABComplex/AP, DAKO, Glostrup, Denmark). Again the plates were washed and incubated at room temperature with 100 µL p-nitrophenyl phosphate (Sigma 104 Phosphatase Substrate, Sigma Chemical Co., St Louis, MO, U.S.) in 10% diethanolamide buffer (pH 9.8). The plates were read at 405 nm in a MRX Microplate Reader (Dynatech Laboratories, U.K.).

7.3.7 STATISTICAL ANALYSIS

One-way Anova analysis for multiple comparisons was used and paired groups compared with Student's t-test when there was a Gaussian distribution of the results, and otherwise by non-parametric Mann-Whitney-U test. P values less than 0.05 were considered significant.

7.4 RESULTS

7.4.1 EFFECT OF β-GLUCAN ON THE PLN WEIGHT, CELL NUMBERS, AND PROLIFERATION

On day 26, the mice in the SSG + OA group had significantly increased PLN weight and cell numbers compared with the mice given OA or SSG alone (Figure 7.3). SSG + OA increased the PLN weight 1.7-fold and the cell number 2.5-fold compared to OA alone ($p \le 0.02$). Compared to SSG alone, SSG + OA increased the PLN weight and cell number 1.5-fold ($p \le 0.05$) and 1.3-fold, respectively. However, the cell proliferation was not significantly increased. The PLN indices in the SSG + OA group did not increase significantly, neither on day 20 before the OA-boosting, nor on day 33. In contrast with SSG, β-glucans BG or MG showed no significant effects on the PLN scores.

7.4.2 EFFECT OF β-GLUCAN ON THE ANTIBODY RESPONSE TO OA

SSG + OA was found to increase the anti-OA IgE levels 2-, 19-, and 9-fold ($p \le 0.038$), on days 20, 26 (Figure 7.4), and 33, respectively, compared to OA alone. The IgG1 levels increased 30- and 180-fold ($p < 0.001$) compared to OA alone on days 26 (Figure 7.4) and 33, respectively. On day 20 the levels of IgG1 were not detectable in mice treated with OA alone. A similar increase was not found for anti-OA IgG2a (Figure 7.4). These results have been reproduced in a second study. Similar results were found with MG. On day 26 MG + OA increased the anti-OA IgE and IgG1, but not IgG2a, levels more than 10-fold relative to OA alone ($p \le 0.038$) (Figure 7.5).

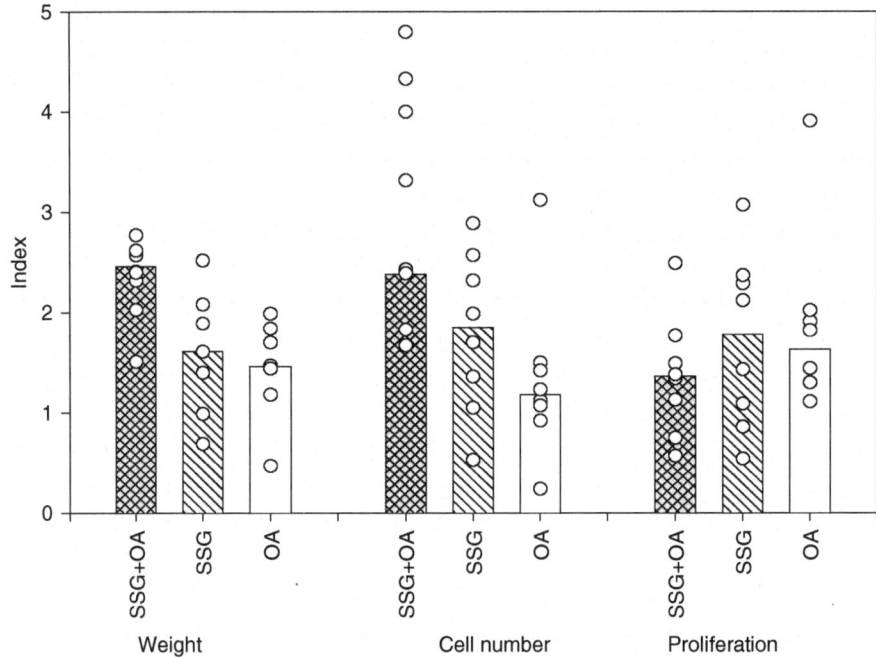

FIGURE 7.3 Indices for PLN weight, cell numbers, and proliferation 26 days after immunization with 100 μg SSG + 50 μg OA, 100 μg SSG, or 50 μg OA. A booster dose of OA (50 μg) was given to all groups on day 21. Values for individual mice and median values for groups of eight mice (columns) are given.

BG had less effect than SSG and MG. However, addition of BG with OA doubled the anti-OA IgE level ($p \leq 0.03$) at days 26 and 33 (Figure 7.6). The IgG1 levels did also increase significantly on days 26 and 33, compared with OA (data not shown). In addition, IgG2a did not show any significant increase in any BG group.

7.5 DISCUSSION

The highest level of β-glucan-adjuvated OA-specific IgE was measured on day 26 (5 days after OA boosting), and it declined again through day 33, while the level of OA-specific IgG1 reached the highest value on day 33. The different kinetics of these two antibodies is most probably due to the fact that IgE has a shorter lifetime than IgG1 in mice. In contrast to OA-specific IgE and IgG1, the levels of OA-specific IgG2a was not found to increase significantly. IgE and IgG_1 indicate a Th2 response and IgG_{2a} a Th1 response in mice. These results suggest that β-glucan, when given together with OA, strongly increases the allergic (Th2), but not the nonallergic (Th1) response to OA. Hence, β-glucan has an adjuvant effect on the allergic response to the model allergen, OA, in mice.

Regarding the PLN indices, only one of three β-glucans tested, SSG, induced a significant increase in PLN weight and cell number, while not affecting cell proliferation

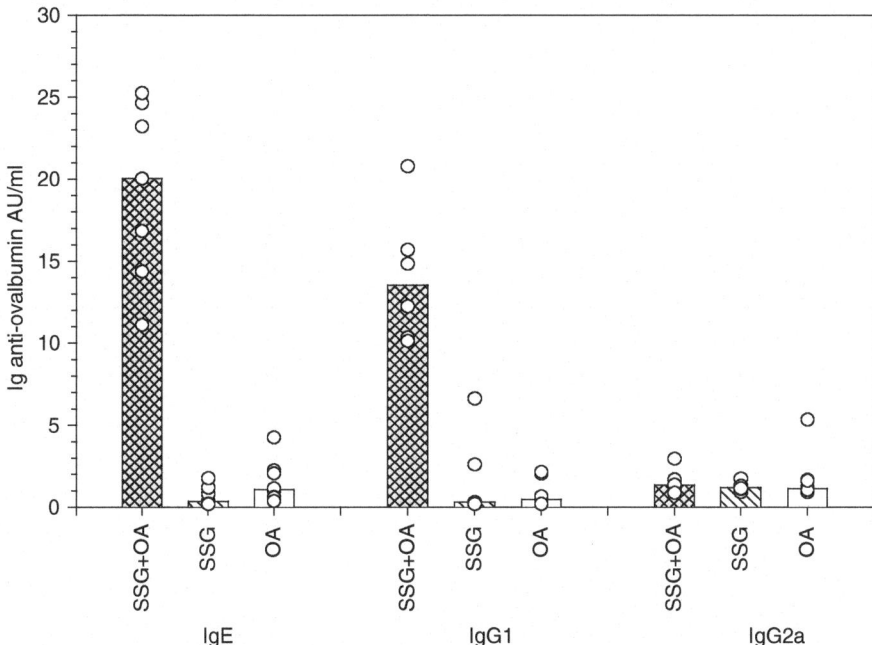

FIGURE 7.4 Levels of specific IgE, IgG1, and IgG2a anti-OA measured in serum 26 days after immunization with 100 μg SSG + 50 μg OA, 100 μg SSG, or 50 μg OA. A booster dose of OA (50 μg) was given to all groups on day 21. Values for individual mice and median values for groups of eight mice (columns) are given.

(thymidine incorporation). These three parameters were used to measure the local cellular response in PLN. One possible explanation for the findings is that the proliferation had already returned to a normal level by day 26, 5 days after the OA boosting. However, the increased cell number on day 26 is most probably the result of an earlier increased proliferation. The time-points used in the present experiment were selected because they were considered optimal in relation to measurement of specific IgE (Løvik et al., 1997).

Much less pronounced effects than with SSG or MG were found with a β-glucan from barley (BG). In contrast to SSG and MG, BG is a mixture of approximately 70% β-1,4-glucan and 30% 1→3-β-D-glucan (Sigma Chemical Co.). The reason for the apparent lesser adjuvant effect of this β-glucan is most likely the lower content of the effective molecule 1→3-β-D-glucan in BG. Other aspects that may play a role are different chain branching and/or the molecular weight of these two polysaccaride molecules. It has previously been found that the branching ratio and molecular weight of 1→3-β-D-glucan are important for their immunological activity (Suzuki et al., 1992; Okazaki et al., 1995).

Results similar to those found systemically in the mice using SSG were obtained with MG, a soluble 1→3-β-D-glucan from baker's yeast. The β-glucan SSG (from the fungus *Sclerotinia sclerotiorum*) is gel-forming and highly branched, branching with β-1,6-linkages at every other glycosyl main chain (Ohno et al., 1986). It has a

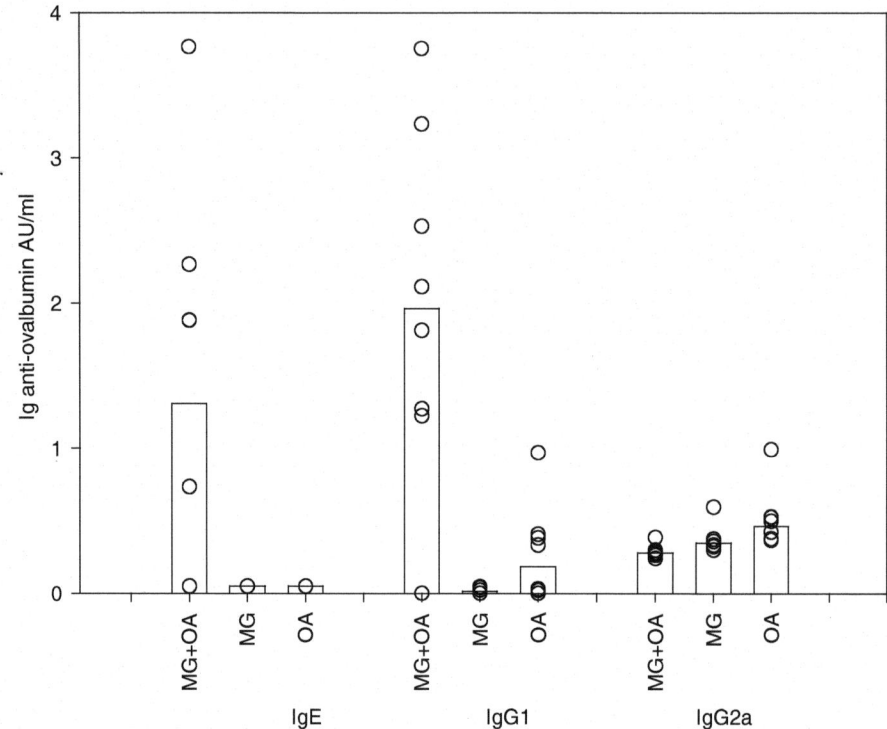

FIGURE 7.5 Levels of specific IgE, IgG1, and IgG2a anti-OA measured in serum 26 days after immunization with 100 μg of soluble MG + 50 μg OA, 100 μg MG, or 50 μg OA. A booster dose of OA (50 μg) was given to all groups on day 21. Values for individual mice and median values for groups of eight mice (columns) are given.

high molecular weight (> 5×10^6) and is more viscous than other 1→3-β-D-glucans (Ohno and Yadomae, 1987). Whereas SSG has one monoglucosyl unit in the side chains, MG from baker's yeast has three, which is believed to be a prerequisite for an optimal engagement of the β-glucan receptor (Engstad and Robertsen, 1995). However, with regard to the local inflammatory response in the PLN, the viscous SSG seemed to induce more adjuvant effects than soluble MG. We have seen this previously in *Mycobacterium tuberculosis*-infected macrophage cultures, where SSG, in contrast with soluble MG, had protective anti-infection effects (Hetland and Sandven, 2002). On the other hand, particulate MG was more effective than SSG in these cell cultures. In contrast, studies investigating the effect of inhaled glucans on respiratory tract inflammation in guinea pigs have shown that the activity of glucans depends on whether they are soluble or not. Only soluble glucan produced a significant increase in the number of PMN in the lungs (Rylander, 1994b; 1994c). However, opposing evidence for effects of particulate 1→3-β-D-glucans on the inflammatory process have been reported (Young et al., 1998). This clearly points out that the "physiological state" of the glucan during exposure seems to be important in dictating the resulting immunological reaction. Hence, the three-dimensional

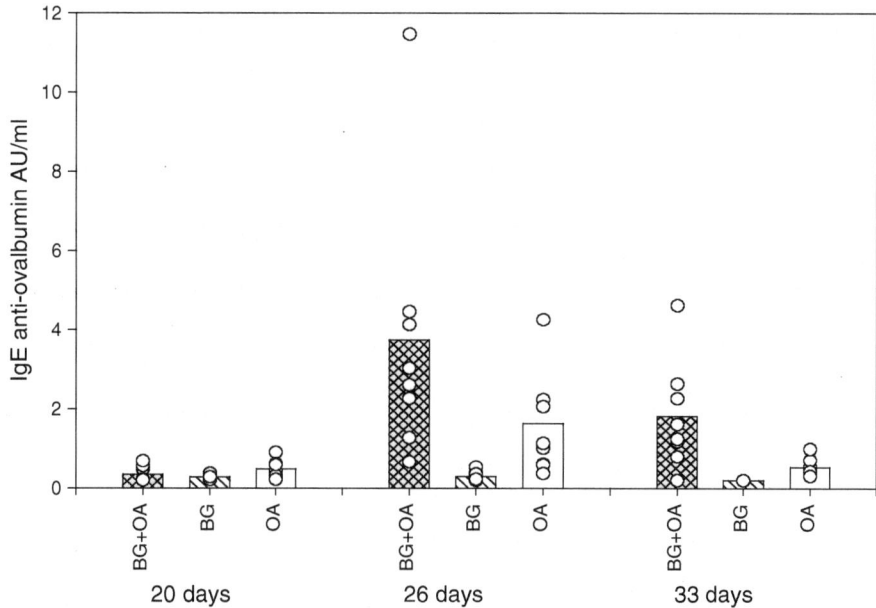

FIGURE 7.6 Levels of specific IgE anti-OA measured in serum 20, 26, and 33 days after immunization with 200 μg BG + 50 μg OA, 200 μg BG, or 50 μg OA. A booster dose of OA (50 μg) was given to all groups on day 21. Values for individual mice and median values for groups of eight mice (columns) are given.

structure of the β-glucan may prove to be more important for its biological effect than the composition of the side chains.

At the present time, we can only speculate about the immunological mechanism involved in the adjuvant activity of β-glucan on the allergic response. It has earlier been shown that the α-M-β-2-integrin CR3 (Mac-1, CD11b/CD18), serves as the β-glucan receptor through one or more lectin binding sites (Thornton et al., 1996). Ross et al. (1987) found that PMN or monocyte ingestion of either yeast or yeast-derived β-glucan particles was blocked by monoclonal anti-CR3. Most probably, binding of SSG and MG (and BG) to CR3 is involved in the adjuvant activity on the allergic response to OA observed in the present study. However, effects mediated via β-glucan binding to the dectin-1 receptor must also be taken into consideration. In our study of the protective effect of MG against *M. tuberculosis* infection of macrophages (Hetland and Sandven, 2002), anti-CD11b blocked the effect of serum-opsonized particulate MG, but not that of native particulate MG. We have also observed lack of inhibition of native β-glucan-mediated adjuvanticity in the present mouse model for allergy (data not shown). Since β-glucan activates the alternative pathway of complement (Czop and Austen, 1985), one can assume that the serum treated β-glucan (MG) particles above were covered with complement activation product C3b/iC3b. Hence, the anti-CD11b antibody most probably inhibited the iC3b-ligand-mediated binding of MG to CR3 in these experiments, but not the β-glucan-mediated binding of native MG to the cells. The latter could partly have occurred via the dectin-1 receptor.

Furthermore, it has previously been found that β-glucan may enhance the activation state of both monocytes (Czop and Austen, 1985) and PNM or NK cells (Vetvicka et al., 1996). If we assume that the PLN response observed in this study is a specific immune response leading to an allergic response, this means that β-glucan may enhance the activation state of immunological cells in the PLN, thus functioning as antigen presenting cells (APCs) for the OA antigen. Generally, upon stimulation by an appropriate APC, a Th cell is activated to produce cytokines and start proliferating. The type of response arising is defined by the cytokines produced by the Th cell. The differentiation of T-helper (CD4$^+$) cells into two subsets, Th1 and Th2, each with a characteristic profile of cytokine production, is central to the understanding of the pathogenetic mechanisms of allergy. Th2 cells play a key role in allergic immune response, as they produce IL-4 (reviewed in Belardelli, 1995; Borish and Rosenwasser, 1997), which is essential for IgE production (Del Prete et al., 1988). Our results show that SSG and MG (and BG) provide a clear Th2-dependent antibody response to OA, indicated by elevated levels of IgE and IgG1 and not IgG2a. We presume that the β-glucan must in some way give rise to the differentiation of T-helper (CD4$^+$) into Th2 cells.

Factors that may play a role in the development of naive CD4$^+$ T cells into Th1 or Th2 dominated populations are the type of antigen presenting cells, the nature and amount of antigen and other micro-environmental factors, such as hormones and cytokines (Romagnani, 1992; Rook et al., 1994). The most definitive studies highlight a primary role for cytokines themselves (Seder et al., 1992; Maggi et al., 1992; Manetti et al., 1993). IL-4 appears to be essential for the development of Th2 cells, whereas IL-12 favors the development of Th1 cells (Romagnani, 1995). Thus, it is possible that SSG and MG (and BG) in some way stimulates the APC involved to produce IL-4. The mechanism involved in the adjuvant activity on the allergic response of β-glucans has to be studied in more detail.

In conclusion, our results show that 1→3-β-D-glucan provides a clear Th2-dependent antibody response to OA, indicated by elevated levels of IgE and IgG1 but not IgG2a. In other words, 1→3-β-D-glucan has an adjuvant activity on the allergic immune response in the mouse model used. In relation to humans, this could mean that 1→3-β-D-glucan may alter the susceptibility to other environmental allergens. If this is the case, allergic individuals would suffer more by living or working in moldy environments, as their allergy may worsen due to presence of 1→3-β-D-glucan. Finally, our results indicate that 1→3-β-D-glucan may increase the induction of allergy to substances in the indoor environment, by shifting the Th1/Th2 balance of the immune response towards a Th2 polarization and the production of specific IgE.

ACKNOWLEDGMENTS

This work was supported by grants from the Norwegian Research Council and the Norwegian Red Cross. The authors are grateful to Dr. Naohito Ohno and Professor

Jan Raa for kindly supplying SSG and MG, respectively. We also thank our co-authors of the reports which this review was based upon, and the technical staff at the Department of Environmental Immunology, Norwegian Institute of Public Health, Oslo.

REFERENCES

Amino, M., Noguchi, R., Yata, J., Matsumura, J., Hirayama, R., Abe, O., Enomoto, K., and Asato, Y. (1983). Studies on the effect of lentinan on human immune system. II. *In vivo* effect on NK activity, MLR induced killer activity, and PHA induced blastic response of lymphocytes in cancer patients. *Jpn. J. Cancer Chemother.* 10, 2000–2006.

Belardelli, F. (1995). Role of interferons and other cytokines in the regulation of the immune response. *APMIS* 103, 161–179.

Bøgwald, J., Gouda, I., Hoffman J., Larm, O., Larsson, R., and Seljelid, R. (1984). Stimulatory effect of immobilized glucans on macrophages *in vitro*. *Scand. J. Immunol.* 20, 355–360.

Borish, L. and Rosenwasser, L. (1997). Th1/Th2 lymphocytes: doubt some more? *J. Allergy Clin. Immunol.* 99, 161–164.

Boushey, H.A. and Fahy, J.V. (1995). Basic mechanisms of asthma. *Environ. Health Perspect.* 103, 229–233.

Breitman, T.R., Hemmi, H., and Imaizumi, M. (1986). Induction by physiological agents of differentiation of the human leukemia cell line HL-60 to cells with functional characteristics. *Prog. Clin. Biol. Res.* 226, 215–233.

Brown, G.D. and Gordon, S. (2001). Immune recognition. a new receptor for beta-glucans. *Nature* 413, 36–37.

Cook, J.A., Holbroook, T.W., and Dougherty, W.J. (1982). Protective effect of glucan against visceral leishmaniasis in hamsters. *Infect. Immun.* 37, 1261–1269.

Czop, J.K. and Austen, K.F. (1985). Properties of glucans that activate the human alternative complement pathway and interact with the human monocyte β-glucan receptor. *J. Immunol.* 133, 3388–3393.

Czop, J.K. and Kay, J. (1991). Isolation and characterization of beta-glucan receptors on human mononuclear phagocytes. *J. Exp. Med.* 173, 1511–1520.

Del Prete, G., Maggi, E., Parronchi, P., Chretién, I., Tiri, A., Macchia, D., Ricci, M., Bancherau, J., De Vries, J., and Romagnani, S. (1988). IL-4 is an essential factor for the IgE synthesis induced *in vitro* by human T cell clones and their supernatants. *J. Immunol.* 140, 4193–4198.

Di Luzio, N.R., McNamee, R., Jones, E., Cook, J.A., and Hoffmann, E.O. (1976). The employment of glucan and glucan activated macrophages in the enhancement of host resistance to malignancies in experimental animals. In *The Macrophage in Neoplasia*, M.A. Fink, ed., Academic Press, New York, pp. 181–198.

Engstad, R.E. and Robertsen, B. (1995). Effect of structurally different β-glucans on immune responses in Atlantic salmon (*Salmo salar* L.). *J. Marine Biotech.* 3, 203–207.

Hetland, G., Løvik, M., and Wiker, H.G. (1998). Protective effect of β-glucan against *Mycobacterium bovis*, BCG infection in Balb/c mice. *Scand. J. Immunol.* 47, 548–553.

Hetland, G., Ohno, N., Aaberge, I.S., and Løvik, M. (2000). Protective effect of β-glucan against systemic *Streptococcus pneumoniae* infection in mice. *FEMS Immunol. Med. Microbiol.* 27, 111–116.

Hetland, G. and Sandven, P. (2002). β-1,3-glucan reduces growth of Mycobacterium tuberculosis in macrophage cultures. *FEMS Immunol. Med. Microbiol.* 33, 41–45.

Instanes, C., Ormstad, H., Rydjord, B., Wiker, H.G., and Hetland, G. Mould extracts increase the allergic response to ovalbumin in mice. *Clin. Exp. Allergy* 34, 1634–1641.

Jarvis, D. and Burney P. (1998). ABC of allergies. The epidemiology of allergic disease. *Brit. Med. J.* 316, 607–610.

Kokoshis, P.L., Williams, D.L., Cook, J.A., and Di Luzio, N.R. (1978). Increased resistance to *Staphylococcus aureus* infection and enhancement in serum lysozyme activity by glucan. *Science* 199, 1340–1342.

Leikauf, G.D., Kline, S., Albert, R.E., Baxter, C.S., Bernstein, D.I., Bernstein, J., and Buncher, C.R. (1995). Evaluation of a possible association of urban air toxics and asthma. *Environ. Health Perspect.* 103, 253–271.

Løvik, M., Høgseth, A.K., Gaarder, P.I., Hagemann, R., and Eide, I. (1997). Diesel exhaust particles and carbon black have adjuvant activity on the local lymph node response and systemic IgE production to ovalbumin. *Toxicol.* 121, 165–178.

Maggi, E., Parronchi, P., Manetti, R., Simonelli, C., Piccinni, M.-P., Rugiu, F.S., De Carli, M., Ricci, M., and Romagnani, S. (1992). Reciprocal regulatory effects of IFN-γ and IL-4 on the *in vitro* development of human Th1 and Th2 clones. *J. Immunol.* 148, 2142–2147.

Manetti, R., Parronchi, P., Guidizi, M.G., Piccinni, M-P., Maggi, E., Trinchieri, G., and Romangani, S. (1993). Natural killer cell stimulatory factor (interleukin 12 (IL-12)) induces T helper type 1 (Th1)-specific immune responses and inhibits the development of IL-4-producing Th cells. *J. Exp. Med.* 177, 1199–1204.

Morikawa, K., Takeda, R., Yamazaki, M., and Mizuno, D. (1985). Induction of tumoricidal activity of polymorphonuclear leukocytes by a linear β-1,3-D-glucan and other immunomodulators in murine cells. *Cancer Res.* 45, 1496–1501.

Mushinski, M. (1997). Average hospital charges for asthma treatment: United States, 1995. *Stat. Bull. Metrop. Insur. Co.* 87, 26–32.

Nathan, C.F. (1987). Secretory products of macrophages. *J. Clin. Invest.* 79, 319–326.

Newman-Taylor, A. (1995). Environmental determinants of asthma. *Lancet* 345, 296–299.

Ohno, N., Suzuki, I., and Yadomae T. (1986). Structure and antitumor activity of a beta-1,3-glucan isolated from the culture filtrate of Sclerotinia sclerotiorum IFO 9395. *Chem. Pharm.* 34, 1362–1365.

Ohno, N., Kurachi, K., and Yadomae, T. (1987). Antitumor activity of highly branched (1→3)beta-D-glucan, SSG, obtained from Sclerotinia sclerotiorum IFO 9395. *J. Pharmacobiodyn.* 10, 478–486.

Ohno, N. and Yadomae, T. (1987). Two different conformations of the antitumor beta-D-glucan produced by Sclerotinia sclerotiorum IFO 9395. *Carbohydr. Res.* 159, 293–302.

Ohno, N., Uchiyama, M., Tsuzuki, A., Tokunaka, K., Miura, N.N., Adachi, Y., Aizawa, W., Tamura, H., Tanaka, S., and Yadomae, T., (1999). Solubilization of yeast cell-wall beta-(1→3)-D-glucan by sodium hypochlorite oxidation and dimethyl sulfoxide extraction. *Carbohydr. Res.* 316, 161–172.

Okazaki, M., Adachi, Y., Ohno, N., and Yadomae, T. (1995). Structure-activity relationship of (1→3)-beta-D-glucans in the induction of cytokine production from macrophages, in vitro. *Biol. Pharm. Bull.* 18, 1320–1327.

Ormstad, H., Gaarder, P.I., and Johansen, B.V. (1997). Quantification and characterisation of suspended particulate matter in indoor air. *Sci. Total Environ.* 193, 185–196.

Ormstad, H., Gaarder, P.I., Johansen, B.V., and Løvik, M. (1998a). Airborne house dust elicits a local lymph node reaction and has an adjuvant effect on specific IgE production in the mouse. *Toxicol.* 129, 227–236.

Ormstad, H., Johansen, B.V., and Gaarder, P.I. (1998b). Airborne house dust particles and diesel exhaust particles as allergen carriers. *Clin. Exp. Allergy* 28, 702–708.

Ormstad, H., Groeng, E.-C., Løvik, M., and Hetland, G. (2000). The fungal cell wall component β-1,3-glucan has an adjuvant effect on the allergic response to ovalbumin in mice. *J. Toxicol. Environ. Health Part A* 61, 55–67.

Platts-Mills, T.A.E. (1995). Is there a dose-response relationship between exposure to indoor allergens and symptoms of asthma? *J. Allergy Clin. Immunol.* 96, 435–440.

Riggi, S. and Di Luzio, N.R. (1961). Identification of a RE stimulating agent in zymosan. *Am. J. Physiol.* 200, 297–300.

Robertsen, B., Engstad, R.E., and Jørgensen, T. (1994). β-glucans as immunostimulants in fish. In *Modulators of Fish Immune Responses*, J.S. Stolen and T.C. Fletcher, eds., SOS Publications, Fair Haven, NJ, pp. 1, 83–99.

Romagnani, S. (1992). Induction of Th1 and Th2 responses: a key role for the "natural" immune response. *Immunol. Today* 13, 379–381.

Romagnani, S. (1995). Th1 and Th2 cells: in human disorders. In *XVI European Congress of Allergy and Clinical Immunology. ECACI'95*, A. Basomba and J. Sastre, eds., Mondussi Editore, Bologna, pp. 5–11.

Rook, A.W., Hernandez-Pando, R., and Lightman, S.L. (1994). Hormones, peripherally activated prohormones and regulation of the Th1/Th2 balance. *Immunol. Today* 15, 301–303.

Ross, G.D., Cain, J.A., Myones, B.L., Newman, S.L., and Lachmann, P.J. (1987). Specificity of membrane complement receptor type three (CR3) for beta-glucans. *Complement* 4, 61–74.

Rylander, R., Persson, K., and Goto, H. (1992). Airborne β-1,3-glucan may be related to symptoms in sick buildings. *Indoor Environ.* 1, 263–267.

Rylander, R., Hsieh, V., and Courteheuse, C. (1994a). The first case of sick building syndrome in Switzerland. *Indoor Environ.* 3, 159–162.

Rylander, R. (1994b). Special envivironment: office and domestic environment. In *Organic Dusts: Exposure, Effects and Prevention*, R. Rylander, and R.R. Jacobs, eds., CRC Press, Inc., Boca Raton, FL, pp. 247–256.

Rylander, R. (1994c). Pulmonary toxicity of inhaled (1-3)-D-glucan. In *Cotton Dust. Proceedings of the 18th Cotton Dust Reseach Conference. Beltwide Cotton Conferences*, R.R. Jacobs, P.J. Wakelyn, and R. Rylander, eds., National Cotton Council, Memphis, TN, pp. 347–349.

Rylander, R. (1996). Airway responsiveness and chest symptoms after inhalation of endotoxin or (1-3)-b-D-glucan. *Indoor Built. Environ.* 5, 196–111.

Rylander, R., Norrhall, M., Engdahl, U., Tunsater, A., and Holt, P.G. (1998). Airways inflammation, atopy, and (1→3)-beta-D-glucan exposures in two schools. *Am. J. Respir. Crit. Care Med.* 158, 1685–1687.

Rylander, R. and Lin, R.-H. (2000). (1→3)-beta-D-glucan: relationship to indoor air-related symptoms, allergy and asthma. *Toxicol.* 152, 47–52.

Sears, M.R. (1997). Epidemiology of childhood asthma. *Lancet.* 350, 1015–1020.

Seder, R.A., Paul, W.E., Davis, M.M., and Fazekas de St. Groth, B. (1992). The presence of interleukin 4 during *in vitro* priming determines the lymphokine-producing potential of CD4+ T cells from T cell receptor transgenic mice. *J. Exp. Med.* 176, 1091–1098.

Shy, M.R. (1999). Changing prevalence of alleric rhinitis and asthma. *Ann. Allergy Asthma Immunol.* 82, 233–252.

Stone, B.A. and Clarke, A.E. (1992). *Chemistry and Biology of (1→3)-β-$_D$-Glucans*, La Trobe University Press, Victoria, Australia.

Suda, M., Ohno. N., and Yadomae, T. (1995). Modulation of the antitumor effect and tissue distribution of highly branched (1→3)-β-D-glucan, SSG, by carrageenan. *Biol. Pharm. Bull.* 18, 772–775.

Suzuki, I., Hashimoto, K., Ohno, N., Tanaka, H., and Yadomae, T. (1989). Immunomodulation by orally administered beta-glucan in mice. *Int. J. Immunopharm.* 11, 761–769.

Suzuki, T., Ohno, N., Saito, K., and Yadomae, T. (1992). Activation of the complement system by (1→3)-beta-D-glucans having different degrees of branching and different ultrastructures. *J. Pharmacobiodyn.* 15, 277–285.

Taguchi, T., Furue, H., Kimura, T., Kondo, T., Hattori, T., and Ogawa, N. (1983). Clinical efficacy of lentinan on neoplastic diseases. *Adv. Exp. Med. Biol. Biotherapy* 166, 181–187.

Thorne, J. and Rylander, R. (1998). Airways inflammation and glucan in a rowhouse area. *Am. J. Respir. Crit. Care Med.* 157, 1798–1803.

Thornton, B.P., Vetvicka, V., Pitman, M., Goldman, R.C., and Ross, G.D. (1996). Analysis of the sugar specificity and molecular location of the beta-glucan-binding lectin site of complement receptor type 3 (CD11b/CD18). *J. Immunol.* 156, 1235–1246.

Vetvicka, V., Thornton, B.P., and Ross, G.D. (1996). Soluble beta-glucan polysaccharide binding to the lectin site of neutrophil or natural killer cell complement receptor type 3 (CD11b/CD18) generates a primed state of the receptor capable of mediating cytotoxicity of iC3b-opsonized target cells. *J. Clin. Invest.* 98, 50–61.

Wan, G.H. and Li, C.S. (1999). Indoor endotoxin and glucan in association with airway inflammation and systemic symptoms. *Arch. Environ. Health* 54, 172–179.

Wan, G.H., Li, C.S., Guo, S.P., Rylander, R., and Lin, R.H. (1999). An airbone mold-derived product, beta-1,3-D-glucan, potentiates airway allergic responses. *Eur. J. Immunol.* 29, 2491–2497.

Wilson, M.E. and Morrison, D.C. (1982). Evidence for different requirements in physical state for the interaction of lipopolysaccharides with the classical and alternative pathways of complement. *Eur. J. Biochem.* 128, 137–141.

Williams, D.L., Cook, J.A., Hoffmann, and Di Luzio, N.R. (1978). Protective effect of glucan in experimentally induced candidiasis. *J. Reticuloendothel. Soc.* 23, 479–490.

Williams, D.L., Mueller, A., Raptis, J., and Rice, P. (1997). Binding of fungal and plant glucans to the human macrophage receptor. In *Proceedings of the 21st Cotton and Other Organic Dust Reseach Conference. Beltwide Cotton Conferences*, P.J. Wakelyn, R.R. Jacobs, and R. Rylander, eds., National Cotton Council, Memphis, TN, pp. 169–171.

Young, R.S., Jones, A.M., and Nicholls, P.J. (1998). Something in the air: endotoxins and glucans as environmental troublemakers. *J. Pharm. Pharmacol.* 50, 11–17.

8 Endogenous Septic Shock by Combination of β-Glucan and NSAIDs

Naohito Ohno

CONTENTS

8.1 Introduction 143
8.2 Expression of Lethal Toxicity by Concomitant Administration
    of Microbial Components and NSAIDs 145
8.3 Changes in Inflammatory and Immune Parameters during
    Concomitant Administration of β-Glucan and Indometacin 149
8.4 Increased Sensitivity to Endotoxin Due to Concomitant
    Administration of β-Glucan and Indometacin 151
8.5 Effects of Nitric Oxide in the Appearance of Lethal Side Effects
    Caused by β-Glucans 152
8.6 Strain Differences in Response to β-Glucans 153
8.7 Conclusion 154
References 156

8.1 INTRODUCTION

Nonsteroidal antiinflammatory drugs (NSAIDs) is the generic term for anti-inflammatory drugs that reduce pain and inflammation by suppressing the production of prostaglandins (PGs) as a result of cyclooxygenase (COX) inhibition (Dubois et al., 2004; Brooks, 2003; Mahmud et al., 2004; Kamat, 2003; Smithard, 2003; Swartz, 2003; Shah, 2003; FitzGerald, 2003; Weinberg, 2003; Mayrink et al., 2003). More than 30 million people around the world are using NSAIDs every day. Some 35 million prescriptions are written each year in the U.S. alone, and if single administrations are included, it is estimated that approximately 1.5% of the U.S. population uses NSAIDs. Moreover, some surveys in the U.S. and U.K. have revealed that NSAIDs account for approximately 5% of all prescriptions. It was first reported in the 1990s that there are two isotypes of COX that are inhibited by NSAIDs, namely COX1 and COX2, and that drugs selective for those two isotypes have been targeted in drug development (Gomez Cerezo et al., 2003; Mengle Gaw and Schwartz, 2002).

Normally, COX2 is hardly found in any cells, and its expression is induced in inflammatory cells and cancer cells by stimulation with such substances as cytokines, hormones, and carcinogenic promoters. For this reason, PGs induced in this manner have been clearly demonstrated to be involved in inflammation, cell growth, vascularization, carcinogenesis, ovulation, and childbirth. On the other hand, PGs induced by the constitutive enzyme COX1 are thought to protect the body by protecting the gastric mucosa and maintaining normal kidney function.

Despite the fact that a large number of NSAIDs are in use as indicated above, numerous adverse effects have been reported (Johson and Day, 1991; Ng, 1992; Levi and Shaw-Smith,1994; Wright, 1995; Lee et al., 2003; Makins and Ballinger, 2003; de Jong et al., 2003; Gomez Cerezo et al., 2003; van Gelder et al., 2003). The major side effect of NSAIDs is digestive tract disorder, and this is said to be mainly due to the inhibition of COX1 (Makins and Ballinger, 2003). In the U.S., roughly 25% of all side effects attributable to drugs are said to be caused by NSAIDs, and the digestive tract disorders caused by NSAIDs number more than 15,000 according to reports of the cause of death by disease in 1997. Other related side effects include liver disorders, Reye's syndrome, and Stevens-Johnson syndrome. In addition, the use of diclofenac, an NSAID, for the treatment of influenza encephalopathy has been prohibited in Japan. However, as there are also reports that NSAIDs are effective against Alzheimer's disease and colon cancer, the NSAIDs will undoubtedly be one of the major drugs used in the 21st century. Moreover, there are reports that not only does the mechanism of action of NSAIDs involve COX inhibition, but also NSAIDs are agonists of peroxisome proliferator-activated receptor-γ (PPAR-γ), which acts through IκB kinase β to decrease signal transduction and the expression of cytokines and other inflammation-inducing molecules, indicating that NSAIDs have other hitherto unknown potentials (Yamamoto et al., 1999; Jaradat et al., 2001; Naito et al., 2001; Paccani et al., 2002; Kojo et al., 2003; Sastre et al., 2003; Bishop Bailey and Warner, 2003; Kim et al., 2002). A detailed analysis of the mechanism of expression of the side effects is essential for preventing the occurrence of side effects in advance and ensuring safer use.

We have conducted intensive analyses of the structure and biological activity of branched 1,3-β-glucans (hereafter referred to as β-glucans) from fungi (Ohno, 2003; Harada et al., 2003; Sugawara et al., 2003; Miura et al., 2003; Nakagawa et al., 2003; Harada et al., 2002; Kikuchi et al., 2002; Ishibashi et al., 2002; Moriya et al., 2002; Suzuki et al., 2002). From the perspective of application to cancer immunotherapy, the development of β-glucans seemed to have reached a peak in the 1980s with the clinical application of lentinan and sizofiran. However, starting in the latter half of the 1990s, they have again become the subject of considerable attention for potential application to alternative therapies and cell therapy. In addition, the measurement of blood β-glucan levels has become commonplace for early auxiliary diagnosis of deep mycoses. However, it cannot be denied that fundamental research on β-glucans is still inadequate compared with other active components originating in microorganisms. One of the reasons behind this is the extremely diverse nature of β-glucans in terms of both structure and activity. For example, although particulate β-glucans act on macrophages to markedly enhance active oxygen production, arachidonic acid release, and inflammatory cytokine production, those actions are extremely weak in the case of soluble β-glucans. In addition, although single-helix β-glucans strongly induce nitric oxide (NO) production from macrophages, the triple-helix forms do not exhibit such an activity. Conclusions have yet to be reached in the research of β-glucan receptors as well, and this is also one of the reasons for the lack of a general understanding of those substances. As β-glucans also interact with the complement system and the coagulation system *in vivo*, the receptors of these biocomponents and activated fragments are also involved in the activity. In addition, after reaching a host organ, β-glucans accumulate for a long period of time due to the absence of an aggressive elimination system. Although this accumulation and residual presence is thought to be one of the factors that lead to the harmful effects of β-glucans, due to the lack of definitive evidence a consensus regarding the harmful nature of β-glucans has yet to be obtained. In addition, in terms of environmental hygiene, fungal β-glucans have been proposed as the exacerbating factors of sick building syndrome and asthma (Wan et al.; 1999; Korpi et al., 2003; Beijer et al., 2002; Thorn et al., 2001; Fogelmark et al., 1997 and 2001; Rylander, 1997; Rylander and Lin, 2000).

When we attempted the concomitant administration of β-glucan and indometacin to tumor-bearing mice as part of our studies on the cancer immunotherapy of β-glucans, we accidentally found that marked lethal toxicity was induced (Yoshioka et al., 1998; Takahashi et al., 2001; Moriya et al., 2002). Figure 8.1 shows the results of concomitant administration of the antitumor drug sonifilan (SPG in the figure) and indometacin (IND in the figure) to mice. Although a certain percentage of the mice expired even with the administration of indometacin alone, a definite increase in the number of mice that expired was observed in the concomitant dose group. The fact that the mice expired when dosed with indometacin alone suggests that the increase in mortality due to concomitant administration may be intimately involved with digestive tract disorder, a typical side effect of NSAIDs.

What we would like to emphasize here is that β-glucans may have greatly enhanced this side effect. It is quite common for the side effects of drugs to be influenced considerably by patient background. The results of this study ought to provide a glimpse into understanding this background. Due in part to the fact that β-glucans are also immunostimulators, there are few reports thus far describing their harmful action. During the combined use of immunostimulators and anti-inflammatory drugs, which have completely opposite effects, prominent side effects were observed, which are of much interest to basic as well as medical science. We discuss herein the occurrence of lethal side effects, along with the mechanism underlying their occurrence, based on the interaction between NSAIDs and microbial components, particularly β-glucans.

8.2 EXPRESSION OF LETHAL TOXICITY BY CONCOMITANT ADMINISTRATION OF MICROBIAL COMPONENTS AND NSAIDS

Although a lethal action was discovered after the concomitant administration of β-glucans and indometacin, it is necessary to determine the degree of specificity for

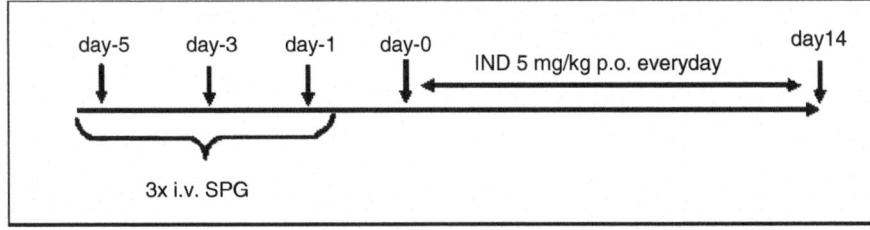

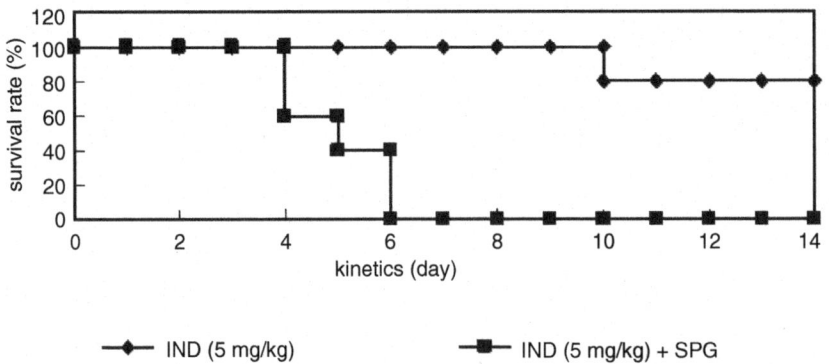

FIGURE 8.1 Immunotoxicity of soluble β-glucans induced by indomethacin treatment. SPG (250 μg per mouse) was intravenously administered 3 times to male ICR mice once every other day (days -5, -3, and -1). IND (5 mg/kg) was orally administered once a day from day 1 to day 14 and survival was monitored.

various substances. Therefore, a comparative study was conducted of the lethal activity from three viewpoints: microbial component, NSAID type, and mouse strain.

Figure 8.2 shows the microorganisms and the microbial components used. All the microorganisms were attenuated. Screening revealed that numerous microbial components other than β-glucans enhanced lethal activity, as shown in Figure 8.3, thereby indicating the possibility that this action is universally caused by microorganisms. However, of note is that whereas Gram-negative microorganisms demonstrated this enhancing action, LPS did not (Figure 8.4). Moreover, there were no increases in lethal activity when the origin of LPS, the LPS extraction method, the LPS administration method, or the dosage of LPS was changed. Although LPS is believed to exhibit a variety of biological activities, it appears that it does not exhibit lethal activity in combination with indometacin. In addition, lethal activity was also not observed for mycelial fungi. The fact that the mycelial type typically contains lower β-glucan levels than the yeast type may be responsible for the difference in activity. However, the detailed reason behind this difference remains unclear. Figure 8.5 shows the types of NSAIDs tested. The experiment was conducted using the highly branched β-glucan SSG, derived from the ascomycete *Sclerotinia sclerotiorum* IFO9395 (Sakurai et al., 1995; Suda et al., 1994; Sakurai et al., 1992). With the exception of nabumetone, many of the NSAIDs exhibited an increase in lethal toxicity when administered in combination with SSG. Because nabumetone is a

Tested Materials

Bacteria

Escherichia coli (E.coli) K-12
E.coli acetone powder (ATCC 11246)
Staphylococcus aureus (Newman D_2C)
Mycobacterium butyricum
Mycobacterium tuberculosis H37 Ra

Lipopolysaccharide

- from *E. coli* O111:B4, phenol-water ext
- from *E. coli* O111:B4 TCA extract
- from *E. coli* O127:B8 butanol extract
- from *Salmonella minnesota* Re 595
- from *Salmonella minnesota* phenol extract
- from *Pseudomonas aeruginosa* Serotype 10
- from *Serratia marcescens*

Fungi/yeast

Candida albicans
- Defatted whole cell
- NaClO-oxidized cell (OX-CA)
- Hot NaOH-extract
- Cold NaOH-extract
- Zymolyase solubilized fraction (Z-CA)
- Solubilized cell wall β-glucan (CSBG)
- Soluble polysaccharide fraction (CAWS)

Aspergillus fumigatus
Mucor racemosus

Saccharomyces cerevisiae
- Zymosan (ZYM)
- Zymocel (ZYC)
- NaClO-oxidized ZYM
  (OX-ZYM particle and DMSO solubilized)

β-glucan from *Sclerotinia sclerotiorum* (SSG)

FIGURE 8.2 List of materials tested for lethal toxicity.

Materials enhancing lethal toxicity

E. coli K-12, *E.coli* ATCC, *S aureus*, *M. butyricum*, *M. tuberculosis* H37 Ra
CA, YPG-CA, OX-CA, Hot NaOH-extract, Z-CA, CSBG, CAWS
ZYM, ZYC, OX-ZYM (particle, single helix, triple helix), SSG

FIGURE 8.3 Summary of materials enhancing the side effects of indomethacin. The lethal toxicity of various materials shown in Figure 8.2 were tested by the protocol in Figure 8.1. List of materials showing enhanced lethal toxicity.

drug that is associated with a low level of digestive tract disorder, the lack of lethal toxicity is most likely related to this. Figure 8.6 shows the differences among mouse strains. Lethal activity was increased in nude mice and other mutant mouse strains, suggesting that this action is not governed by specific cells, such as T cells and mast cells. In addition, a closer examination of the data revealed that there were differences in the expression of lethal activity among strains during the administration of indometacin alone. As there were also differences in the average survival time among strains during concomitant administration, the involvement of multiple factors was suggested. DBA/2 mice exhibited strong sensitivity equivalent to that of ICR mice, and the sensitivity of C3H/HeJ mice, which exhibit low sensitivity to LPS, tended

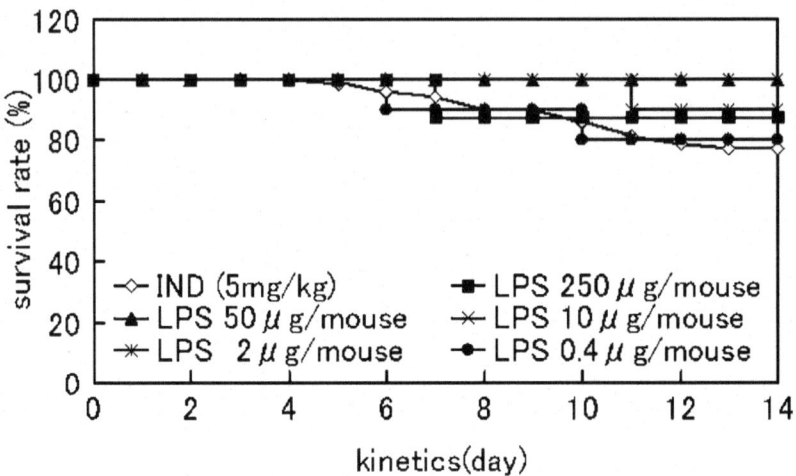

FIGURE 8.4 Survival of lipopolysaccharide-treated mice in response to indomethacin. The protocol shown in Figure 8.1 was used. Indicated quantities of lipopolysaccharide (LPS) were administered instead of SPG.

^{b} p<0.05 ^{c} p<0.01

NSAID	treatment NSAID mg/kg	SSG μg/mouse	life span (days, mean±SD)	number of mice dead/total on day 14	mortality (%)
indomethacin	5		12±4.5	1/5	20
	5	250	2.8±1.8 ^{c}	5/5	100
aspirin	600		14±0	0/5	0
	600	250	8.6±5.3 ^{b}	3/5	60
diclofenac	20		14±0	0/5	0
	20	250	6.6±4.6 ^{b}	5/5	100
sulindac	60		11.2±3.9	2/5	40
	60	250	4±1.9 ^{c}	5/5	100
nabumetone	320		14±0	0/5	0
	320	250	14±0	0/5	0

FIGURE 8.5 Effect of various NSAIDs on mortality of SSG-administered mice. SSG (250 μg per mouse) was administered to these mice on days 5, 3, and 1, and the indicated dose of various kinds of NSAIDs was orally administered to these mice daily from day 0 and lethality monitored.

to be somewhat low. These findings are discussed later. On the basis of these findings, it was determined that the enhancement of lethal activity may be induced by β-glucans and other microbial components.

	treatment		life span		treatment		life span
strain	IND mg/kg	SSG μg/mouse	(days, mean ± SD)	strain	IND mg/kg	SSG μg/mouse	(days, mean ± SD)
ICR	5		12±4.5	DBA/2	5		13±2.3
	5	250	2.8±1.8 c		5	250	3.2±0.5 b
C3H/HeJ	5		14±0	A/J	5		12.4±2.1
	5	250	10.8±3.7		5	250	6±4.2 b
C3H/HeN	5		13±2.2	BALB/c nu/nu	3		13.2±1.8
	5	250	6.8±2.3 c		3	250	6.6±1.8 c
AKR/N	5		13.6±0.9	KSN	5		7.2±2.3
	5	250	4.6±0.6 c		5	250	3.6±0.6 b
BALB/c	5		14±0	WBB6F1	5		9.6±1.8
	5	250	7.2±4.3 b		5	250	4.8±2.2 c
C57BL/6	3		14±0				
	3	250	5.6±4.9 c				

b $p<0.05$ c $p<0.01$

FIGURE 8.6 Strain difference on mortality of SSG and IND administered mice. SSG (250 μg per mouse) was administered to these mice on days -5, -3, and -1 and an indicated dose of IND was orally administered to these mice daily from day 0 and lethality monitored.

8.3 CHANGES IN INFLAMMATORY AND IMMUNE PARAMETERS DURING CONCOMITANT ADMINISTRATION OF β-GLUCAN AND INDOMETACIN

Next, a study was conducted from various perspectives regarding the manner in which changes are induced in the host during the course of the appearance of the lethal side effects of concomitant administration of β-glucan and indometacin. Because it was necessary to conduct measurements prior to death, data were collected on day 2 of indometacin administration. As a result, decreased liver weight, increased GOT and GPT levels, splenomegaly, increased production of IFN-γ and IL-6 in the blood, increased ceruloplasmin activity, increased colony-forming activity, increased NO production by spleen cells, and changes in the electrophoretic patterns of serum proteins and glycoproteins were observed. Moreover, serum IFN-γ levels were increased 6 hours after indometacin administration, whereas body temperature and blood sugar were subsequently decreased rapidly approximately 24 hours after indometacin administration. In addition, leukocytes markedly infiltrated the liver, the spleen, the lungs, and the peritoneal cavity, and they consisted primarily of granulocytes and macrophages fractions. The infiltrating leukocytes exhibited strong peroxidase activity, and because the expression of L-selectin was decreased, all were considered to be in the activated state. As shown in Figures 8.7 and 8.8, FACS analysis of cells that had infiltrated the lungs due to the concomitant administration of *Candida* (CA in the figures) and indometacin revealed infiltration by granulocytes

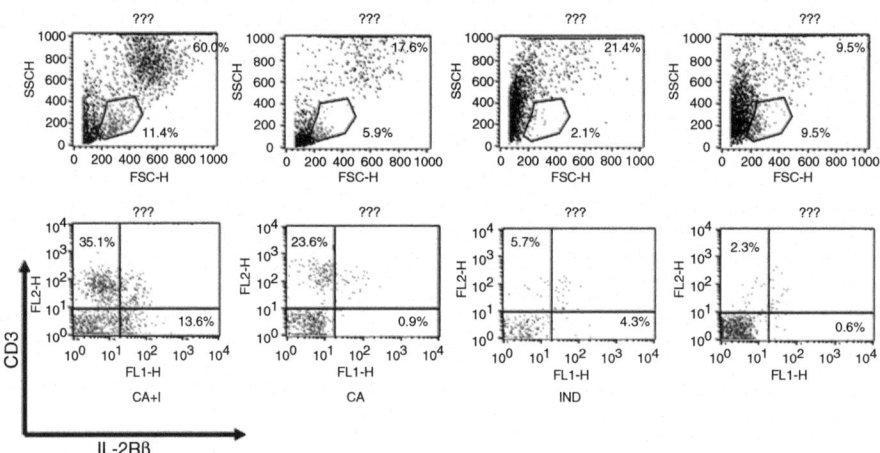

FIGURE 8.7 (See color insert following page 20.) Lymphocyte phenotype in lung (1). Cells were collected from BAL of mice which were administered CA/IND. The surface phenotypes of these cells were analyzed using a two-color immunofluorescence test. FITC anti-mouse CD122 (IL-2 Receptor β-chain) mAbs and R-PE anti-mouse CD3e (CD3 ε–chain) mAbs were used.

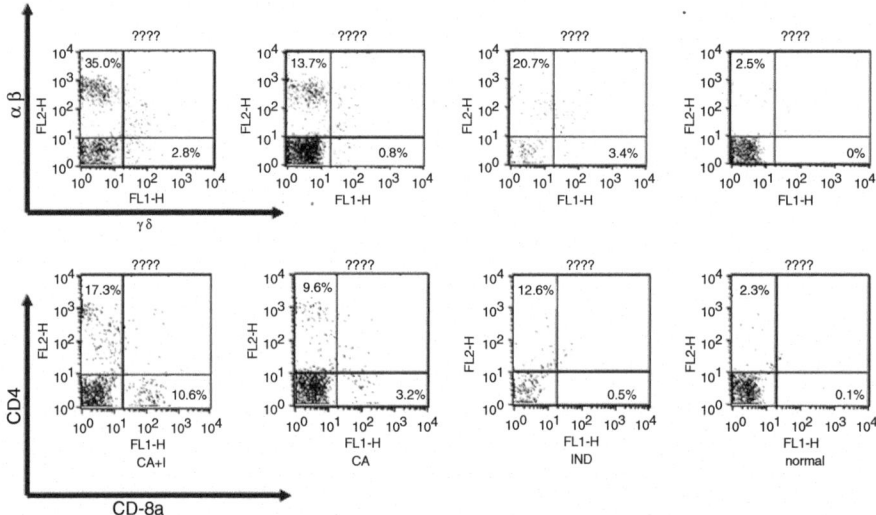

FIGURE 8.8 (See color insert following page 20.) Lymphocyte phenotype in lung (2). Cells were collected from BAL of mice which were administered CA/IND. The surface phenotypes of these cells were analyzed using a two-color immunofluorescence test. FITC anti-mouse δ T cell receptor mAbs and R-PE anti-mouse TCR chain (α) mAbs (upper side), and FITC anti-mouse CD8a mAbs and R-PE anti-mouse CD4 mAbs (lower side) were used.

as well as cells thought to be mature T cells, NK cells (CD3-NK1.1+), and NKT cells (CD3+NK1.1+). In addition, Gram-positive and Gram-negative organisms were detected from peritoneal washings, thereby indicating the possibility of the occurrence of translocation of intestinal bacteria. When indometacin was added to an *in vitro* culture of peritoneal exudate cells induced with β-glucans, IFN-γ production was increased. As this increase was due to nonadhering cells, the interaction between macrophages and other cells was suggested to be important for the increase in IFN-γ production by β-glucans and indometacin. The changes in these inflammatory and immune parameters created a situation that resembled the so-called systemic inflammatory response syndrome (SIRS).

8.4 INCREASED SENSITIVITY TO ENDOTOXIN DUE TO CONCOMITANT ADMINISTRATION OF β-GLUCAN AND INDOMETACIN

As previously mentioned, a state of systemic inflammation is achieved prior to death. Examination of the organ tissues of expired mice revealed that the organ that exhibited the most prominent changes was the liver, and liver failure was suggested to be one cause of death. These changes resembled the SIRS-MOF (multiple organ failure) observed in septic shock. Therefore, an investigation was conducted to determine whether or not an increase in sensitivity to LPS occurs as a result of such changes. The sensitivity to extrinsic LPS in the *Candida* and indometacin dose groups is shown in Figure 8.9. Whereas a considerable number of mice survived even when 1 mg of extrinsic LPS was administered, the sensitivity to LPS was increased by the administration of *Candida* and all the mice expired when 100 μg of LPS was administered. On the other hand, all the mice in the concomitant dose group expired within 4 days after administering 1 μg of LPS, and nearly all of the mice expired within 7 days even in the 0.1 μg dose group, thus demonstrating a considerable increase in sensitivity. On the basis of these findings, the sensitivity to extrinsic LPS in the concomitant dose group was found to be 10,000 times higher than that of untreated mice, and approximately 1000 times higher than that of the *Candida*-dosed mice. As was previously indicated, as microorganisms are detected in the abdominal cavity, there is a possibility that as a result of a digestive tract disorder, intestinal bacteria were translocated, triggering an excessive reaction to the LPS of these bacteria that eventually led to death. In order to further examine this point, survival rates were compared after sterilizing the digestive tract by continuous administration of antibiotics. As a result of antibiotic treatment, the survival rates were improved, suggesting even more strongly the involvement of intrinsic LPS, as shown in Figure 8.10. As was indicated at the outset, LPS alone did not enhance the lethality of indometacin. However, an increase in sensitivity to LPS may lead to death, and although these findings may appear contradictory at a glance, this contradiction is thought to have occurred due to differences in the disease stage of the host due to the concomitant administration of β-glucan.

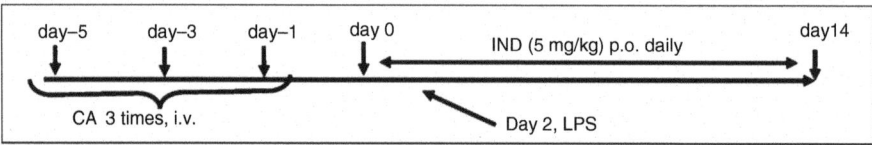

CA dose ×3 (μg/mouse)	IND (5 mg/kg)	LPS dose, (μg/mouse)	route	dead/sum	life span (days, mean)
500	O	–	–	14/15	7.7
500	O	100	i.v.	4/5	5.2
	X		i.v.	4/5	4.4
500	O	50	i.v.	5/5	3.0
	X		i.v.	1/5	11.8
500	O	10	i.v.	5/5	3.0
	X		i.v.	1/5	12.0
500	O	1	i.v.	10/10	4.5
	X		i.v.	3/10	10.7
500	O	0.1	i.v.	9/10	7.6
	X		i.v.	0/10	14.0
500	O	50	i.p.	3/5	9.0
	X		i.p.	1/5	11.8
500	O	10	i.p.	5/5	5.6
	X		i.p.	1/5	11.8
–	O	1	i.v.	0/5	14.0

FIGURE 8.9 Sensitivity of CA/IND administered mice to LPS. Aceton-dried cells of *Candida albicans* (250 μg per mouse) was administered to these mice on days -5, -3, and -1 and IND was orally administered to these mice daily from day 0. Various concentrations of lipopolysaccharide were administered on day 2 via the intravenous or the i.p. route, and lethality monitored. O indicates the presence of indometacin, while X indicates its absence.

8.5 EFFECTS OF NITRIC OXIDE IN THE APPEARANCE OF LETHAL SIDE EFFECTS CAUSED BY β-GLUCANS

In addition to its involvement in blood pressure regulation, nitric oxide is also an important biological defense factor. Isozymes exist for enzymes that produce nitric oxide, and the iNOS produced by macrophages is known to play an important role in microbial defense. Nitric oxide demonstrates potent cell damaging action and contributes to the elimination of infecting bacteria. On the other hand, nitric oxide produces excess damage to the host tissue. Therefore, a study was conducted to elucidate the significance of nitric oxide production in this model. The survival rate was lowered considerably by the addition of an inhibitor of iNOS, indicating a strong possibility that nitric oxide demonstrates an antimicrobial action in this model. This suggests that nitric oxide plays an important role in the early stage of eliminating microorganisms that have invaded as a result of translocation.

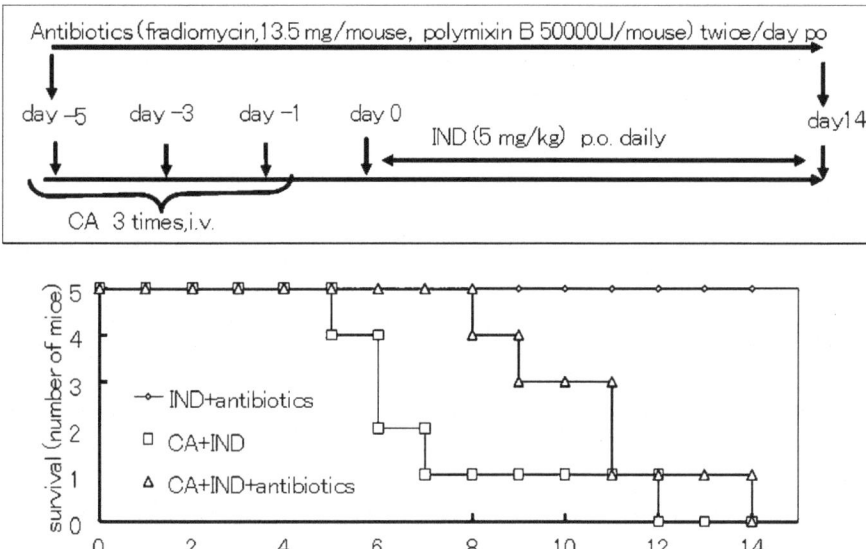

FIGURE 8.10 Effect of antibiotics on lethal toxicity of CA/IND. Aceton-dried cells of *Candida albicans* (250 μg per mouse) was administered to these mice on days -5, -3, and -1. A mixture of antibiotics (fradiomycin and polymyxin B) were orally administered to mice twice a day from day 5. IND was orally administered to these mice daily from day 0 and lethality monitored.

8.6 STRAIN DIFFERENCES IN RESPONSE TO β-GLUCANS

β-Glucans exhibit various biological activities in various organisms. The SNPs of drug-metabolizing enzymes represented by P450 have been found to be intimately involved in drug absorption, metabolism, and excretion. Gender differences and HLA differences are suggested to be involved in the frequency of onset of various autoimmune diseases. Strain differences are predicted to be involved in the response to β-glucans as well. Although antibodies against β-glucans have hitherto been considered to be difficult to produce, high β-glucan antibody titers were detected in strains DBA/1 and DBA/2 (Harada et al., 2003; Uchiyama et al., 2000). The high titers of these antibodies in the nonimmune state are thought to reflect differences in the manner of spontaneous sensitization by substances containing β-glucans in the environment. In addition, differences among strains were observed in the response of spleen cells to β-glucans, and strain DBA/2, which demonstrated a high level of sensitivity in this model, showed a high response in this system as well. Although the causes of these strain differences remain unknown, continuing studies will lead to further insights into the mechanism of occurrence in this model.

8.7 CONCLUSION

The lethal side effects induced by the concomitant administration of β-glucans and indometacin may have been a result of shock caused by the transmigration of bacteria from intestine damaged by the co-exposure to both drugs. As was indicated at the outset, LPS did not replace β-glucan in this enhanced lethality model, and C3H/HeJ mice having decreased sensitivity to LPS also expired. Therefore, the involvement of intrinsic endotoxins was not initially concluded. However, it appears that once concomitant β-glucan and indometacin have damaged the intestine and bacteria have entered the peritoneal cavity, enhanced LPS sensitivity becomes a factor.

As the frequency of occurrence of digestive tract disorders as the side effects of NSAID administration is extremely high, the development of new drugs has been pursued and very recently, drugs that are highly selective for COX2 have become available commercially. Although clinical data are awaited to determine the extent of the side effects of these drugs, they are, at least, expected to have fewer side effects than conventional drugs. The possibility of long-term administration will offer new potential for overcoming diseases.

Although the action of NSAIDs has long been considered to involve COX inhibition, a new action has recently attracted attention (Kojo et al., 2003; Sastre et al., 2003; Bishop Bailey and Warner, 2003; Kim et al., 2002). For example, although aspirin suppresses prostaglandin (PG) production even at low concentrations, no anti-inflammatory action can be observed at such low concentrations in the blood. In addition, although PGE is known to cause inflammation, it has been reported to demonstrate antiinflammatory action in some animal inflammation models. Therefore, several reports have suggested that the mechanism of action of NSAIDs cannot be fully explained by COX inhibition alone. Because new drug development has proceeded focusing primarily on selectivity for COX, it will be necessary to examine the manner in which new compounds behave in response to this action. One new important action involves the intranuclear receptor PPAR. Although PPAR was initially discovered to be a receptor that increases the number of liver peroxisomes, it was subsequently determined to have three types in humans (α, β, and γ), and PPAR-γ was discovered to have an important function in fat cell differentiation, bone metabolism, vascularization, and other physiological and pathological reactions. In addition, attention has also been focused on its intrinsic ligand, 15-deoxy-12,14-PGJ2. During the course of research on this ligand, it was determined that arachidonic acid and NSAIDs involved in PG production also serve as ligands of PPAR-γ, and that suppression of the production of inflammatory cytokines is mediated by the activation of PPAR. In addition, certain types of NSAIDs have been found to suppress the activity of IκB kinase β, which is important for the migration into the nucleus of NF-κB, a transcription factor intimately involved in cytokine production. New attention is also being focused on the involvement of this pathway in the inhibition of lymphocyte function. Although an analysis of changes at the molecular level which accompany the lethal side effects discovered in this study has yet to be completed, the actions of NSAIDs on PPAR and IκB kinase β are all antiinflammatory or immunosuppressive, and those actions are the reciprocal of the systemic inflammatory symptoms observed in this study. It is therefore necessary to conduct detailed studies on not only the pathways mediated by COX inhibition, but also the control failure mediated by these pathways.

β-Glucans are the major components of the fungal cell wall. It has recently been reported that *Candida*, which is a typical causative microorganism of deep mycoses, possesses a COX-like enzyme and produces a PG-like substance (Noverr et al., 2001). Other microorganisms have also been reported to produce PG-like substances (Lamacka and Sajbidor 1995; Kock, 1997; Botha, 1997). These microbial PGs have been shown to demonstrate pharmacological actions on host cells. In the relationship between microorganisms and their hosts, not only do the PGs act as growth factors for the microorganisms, but they also create a favorable growing environment for the microorganisms by acting as immunosuppressants of the host. As indometacin is considered to act in a cytostatic manner by suppressing microbial COX activity, it is thought to create a suitable environment for the host. However, the results of this study appear to contradict the aforementioned reports in that the combined use of fungal components and indometacin creates a situation that is extremely unsuitable for the host. This most likely indicates that the response varies with the difference in the disease stage of the host.

The administration of endotoxin to galactosamine- or carrageenan-injected mice is commonly used as an animal model for analyzing endotoxin shock. We also used those models to analyze the pathological state and the detoxification process. As those models have a high degree of universality and are used by numerous research groups throughout the world, they have a high degree of reliability. However, they are continuously confronted with the problem of the animals expiring soon after direct administration of large doses of endotoxin, thereby creating a gap between those models and septic shock in humans. In the process of the appearance of lethal side effects discovered in this study, it is highly likely that the sensitivity to endotoxin is increased gradually, with shock ultimately being induced by the intrinsic endotoxin. Although numerous aspects of this process remain for future studies, including at what point in time the endotoxin entered the body and what triggered the shock, as well as analyses at the molecular, cellular, and tissue levels, since the pathological state can be induced with good reproducibility, this model is considered to be applicable for the analysis of various parameters and drug development.

In this paper, we did not discuss toxicity directly resulting from β-glucans. This can be easily understood from the fact that β-glucans are not inherently recognized as "toxins." In addition, two types of β-glucans, namely, lentinan and sonifilan, are available commercially as cancer immunotherapy drugs in Japan. Although high concentrations of β-glucans are detected in the blood in the case of deep mycoses, there would be little likelihood of an increase in blood concentration if prominent toxicity were demonstrated as a result of directly stimulating receptors. In addition, β-glucans distributed in organs are known to accumulate in a state that is resistant to decomposition over a long period of time. All these facts suggest that the toxicity of β-glucans is not considered to be extremely potent. However, β-glucans have been found to significantly alter the toxicity of various microbial components routinely encountered by people in their normal living environment. Certain mouse strains have been found to be spontaneously sensitized by β-glucans. In addition, antibodies against β-glucans have been determined to exist universally in human blood. These findings indicate that humans are constantly subjected to a high frequency of exposure to β-glucans. Altogether, these findings indicate the possibility that β-glucans are involved in human diseases in various forms. As basic and applied research of β-glucans is still inadequate, it is hoped that further progress will be made in the future.

REFERENCES

Beijer, L., Thorn, J., and Rylander, R. (2002). Effects after inhalation of (1→3)-beta-D-glucan and relation to mould exposure in the home. *Mediators Inflam.* 11, 149–153.

Bishop Bailey, D., and Warner, T.D. (2003). PPARgamma ligands induce prostaglandin production in vascular smooth muscle cells: indomethacin acts as a peroxisome proliferator-activated receptor-gamma antagonist. *FASEB J.* 17, 1925–1927.

Botha, A., Kock, J.L., and Nigam, S. (1997). The production of eicosanoid precursors by mucoralean fungi. *Adv. Exp. Med. Biol.* 433, 227–229.

Brooks, P. (2003). Inflammation as an important feature of osteoarthritis. *Bull. World Health Organ.* 81, 689–690.

de Jong, J.C., van den Berg, P.B., Tobi, H., and de Jong van den Berg, L.T. (2003). Combined use of SSRIs and NSAIDs increases the risk of gastrointestinal adverse effects. *Br. J. Clin. Pharmacol.* 55, 591–595.

Dubois, R.W., Melmed, G.Y., Henning, J.M., and Laine, L. (2004). Guidelines for the appropriate use of non-steroidal anti-inflammatory drugs, cyclo-oxygenase-2-specific inhibitors and proton pump inhibitors in patients requiring chronic anti-inflammatory therapy. *Aliment. Pharmacol. Ther.* 19, 197–208.

FitzGerald, G.A. (2003). COX-2 and beyond: approaches to prostaglandin inhibition in human disease. *Nat. Rev. Drug Discov.* 2, 879–890.

Fogelmark, B., Sjostrand, M., Williams, D., and Rylander, R. (1997). Inhalation toxicity of (1→3)-β-D-glucan. Recent advances. *Mediators Inflam.* 6, 263–265.

Fogelmark, B., Thorn, J., and Rylander, R. (2001). Inhalation of (1→3)-beta-D-glucan causes airway eosinophilia. *Mediators Inflam.* 10, 13–19.

Gomez Cerezo, J., Lubomirov Hristov, R., Carcas Sansuan, A.J., and Vazquez Rodriguez, J.J. (2003). Outcome trials of COX-2 selective inhibitors: global safety evaluation does not promise benefits. *Eur. J. Clin. Pharmacol.* 59, 169–175.

Harada, T., Miura, N.N., Adachi, Y., Nakajima, M., Yadomae, T., and Ohno, N. (2002). IFN-gamma induction by SCG, 1,3-beta-D-glucan from *Sparassis crispa*, in DBA/2 mice in vitro. *J. Interferon Cytokine Res.* 22, 1227–1239.

Harada, T., Nagi Miura, N., Adachi, Y., Nakajima, M., Yadomae, T., and Ohno, N. (2003). Antibody to soluble 1,3/1,6-beta-D-glucan, SCG in sera of naive DBA/2 mice. *Biol. Pharm. Bull.* 26, 1225–1228.

Ishibashi, K., Miura, N.N., Adachi, Y., Ogura, N., Tamura, H., Tanaka, S., and Ohno, N. (2002). Relationship between the physical properties of *Candida albicans* cell wall beta-glucan and activation of leukocytes *in vitro*. *Int. Immunopharmacol.* 2, 1109–1122.

Jaradat, M.S., Wongsud, B., Phornchirasilp, S., Rangwala, S.M., Shams, G., Sutton, M., Romstedt, K.J., Noonan, D.J., and Feller, D.R. (2001). Activation of peroxisome proliferator-activated receptor isoforms and inhibition of prostaglandin H(2) synthases by ibuprofen, naproxen, and indomethacin. *Biochem. Pharmacol.* 62, 1587–1595.

Johson, A.G. and Day, R.O. (1991). The problems and pitfalls of NSAID therapy in the elderly (Part I). *Drugs Aging* 1, 130–143.
Kamat, A.M. (2003). Chemoprevention of superficial bladder cancer. *Expert Rev. Anticancer Ther.* 3, 799–808.
Kikuchi, T., Ohno, N., and Ohno, T. (2002). Maturation of dendritic cells induced by *Candida* beta-D-glucan. *Int. Immunopharmacol.* 2, 1503–1508.
Kim, T.I., Jin, S.H., Kang, E.H., Shin, S.K., Choi, K.Y., and Kim, W.H. (2002). The role of mitogen-activated protein kinases and their relationship with NF-kappaB and PPAR-gamma in indomethacin-induced apoptosis of colon cancer cells. *Ann. N. Y. Acad. Sci.* 973, 241–245.
Kock, J.L., Venter, A., Botha, A., Coetzee, D.J., Botes, P., and Nigam, S. (1997). The production of biologically active eicosanoids by yeasts. *Adv. Exp. Med. Biol.* 433, 217–219.
Kojo, H., Fukagawa, M., Tajima, K., Suzuki, A., Fujimura, T., Aramori, I., Hayashi, K., and Nishimura, S. (2003). Evaluation of human peroxisome proliferator-activated receptor (PPAR) subtype selectivity of a variety of anti-inflammatory drugs based on a novel assay for PPAR delta(beta). *J. Pharmacol. Sci.* 93, 347–355.
Korpi, A., Kasanen, J.P., Kosma, V.M., Rylander, R., and Pasanen, A.L. Slight respiratory irritation but not inflammation in mice exposed to (1→3)-beta-D-glucan aerosols. *Mediators Inflam.* 12, 139–146.
Lamacka, M. and Sajbidor, J. (1995). The occurence of prostaglandins and related compounds in lower organisms. *Prostaglandins Leukot. Essent. Fatty Acids* 52, 357–364.
Lee, J.L., Mukhtar, H., Bickers, D.R., Kopelovich, L., and Athar, M. (2003). Cyclooxygenases in the skin: pharmacological and toxicological implications. *Toxicol. Appl. Pharmacol.* 192, 294–306.
Levi, S. and Shaw-Smith, C. (1994). Non-steroidal anti-inflammatory drugs: how do they damage the gut? *Br. J. Rheumatol.* 33, 605–612.
Mahmud, S., Franco, E., and Aprikian, A. (2004). Prostate cancer and use of nonsteroidal anti-inflammatory drugs: systematic review and meta-analysis. *Br. J. Cancer* 90, 93–99.
Makins, R. and Ballinger, A. (2003). Gastrointestinal side effects of drugs. *Expert Opin. Drug Saf.* 2, 421–429.
Mayrink, M., Mendonca, A.C., and da Costa, P.R. (2003). Soft-tissue sarcoma arising from a tissue necrosis caused by an intramuscular injection of diclofenac. *Plast. Reconstr. Surg.* 112, 1970–1971.
Mengle Gaw, L.J., and Schwartz, B.D. (2003). Cyclooxygenase-2 inhibitors: promise or peril? *Mediators Inflam.* 11, 275–286.
Miura, N.N., Adachi, Y., Yadomae, T., Tamura, H., Tanaka, S., and Ohno, N. (2003). Structure and biological activities of beta-glucans from yeast and mycelial forms of *Candida albicans*. *Microbiol Immunol.* 47, 173–182.
Moriya, K., Ohno N., Miura N.N., Adachi, Y., and Yadomae, T. (2002a). Septic shock induced by microbial products and indomethacin, *Drug Develop. Res.* 55, 139–148.
Moriya, K., Miura, N.N., Adachi, Y., and Ohno, N. (2002b). Systemic inflammatory response associated with augmentation and activation of leukocytes in *Candida*/indomethacin administered mice. *Biol. Pharm. Bull.* 25, 816–822.
Nakagawa, Y., Ohno, N., and Murai, T. (2003). Suppression by *Candida albicans* beta-glucan of cytokine release from activated human monocytes and from T cells in the presence of monocytes. *J. Infect. Dis.* 187, 710–713.

Naito, Y., Takagi, T., Matsuyama, K., Yoshida, N., and Yoshikawa, T. (2001). Pioglitazone, a specific PPAR-gamma ligand, inhibits aspirin-induced gastric mucosal injury in rats. *Aliment Pharmacol. Ther.* 15, 865–873.

Ng, S.C. (1992). Non-steroidal anti-inflammatory drugs—uses and complications. *Singapore Med. J.* 33, 510–513.

Noverr, M.C., Phare, S.M., Toews, G.B., Coffey, M.J., and Huffnagle, G.B. (2001). Pathogenic yeasts *Cryptococcus neoformans* and *Candida albicans* produce immunomodulatory prostaglandins. *Infect. Immun.* 69, 2957–2963.

Ohno, N. (2003). Chemistry and biology of angiitis inducer, *Candida albicans* water-soluble mannoprotein-beta-glucan complex (CAWS). *Microbiol. Immunol.* 47, 479–490.

Paccani, S.R., Boncristiano, M., Ulivieri, C., D'Elios, M.M., Del Prete, G., and Baldari, C.T. (2002). Nonsteroidal anti-inflammatory drugs suppress T-cell activation by inhibiting p38 MAPK induction. *J. Biol. Chem.* 277, 1509–1513.

Rylander, R. (1997). Investigations of the relationship between disease and airborne (1→3)-β-D-glucan in buildings. *Mediators Inflam.* 6, 275–277.

Rylander, R. and Lin, R.H. (2000). (1→3)-beta-D-glucan: relationship to indoor air-related symptoms, allergy and asthma. *Toxicol.* 152, 47–52.

Sakurai, T., Hashimoto, K., Suzuki, I., Ohno, N., Oikawa, S., Masuda, A., and Yadomae, T. (1992). Enhancement of murine alveolar macrophage functions by orally administered beta-glucan. *Int. J. Immunopharmacol.* 14, 821–830.

Sakurai, T., Ohno, N., Suzuki, I., and Yadomae, T. (1995). Effect of soluble fungal (1→3)-beta-D-glucan obtained from *Sclerotinia sclerotiorum* on alveolar macrophage activation. *Immunopharmacol.* 30, 157–166.

Sastre, M., Dewachter, I., Landreth, G.E., Willson, T.M., Klockgether, T., van Leuven, F., and Heneka, M.T. (2003). Nonsteroidal anti-inflammatory drugs and peroxisome proliferator-activated receptor-gamma agonists modulate immunostimulated processing of amyloid precursor protein through regulation of beta-secretase. *J. Neurosci.* 23, 9796–9804.

Shah, S. (2003). Should a prolonged or short course of indomethacin be used in preterm infants to treat patent ductus arteriosus? *Arch. Dis. Child* 88, 1132–1133.

Smithard, D.G. (2003). Management of stroke: acute, rehabilitation, and long-term care. *Hosp. Med.* 64, 666–672.

Suda, M., Ohno, N., Adachi, Y., and Yadomae, T. (1994). Relationship between the tissue distribution and antitumor activity of highly branched (1→3)-beta-D-glucan, SSG. *Biol. Pharm. Bull.* 17, 131–135.

Sugawara, T., Sato, M., Takagi, T., Kamasaki, T., Ohno, N., and Osumi, M. (2003). In situ localization of cell wall alpha-1,3-glucan in the fission yeast *Schizosaccharomyces pombe*. *J. Electron Microsc.* (Tokyo) 52, 237–242.

Suzuki, T., Tsuzuki, A., Ohno, N., Ohshima, Y., Adachi, Y., and Yadomae, T. (2002). Synergistic action of beta-glucan and platelets on interleukin-8 production by human peripheral blood leukocytes. *Biol. Pharm. Bull.* 25, 140–144.

Swartz, E.N. (2003). Is indomethacin or ibuprofen better for medical closure of the patent ductus arteriosus? *Arch. Dis. Child* 88, 1134–1135.

Takahashi, H., Ohno, N., Adachi, Y., and Yadomae, T. (2001). Association of immunological disorders in lethal side effect of NSAIDs on beta-glucan-administered mice. *FEMS Immunol. Med. Microbiol.* 31, 1–14.

Thorn, J., Beijer, L., and Rylander, R. (2001). Effects after inhalation of (1→3)-beta-D-glucan in healthy humans. *Mediators Inflam.* 10, 173-178.

Uchiyama, M., Ohno, N., Miura, N.N., Adachi, Y., Tamura, H., Tanaka, S., and Yadomae, T. (2000). Solubilized cell wall beta-glucan, CSBG, is an epitope of *Candida* immune mice. *Biol. Pharm. Bull.* 23, 672–676.

van Gelder, T., ter Meulen, C.G., Hene, R., Weimar, W., and Hoitsma, A. (2003). Oral ulcers in kidney transplant recipients treated with sirolimus and *mycophenolate mofetil*. *Transplantation* 75, 788–791.

Wan, G.H., Li, C.S., Guo, S.P., Rylander, R., and Lin, R.H. (1999). An airbone mold-derived product, beta-1,3-D-glucan, potentiates airway allergic responses. *Eur. J. Immunol.* 29, 2491–2497.

Weinberg, J.M. (2003). An overview of infliximab, etanercept, efalizumab, and alefacept as biologic therapy for psoriasis. *Clin. Ther.* 25, 2487–2505.

Wright, V. (1995). Historical overview of non-steroidal anti-inflammatory drugs. *Br. J. Rheumatol.* 34 (Suppl. 1), 2–4.

Yadomae, T. and Ohno, N. (1999). Structure-activity relationship of immunomodulating (1→3)-β-D-glucans, *Recent Res. Devel. Chem. Pharm. Sci.* 1, 23–33.

Yamamoto, Y., Yin, M.J., Lin, K.M., and Gaynor, R.B. (1999). Sulindac inhibits activation of the NF-kappaB pathway. *J. Biol. Chem.* 274, 27307–27314.

Yoshioka, S., Ohno, N., Miura, T., Adachi, Y., and Yadomae, T. (1998). Immunotoxicity of soluble beta-glucans induced by indomethacin treatment. *FEMS Immunol. Med. Microbiol.* 21, 171–179.

9 Particulate and Soluble β-Glucans from *Candida albicans* Modulate Cytokine Release from Human Leukocytes

Ken-ichi Ishibashi, Yukari Nakagawa, Naohito Ohno, and Toshimi Murai

CONTENTS

9.1 Summary 161
9.2 Introduction 162
9.3 Preparation and Biological Activity of *Candida* Cell Wall β-Glucans 164
9.4 Higher Order Structure of Glucans 166
9.5 Anti-CSBG Antibody 167
9.6 Analysis of Gene Expression in Leukocytes Activating *Candida* Cell Wall β-Glucans Using the DNA Microarray Method 169
9.7 Conclusion 175
References 176

9.1 SUMMARY

β-Glucans are widely distributed in nature, including fungi, bacteria, and plants, and humans are constantly in contact with them. Although β-glucans exhibit actions that modify the biodefence system in various ways, those actions vary considerably according to not only the primary structure, but also the conformation and solubility. In the case of deep mycoses, it was found that β-glucans are released into the blood, and that the human body interacts with insoluble β-glucans present as components of the cell wall as well as soluble β-glucans released by the cells. We prepared the particulate form, OX-CA, and the soluble form, CSBG, from *Candida albicans* using the NaClO-DMSO method. Although they have different solubilities, their primary structures are the same. Comparison of the abilities of OX-CA and CSBG to induce the production of IL-8 and TNF-α using human peripheral blood mononuclear cells (PBMCs) revealed that although they both induced the production of the two cytokines in the presence of autologous plasma, only OX-CA demonstrated cytokine production when heat-inactivated autologous plasma was used. Anti-CSBG antibody was detected in human blood. Similar differences were also observed for mouse macrophages. When gene expression in PBMCs stimulated with OX-CA or CSBG was examined by comprehensive gene expression analysis using a microarray, 114 genes having different expression forms were found. On the basis of those findings, the biological activity of β-glucans was clearly demonstrated to be influenced considerably by their physical properties. As regards the toxicity of β-glucans, it is important to clearly define the structural and physical characteristics of β-glucan used for these investigation.

9.2 INTRODUCTION

β-Glucans are widely distributed in nature, including fungi, bacteria, and plants. In addition to such biological functions as a cell wall component in various organisms, β-glucans have attracted attention because of the many roles they play in organisms (e.g., stimulating actions on the biodefense mechanisms) as well as their existence as potentially important industrial materials. The structures of many β-glucans have been analyzed and comparisons of their detailed structures have revealed differences. However, as β-glucans are of biological (microbiological) origin, conclusions have yet to be reached as to whether those differences are truly significant, and as to whether a common biological activity exists among β-glucans when considering individual differences in the materials used, subtle differences in the extraction methods (including partial degradation in the extraction process), as well as methods of analysis and their accuracies. In addition, for instance, although endotoxins have a well-defined pathogenesis of septic shock, such a property of β-glucans has yet to be fully understood, and this too has contributed to ambiguity regarding the nature of these substances.

With progress in advanced therapy, it has become clear that the number of deep mycoses has been increasing primarily in high-risk patients (Rubin, 1993). β-Glucans are detected at the early stage of deep mycoses (Obayashi et al., 1995). Because the concentration of β-glucans in the blood of healthy subjects is below the detection limit, this detection method has been established for the early diagnosis of deep mycoses. Blood β-glucan concentration may increase to as high as approximately 1 ng/mL. Although it is presumed that β-glucans in blood are the released fraction from fungal cell walls, details of their structure remain unknown.

In general, the biological activities of particulate and soluble β-glucans are quite different. We have previously reported that although particulate β-glucans strongly promote greater activation of the alternative pathway of complement, TNF-α production by macrophages, H_2O_2 production, production and release of arachidonic acid, and vascular permeability compared with soluble β-glucans (Suzuki et al., 1992; Adachi et al., 1993; Ohno et al., 1996; Daum and Rohrbach, 1992), their ability to induce serum IFN-γ production and their antitumor effects are comparatively weak (Suzuki et al., 1996). In addition, the comparison of the activity of zymosan (ZYM), which originates from *Saccharomyces cerevisiae* and is commonly used as a particulate β-glucan, with that of soluble β-glucans, has also been carried out by other researchers (Mork et al., 1998; Gallin et al., 1992; Janusz et al., 1987). However, as the purity of commercially available ZYM is extremely low and preparation methods differ among researchers, it is extremely unreliable to discuss the biological reactions of ZYM in terms of β-glucan activity. In addition, as particulate β-glucans and soluble β-glucans used in typical comparative studies have different basic structures, such studies cannot be said to provide comparisons that are related to solubility alone.

We have reported that the particulate β-glucan, OX-CA (NaClO-oxidized *Candida* cells), can be obtained by treating acetone dried cells of *Candida albicans* with hypochlorous acid, and that soluble β-glucan can be obtained by treating these particles with dimethylsulfoxide (DMSO) (Ohno et al., 1999). In addition, we have also previously reported that this *Candida* solubilized β-glucan (CSBG; structure shown in Figure 9.1) exhibits such *in vitro* activities as *Limulus* factor G reactivity and complement activation, as well as such *in vivo* activities as the promotion of vascular permeability, antitumor activity, and adjuvant effects (Tokunaka et al., 2000 and 2002). With this in mind, the present discussion focuses on the results of a study in which soluble and insoluble β-glucans having the same primary structures were prepared and their leukocyte-activating action were examined (Figure 9.1).

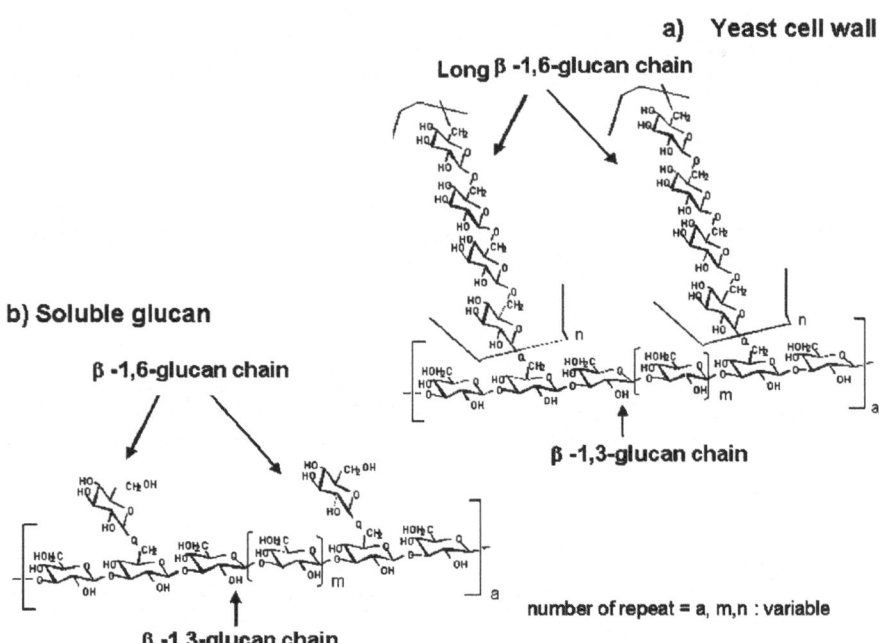

FIGURE 9.1 Structure of fungal β-glucan.

9.3 PREPARATION AND BIOLOGICAL ACTIVITY OF *CANDIDA* CELL WALL β-GLUCANS

Particlulate and soluble β-glucans having the same primary structures were prepared by the NaClO-DMSO method. *Candida* cells were oxidized with hypochlorous acid at various concentrations using available chlorine concentration as an indicator for preparing OX-CA having different degrees of oxidation and compared in terms of IL-8 production in order to determine the biological significance of the solubility difference. As shown in Fig. 2, although IL-8 production was gradually increased to reach a maximum with the progress of oxidation, the IL-8 production disappeared in the strongly oxidized OX-CA (effective chlorine concentration of 1% or higher). In addition, the strongly oxidized OX-CA inhibited IL-8 production and demonstrated behavior resembling that of soluble β-glucan. In order to verify the above, after the strongly oxidized OX-CA was dissolved in DMSO, lyophilized and resuspended to prepare renatured OX-CA, its prominent activity was reproduced. Therefore, the activity of particulate β-glucans is thought to differ from soluble β-glucans because of their physical properties rather than their primary structures. The results strongly suggest that the cross-linking of receptors that recognize β-glucans is required for signal transduction (Figure 9.2).

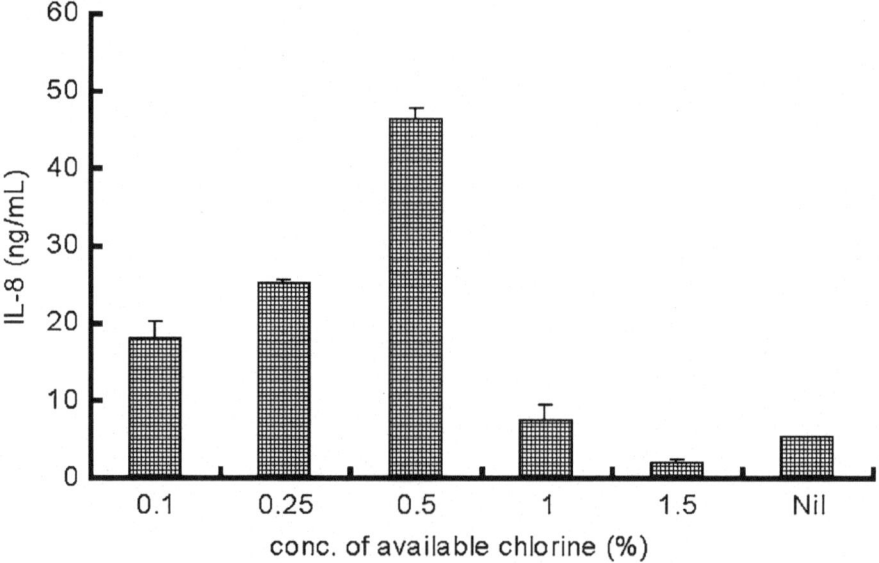

FIGURE 9.2 IL-8 production by human PBMC stimulated with various OX-CA. Acetone-dried cells of *C. albicans* were oxidized with NaClO at various available chlorine concentrations (0.1, 0.25, 0.5, 1.0, 1.5%) for 1 d at 4°C. PBMC obtained from the peripheral blood of healthy donors were adjusted to a concentration of 2×10^6 cells/ml in RPMI1640 medium containing 10% heat-inactivated autologous plasma and cultured with various OX-CA particles 100 μg/ml for 12 h in a 5% CO_2 incubator. Subsequently, the culture supernatants were collected, and IL-8 was measured.

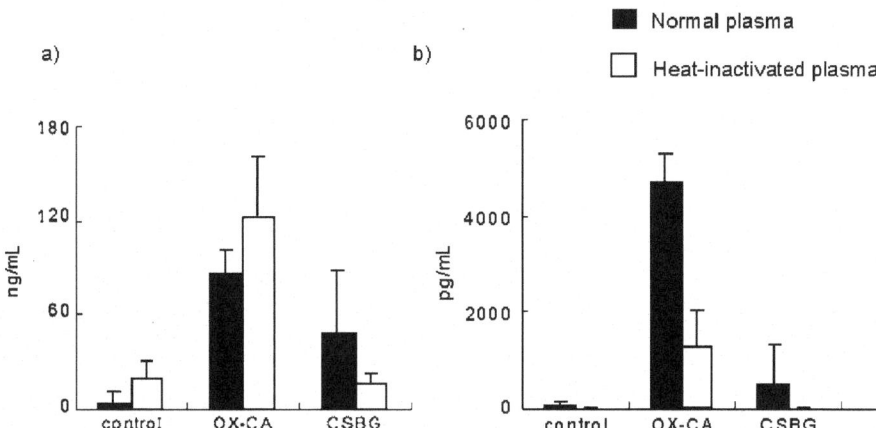

FIGURE 9.3 Activation of leukocytes by OX-CA and CSBG. (a) IL-8 production and (b) TNF-α production by human PBMCs. PBMCs were obtained from the peripheral blood of healthy donors and adjusted to a concentration of 2×10^6 cells/ml in RPMI 1640 medium containing 10% normal or heat-inactivated autologous plasma. Cells were then cultured with OX-CA particles or CSBG (0.5%) (100 μg/ml) for 12 h in a 5% CO_2 incubator. Subsequently, the culture supernatants were collected, and IL-8 and TNF-α was measured.

A comparative study of the leukocyte-activation by soluble and particulate β-glucans using IL-8 and TNF-α production as indicators was conducted. Under complement inactivated conditions, only OX-CA induced IL-8 and TNF-α production (Figure 9.3) On the other hand, in the presence of normal autologous plasma, both OX-CA and CSBG induced IL-8 production. In addition, in contrast to the fact that the degree of IL-8 production induced by OX-CA was roughly the same regardless of the presence of complement, TNF-α production was strongly enhanced in the presence of complement. IL-8 production in the presence of complement was almost the same for OX-CA and CSBG. The above results suggest that CSBG demonstrates activity that is mediated by the complement system, while OX-CA demonstrates activity that is mediated by pathways that are both dependent and independent of the complement system. In addition, the regulatory mechanisms of IL-8 and TNF-α production appear to differ. Furthermore, OX-CA exhibited stronger activity than CSBG under complement inactivated system for all the evaluation systems tested including TNF-α production by RAW 264.7 cells, and H_2O_2 production by peritoneal exudates cell (PEC) of ICR mouse (Ishibashi et al., 2002). The production of the complement-based anaphylatoxin peptide, C5a, has been observed in culture supernatant during stimulation by both glucans. When this was considered together with the differences in activation between the two glucans, it was suggested that the PBMC activation by CSBG is strongly dependent on anaphylatoxin (Figure 9.3).

In addition, when a study was conducted to determine whether the production of inflammatory cytokines (IL-6, TNF-α) was induced by CSBG in a human peripheral blood culture system in the presence of FCS, although CSBG was confirmed to significantly induce the production of both cytokines, its inducing activity was extremely weak compared with that of endotoxins. Similar findings were obtained

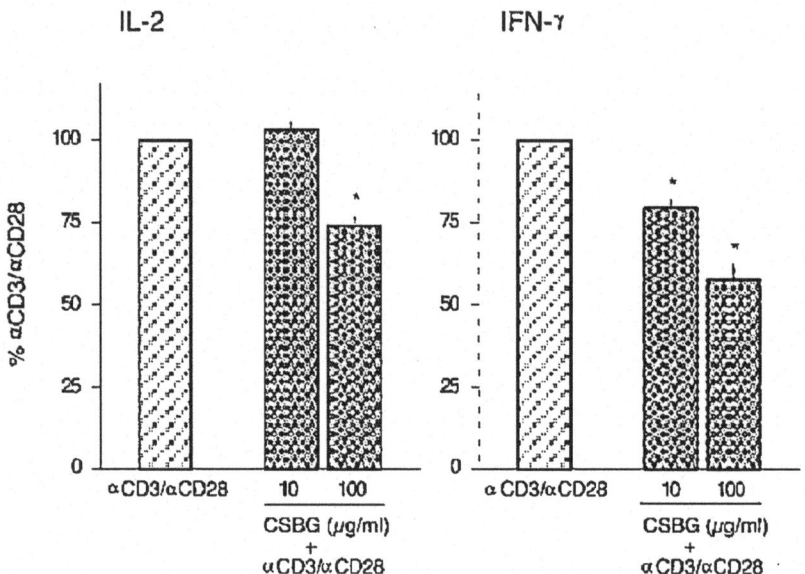

FIGURE 9.4 Suppressive effect of CSBG on αCD3/αCD28-induced IL-2 and IFN-γ production in the PBMC culture. The cells were incubated for 17 h at 37°C with CSBG (10 or 100 μg/mL) in the presence of anti-CD3 (1 μg/mL) and anti-CD28 (5 μg/mL) antibodies. Subsequently, the culture supernatants were collected, and IL-2 and IFN-γ was measured. Data are expressed as a percentage of αCD3/αCD28-induced IL-2 and IFN-γ production (IL-2; 34133.7 pg/mL, IFN-γ; 26943.2 pg/mL) *P<0.05.

for monocytes isolated from human peripheral blood. Examination of the pyrogenic activity of CSBG revealed that pyrexia was not induced in rabbits even when CSBG was administered at a high dose (100 μg/kg, i.v.).

Next, a study was conducted on the effects of CSBG on the IL-2 and IFN-γ producing ability of Th1 cells. CSBG inhibited the αCD3/αCD28-stimulated IL-2 and IFN-γ production by human PBMCs by 70 and 50%, respectively (Figure 9.4). It is interesting to note that the inhibitory effect disappeared upon removing the monocytes from the PBMCs, suggesting that the inhibitory effect of CSBG on Th1 cytokine production is mediated by monocytes. Cellular immunity plays an important role in biodefense reactions against infections caused by *Candida* and other fungi. It has been reported that in the case of decreased Th1 cell function, namely, decreased IL-2 production, the resistance of the host to *C. albicans* decreases and recovery from candidiasis is delayed (Farah et al., 2001 and 2002). Thus, CSBG may be involved in the onset and progressing of candidiasis as a result of impairing the biodefense mechanism against *Candida* infection through inhibition of Th1 cytokine production (Figure 9.4).

9.4 HIGHER ORDER STRUCTURE OF GLUCANS

NMR measurement of the neutral aqueous solutions of high-molecular-weight β1,3-glucans yielded the characteristic result that hardly any signals were detected.

This is consistent with the property of β1,3-glucans being able to adopt a helical structure, namely, high molecular weight β1,3-fragments exhibit strong molecular interaction mediated by hydrogen bonds. Because molecular mobility is significantly inhibited as a result of this molecular interaction, signals of measurable linewidths are not demonstrated by NMR. On the other hand, when the solution is rendered basic, the molecular interaction weakens, resulting in the gradual appearance of signals, and a sharp signal appears in 0.3 N NaOH solution. Under these conditions, a solution of a lower viscosity is formed. The results of high resolution NMR measurement of solution, gel and solid suggest that high molecular weight β1,3-glucans adopt two higher order structures consisting of a single helix and a triple helix (Ohno et al., 1986). In addition, β1,3-glucans having a low molecular weight or a charge exist in the form of random coils.

The extent to which the higher order structure of β1,3-glucans can be controlled is also an important issue in the evaluation of their biological activity. To this end, it is necessary to easily and rapidly measure higher order structures in aqueous solution. The coagulation system of horseshoe crab (G factor system) has been demonstrated to selectively react with single helices and random coils and therefore, use of this system will enable rapid observation of the changes in higher order structure in trace amounts (Nagi et al., 1993). Using this system, all the branched β1,3-glucans that were analyzed were found to have a stable triple helix structure, and even if they were temporarily transformed to the single helix structure, they gradually reverted back to the triple helix structure. Although the transformation from a single helix into a triple helix proceeded even at a temperature of 4°C in the case of the branched β1,3-glucans, grifolan (GRN) and sonifilan (SPG), obtained from *Basidiomycetes*, CSBG only exhibited the transformation from a single helix into a triple helix when autoclaved (Figure 9. 5). In addition, although random coils appeared when an aqueous solution of β1,3-glucans was heated to a temperature of 135°C under pressure, differences occurred in the content of the triple helix structure depending on the cooling conditions.

As has been described above, as the higher order structure of β1,3-glucans differs considerably depending on the conditions, in order to evaluate biological activity, the type of activity that is exhibited by individual higher order structures should be identified to the greatest extent possible (Figure 9.5).

9.5 ANTI-CSBG ANTIBODY

Candida albicans typically exists as normal flora on the mucous membranes of human skin, oral cavity, intestinal tract, and so forth. Considering that the normal flora is intimately involved in the development of mucosal immunity in particular, it is possible that a latent specific immune response to *C. albicans* is evoked through mediation by the intestinal tract. Therefore, in order to assess this possibility, ELISA was conducted using CSBG as the solid phase (Masuzawa et al., 2003). The six-branched β1,3-glucan, GRN, which originates in mushrooms and exhibits antitumor activity, was used as the control antigen, and the reactivities with normal human serum were compared. Differences in the basic structure of the two polysaccharides are shown in Figure 9.1. As a result, anti-CSBG antibody was detected, and when

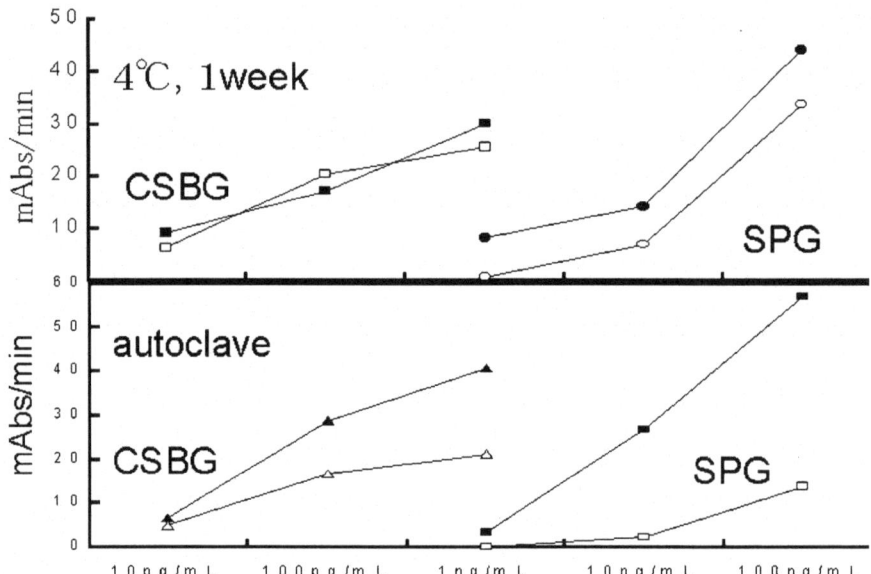

FIGURE 9.5 Conformation of β-1,3-glucan assessed by the Fungitec G test MK. CSBG and SPG were dissolved in 0.5N NaOH, diluted with distilled water, and used for the Fungitec G test MK. ●, ■, ▲ measured after alkaline treatment.

compared on the basis of the dilution factor, it was confirmed to be present at a titer that was more than ten times higher than that of the control GRN (Figure 9.6). In addition, IgG, IgM, IgA isotype classes were also detected. Moreover, this antibody was also detected in a gamma-globulin preparation made from human pooled serum. When a search was conducted by expanding the range to include clinical patients, markedly decreased levels of this antibody were demonstrated in patients with rheumatic disease and vasculitis, and the degree of variation increased among patients with malignant tumors (Figure 9.6).

It is possible that anti-CSBG antibody plays a definite role in the initial immune response and defense against opportunistic infections. In addition, under the conditions in which β-glucans are present in the blood, activation of the classical pathway and signaling mediated by FcR may also occur. It has been reported that in neutrophils, FcR and CR3 act in concert in the processing of foreign substances, and even stronger cellular responses are induced than the case of mediation by either receptor alone. Potent activity may also be induced in monocytes and macrophages as a result of C5aR, CR3, FcR and so forth acting in concert while mediated by complement components in this manner.

When anti-CSBG antibody titer was examined in other animals, high titers were detected in inbred strains DBA1 and DBA2 and in closed colony ICR among mice (Harada et al., 2003). High titers were also observed in pigs and cows. The presence of anti-CSBG antibody in such a wide range of animals alludes to a strong connection between β-glucans and immune function, which is an extremely surprising result. Although analyses of the clinical significance of antibodies in blood have just started,

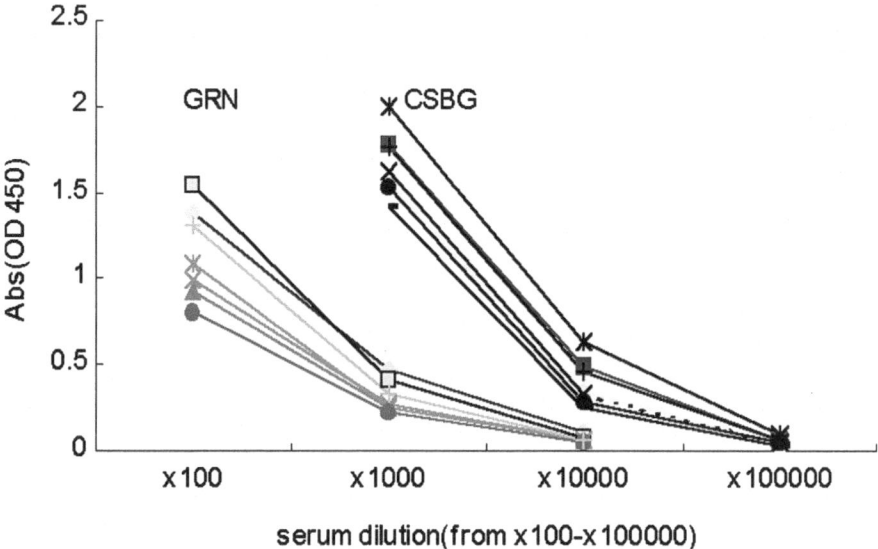

FIGURE 9.6 Reactivity of human sera to CSBG and GRN coated plates. Sera were serially diluted and the plate bound Ig was determined by peroxidase conjugated antihuman IgM+IgG antibody. An ELISA plate was coated with CSBG or GRN (25 μg/mL in carbonate buffer) and blocked by BSA before use. Data from seven volunteers are shown. Each line represents the serum of one volunteer.

we were able to have a glimpse of the specific mechanism by which β-glucans are recognized.

9.6 ANALYSIS OF GENE EXPRESSION IN LEUKOCYTES ACTIVATING *CANDIDA* CELL WALL β-GLUCANS USING THE DNA MICROARRAY METHOD

In order to clarify the differences in the mechanism of action between OX-CA and CSBG, changes in the expression of 1176 genes were examined during β-glucan stimulation using the DNA microarray method. In addition, in order to examine interaction in the body, PBMCs were stimulated for 4 h in the presence of 10% autologous plasma, and this was followed by mRNA extraction. When data that could be evaluated for significant differences in terms of difference and ratio were selected from among the data obtained from the array, 147 genes were selected following OX-CA and CSBG stimulation. Among those genes, 62 exhibited changes induced by both OX-CA and CSBG, 26 exhibited changes specifically induced by OX-CA, and 59 exhibited changes specifically induced by CSBG (Figure 9.7 and Figure 9.8). These included gene expression of a wide range of molecules, including chemokines, cytokines, receptors, adhesion molecules, and signal transducing molecules (Table 9.1). In addition, among the 62 genes that exhibited changes induced by both OX-CA and CSBG, there were those, including MIP-1, TNF-α, and IL-9, that demonstrated differences in the expression level between OX-CA and CSBG.

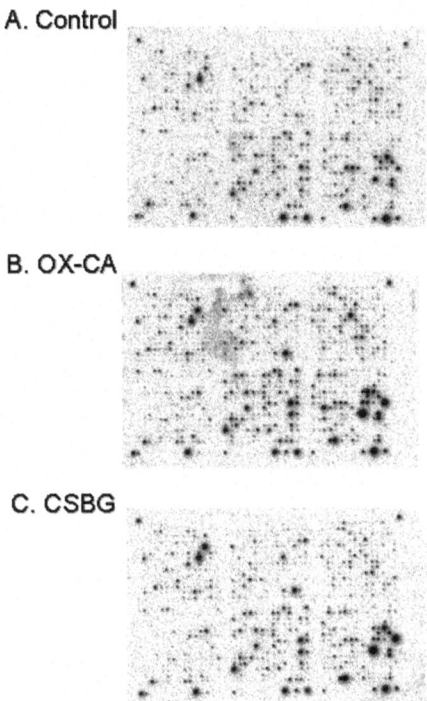

FIGURE 9.7 cDNA microarray analysis of gene expression by PBMCs stimulated with OX-CA and CSBG. Total RNA was extracted from PBMCs (A) unstimulated, or stimulated with 100 μg/mL (B) OX-CA, or (c) CSBG for 4 h and subjected to cDNA expression array assay.

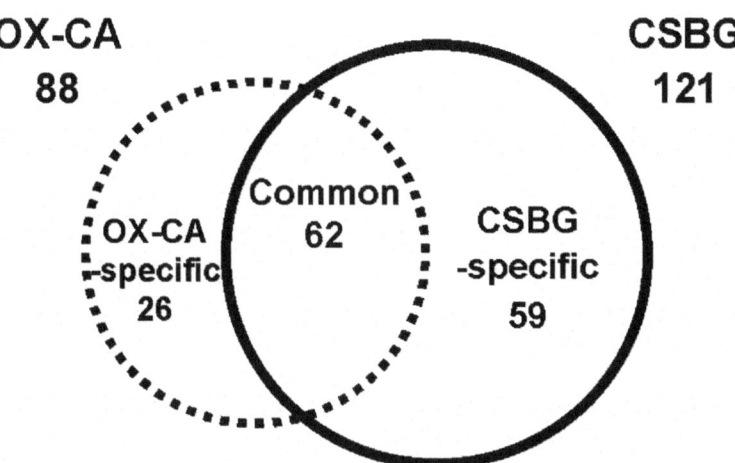

FIGURE 9.8 Influence of solubility on gene expression in human PBMCs stimulated with OX-CA and CSBG. Numbers of genes significantly changed by each stimulation are shown outside of the circle, and stimulant-specific genes or commonly changed genes inside of them.

In this study, although a discussion was attempted by dividing the expressed genes into three groups, it may be necessary to discuss gene expression by grouping in a more complex manner. In addition, it was strongly suggested that the differences in physical properties may bring about changes in cells that differ both qualitatively and quantitatively. The actions of OX-CA and CSBG based on the functions of the gene expression for which changes are discussed below (Figure 9.7; Figure 9.8; Table 9.1).

In the case of stimulation with either OX-CA or CSBG, the levels of such inflammatory cytokines as IL-1, TNF-α, and IL-6 were increased. Those cytokines demonstrate various actions including leukocyte activation, increasing vascular permeability, and induction of acute-phase proteins, as well as symptoms of inflammation such as rubor and fever. In addition, chemokines play the role of releasing and activating immunocompetent cells at local inflammation sites. The levels of various chemokines were increased, including IL-8, MIP-2, and GCP-2, which act on neutrophils, MCP-1, which acts on monocytes, and MIP-1α and MIP-1β, which act on T cell subsets, and they have been suggested to induce chemotactic activity in a wide range of subsets. In addition, increased expression was observed for the CC chemokine receptor gene, which is a G protein coupled receptor, and for the transductin and PLC-δ1 genes, which are involved in signal transduction mediated by GTP-bound protein. Moreover, an increased expression of the G protein coupled receptor HM74 gene, which is a receptor for leukocyte chemotactic factors such as leukotriene, was observed for OX-CA, whereas an increased expression of the C5a receptor gene was observed for CSBG, thus indicating the possibility that cells migrate and are activated by these peptides.

In the case of adhesive molecules, the expression of integrins α5, α7, β1, and β3, which are members of the integrin family and are responsible for adhesion between the extracellular matrix and cells and between cells themselves, as well as the expression of ICAM-1 and catenin was increased. In addition, as integrins are unable to bind in the normal state even if they are expressed in adequate amounts, they are able to adhere as a result of being activated by inside-out signals transduction due to stimulation by chemotactic factors. Because chemokines were adequately induced by stimulation with either OX-CA or CSBG, it is thought that strong adhesion is induced by cross-talk, which in turn leads to induction of cell invasion and other inflammatory reactions.

Increased gene expression of such cytokines as IL-3, IL-5, IL-6, and IL-9, which are produced by activated T cells, accompanying the expression of cytokine receptor genes such as IL-2Rα, IL-3R, IL-4αR, IL-5R, and INF-γR, and the expression of CD27, which play an important role in lymphocyte activation, were observed, indicating activating function of the acquired immunity system. With respect to the B cell activation, increased expression of receptor genes for the growth and differentiation factors of IL-2R, IL-3R, and IL-5R, which are markers for B cell activation, and the increased expression of BASP and PAX-5, which play an important role in the early differentiation of B cells, occur in the case of stimulation by both OX-CA and CSBG.

Many typical gene groups use NF-κB activation that is deeply involved in inflammation, including cytokines, adhesion molecules, NOS, and COX-2. In the

TABLE 9.1
Summary of *Candida* Glucan-Induced Genes in Human PBMCs by cDNA Array[a]

Gene/Protein Name	Chromosomal Location	Relative mRNA Expression Ratio[b]	
		OX-CA	CSBG
a)			
IL-8	4q13-q21	7.0	11.4
MIP-1α	17q11-q21	24.4	3.7
MIP-1β	17q21-q23	16.0	–
MIP-2α	4q21	11.6	9.4
MCP-1	17q11.2-q21.1	–	1.8
GCP-2	Chr.4	–	1.7
TNF-α	6q21.3	32.0	2.4
IL-1α	2q12-q21	5.3	5.4
IL-1β	2q13-q21	5.3	7.8
IL-10	1	2.7	2.1
TGF-β	1q41	2.0	1.7
CC chemokine R1	3p21	2.5	2.7
G-protein-coupled R HM74	Chr.12	9.1	–
C5a R	19q13.3-13.4	–	1.8
IL-1 R1	2q12	–	2.4
Integrin α 5	Chr.2	6.6	
Integrin α 7B	12q13	7.5	9.3
Integrin β1	10q11.2	–	1.9
Integrin β2 (CD11c)		–	2.9
Integrin β3	17q21.32	4.1	3.9
α 1 catenin	5q31	2.2	2.2
ICAM-1	19p13.3-p13.2	2.1	+/–
b)			
IL-3	5q23-q31	6.0	5.2
IL-5	5q23-q31	8.2	
IL-6	7p21-p15	2.6	1.6
IL-9	5q31-q-35	4.9	12.6
IL-2Rα	10p15-p14	2.2	2.4
IL-3R	Xp22.3 ; Yp13.3	4.9	3.9
IL-4Rα	16p11.2-12.1		3.9
IL-5R	3p26-p24	1.9	1.9
IFN-γR	13q34	–	1.7
OX40L	Chr.1	–	8.7
MAL	2cen-q13	1.8	2.1
CD27	12p13	4.0	2.6

(continued)

TABLE 9.1 (CONTINUED)
Summary of *Candida* Glucan-Induced Genes in Human PBMCs by cDNA Array[a]

Gene/Protein Name	Chromosomal Location	Relative mRNA Expression Ratio[b]	
c)			
NF-κB p105 subunit	4q24	2.2	+/–
NF-κB p100 subunit	10q24	–	3.9
MEKK3	17q24	4.9	–
MAPKK 6	17q24.3	8.2	–
neurogranin	11q24	1.9	–
fos-related antigen	11q13	3.3	4.5
Jun proto-oncogene	1p32-p31	2.4	2.5
PRK 1	19p12-p13.1	3.6	2.7
PLC-σ 1	3p22-p21.3	3.3	3.1
transductin β1 subunit	1p36.33	9.3	10.6
cAMP-dependent transcription factor	22q13.1	1.7	3.0
PTP-1B	20q13.1-q13.2	+	–
Rab-7	3q21	2.3	2.5
Gα$_{13}$	Chr 17	–	5.8
Tiam-1	21q22.1	–	2.9
plakoglobin	17q21	–	2.6
Rap-1b	12q14	–	1.9
MLK3	11q13.1-q13.3	+/–	1.8
NF-AT	15	+/–	1.8
lyn	8q13	–	1.9
BASP	9p13	2.1	2.7

[a] cDNA microarrays were analyzed, and gene signals were normalized to the housekeeping gene.
[b] Ratio of the normalized OD gene signal from *Candida* glucan-stimulated cells to the normalized OD gene signal from unstimulated control cells. mRNA expression was assessed at 4 h stimulation.

case of OX-CA stimulation, the gene encoded by NF-κB, p105 subunit was increased, whereas in the case of CSBG stimulation, the gene encoding NF-κB, p100 subunit was increased, thereby suggesting that NF-κB signaling occurs for both OX-CA and CSBG. Changes were also observed in the MAPK cascade. There are four families of the MAPK cascade in mammals, namely, MAPK/ERK, p38, SAPK/JNK and ERK5/BMK. The MAPK cascade is composed of three levels of signals comprising MAPKKK, MAPKK, and MAPK. In the case of OX-CA stimulation, the expression of MEK6, which is a member of the P38 cascade, and that of MEKK3, which is capable of activating both SAPK and ERK, were increased. In the case of CSBG stimulation, the expression of MLK3 was increased, which is capable of activating SAPK/JNK or p38/RK.

As was previously indicated, although OX-CA and CSBG induce the production of the same levels of IL-8 in the presence of complement activation, OX-CA demonstrates markedly higher TNF-α production under identical conditions. At the level of gene expression as well, OX-CA was observed to express more than 15 times more gene products than CSBG. The expression of MIP-1α and MIP-1β was also higher for OX-CA. With respect to MAPK cascade activation, OX-CA was demonstrated to have a stronger potential for activating ERK than CSBG, and to strongly activate p38 and SAPK as well. OX-CA was also associated with an increase in neurogranin, which promotes signal transduction mediated by calmodulin. This difference in signals may be due to the difference in the expression levels of TNF-α and so forth. The stability of the TNF-α mRNA was enhanced by p38, and an intensive expression was required for not only NF-kB but multiple signaling accompanied by MAPK, such as ERK and JNK/p38 (Brook et al., 2000; Zhu et al., 2000). The MAP kinase pathway acts at multiple levels for the full regulation of TNF. OX-CA-stimulated monocytes expressed MAPK-related molecules more strongly than CSBG. Therefore, it was suggested that those molecules have an effect on β-glucan-stimulation of TNF mRNA expression. Assuming that differences occur in the ERK activation according to molecular weight, it is possible that this may be a signal accompanying receptor cross-linking or changes in the cell skeleton. CR3 and CR4, which are β-glucan receptors, are members of the integrin family. Those two receptors do not have any signal transduction regions in their cells; rather, they are known to activate ERK, JNK, and PKC by forming focal adhesion accompanying changes in the cell skeleton caused by the aggregation of those receptors (Alahari et al., 2002; Schlaepfer et al., 1999). Additional evidence for the possibility of OX-CA activating this signal transduction pathway is the increased expression of PTP-1, which interacts with p130Cas that plays an important role in integrin signal transduction.

In the case of CSBG stimulation, the expression of OX40L, which acts as a co-stimulating molecule during T cell activation, was specifically increased. Also, the mRNA expression of IL-9 as produced by activated T cells and NF-AT, a downstream signaling molecule of TCR, was more intensive than in OX-CA. These results show lymphocyte activation is greater in response to CSBG than OX-CA. Besides, it was reported that the expression of OX40L was induced in monocyte-derived dendritic cells (DCs), antigen-presenting cells. Therefore, it was suggested that CSBG induced the differentiation of monocytes to DCs. In addition, the expression of retinoic acid binding protein II (CRABP II) which was induced in human monocyte-derived macrophages (Kreutz et al., 1998; Fritsche et al., 2000) and MAL, which was expressed on the cell membrane during the differentiation of T cells (Liebert et al., 1997), was intensive with CSBG. The results show that CSBG is a powerful activator of differentiation and proliferation. Moreover, the up-regulation of Rho family molecules, such as $G\alpha_{13}$ and Tiam-1, Rap-1b which may suggest the modulation of cytoskeleton in cell division, and lyn, Rac, MLK3 which is related to cell to cell contact, may support the activity of CSBG.

As shown in the above results, a comparison of the mechanisms of action underlying the particulate β-glucan, OX-CA, and the soluble β-glucan, CSBG, was conducted over a broad range of responses. Differences in biological activities, namely, OX-CA exhibiting potent proinflammatory activity and CSBG enhancing cellular interaction and having strong activating effects on lymphocytes, could be examined in greater detail at the molecular level. Because changes were observed in the signal transduction systems, such as changes in MAPK and the cytoskeleton, due to differences in solubility, it is suggested that activity is expressed as a result of different activation mechanisms.

9.7 CONCLUSION

The conclusion that β-glucans exhibit "broad-spectrum" activity has been reached by examining various β-glucans. On the other hand, we found that this broad spectrum is not uniformly observed in all β-glucans by analyzing those having well-defined structures. For example, when NO production by macrophages was analyzed using the clinically used medicine, sonifilan, activity was only observed when a single helix was adopted as the higher order structure (Hashimoto et al., 1997). β-Glucans produced by various organisms have various molecular weights, degrees of branching, solubilities, and higher order structures. Therefore, it is considered to be over-speculative to assume that the β-glucans exhibit similar activity. Moreover, because there are no synthetic products of β-glucans, it is doubtful whether the results obtained thus far were acquired on the basis of analyses conducted after adequately verifying the purities of the β-glucans used. As indicated previously, because it is difficult to acquire commercially available products of high purity, the use of such products coupled with the relative ease of publishing the results has contributed to confusion.

In this study, the structure and biological activity of β-glucans found in the cell wall of *Candida albicans*, a typical causative microorganism of deep mycoses, were analyzed. What should be emphasized here is that both activity and gene expression were compared by changing only the solubility, while using β-glucans having completely identical primary structures. We were able to accomplish this with the NaClO-DMSO method. As a result of comparative studies from various perspectives of the particulate β-glucan, OX-CA, and the soluble β-glucan, CSBG, we demonstrated that the difference in solubility does not simply reflect the extent of activity, but rather clearly exhibits activities that are qualitatively different. This was also verified in the comprehensive analyses of gene expression. Although this difference is attributable to none other than the activation of cells mediated by receptors, the factors involved in this activation most likely include more complex mechanisms than those of typical receptors and so forth. Because these are also high molecular weight polysaccharides, the simultaneous capture of multiple receptors must also be taken into consideration.

The major finding of our research on β-glucans is that considerable changes in activity are observed when different sample dissolution methods are used. We therefore intend to pay close attention to the standardization of the dissolution method in order to obtain reproducibility. Although it is clearly important to determine the activity and the extent to which gene functions are modified, this is unlikely to be possible without an understanding of the basic nature of β-glucans. Attention needs to be focused on the existence of specific antibodies because they are thought to be the parameters that express in the most direct manner possible the importance of β-glucans in immune function.

As indicated, β-glucans have long been known to have a broad spectrum of activities. This chapter was intended to analyze one aspect of that spectrum. In-depth analyses are required in order to fully identify the nature of this broad spectrum.

REFERENCES

Adachi, Y., Ohno, N., and Yadomae, T. (1993) Inhibitory effect of beta-glucans on zymosan-mediated hydrogen peroxide production by murine peritoneal macrophages *in vitro*. *Biol. Pharm. Bull.* **16**, 462–467.

Alahari, S.K., Reddig, P.J., and Juliano, R.L. (2002) Biological aspects of signal transduction by cell adhesion receptors. *Int. Rev. Cytol.* **220**, 145–184.

Brook, M., Sully, G., Clark, A.R., and Saklatvala, J. (2000) Regulation of tumour necrosis factor alpha mRNA stability by the mitogen-activated protein kinase p38 signalling cascade. *FEBS Lett.* **483**, 57–61.

Daum, T. and Rohrbach, M.S. (1992) Zymosan induces selective release of arachidonic acid from rabbit alveolar macrophages via stimulation of a beta-glucan receptor. *FEBS Lett,* **309**, 119–122.

Farah, C.S., Elahi, S., Pang, G., Gothamanos, T., Seymour, G.J., Clancy, R.L., and Ashman, R.B. (2001) T cells augment monocyte and neutrophil function in host resistance against oropharyngeal candidiasis. *Infect. Immun.* **69**, 6110–6118.

Farah, C.S., Elahi, S., Drysdale, K., Gothamanos, T., Seymour, G.J., Clancy, R.L., and Ashman, R.B. (2002) Primary role for CD4(+) T lymphocytes in recovery from oropharyngeal candidiasis. *Infect. Immun.* **70**, 724–731.

Fritsche, J., Stonehouse, T.J., Katz, D.R., Andreesen, R., and Kreutz, M. (2000) Expression of retinoid receptors during human monocyte differentiation *in vitro*. *Biochem. Biophys. Res. Commun.* **270**, 17–22.

Gallin, E.K., Green, S.W., and Patchen, M.L. (1992) Comparative effects of particulate and soluble glucan on macrophages of C3H/HeN and C3H/HeJ mice. *Int. J. Immunopharmacol.* **14**, 173–183.

Harada, T., Miura, N.N., Adachi, Y., Nakajima, M., Yadomae, T., and Ohno, N. (2003) Antibody to soluble 1,3/1,6-beta-D-glucan, SCG in sera of naive DBA/2 mice. *Biol. Pharm. Bull.* **26**, 1225–1228.

Hashimoto, T., Ohno, N., Adachi, Y., and Yadomae, T. (1997) Enhanced production of inducible nitric oxide synthase by β-glucans in mice. *FEMS Immunol. Med. Microbiol.* **19**, 131–151.

Ishibashi, K., Miura, N.N., Adachi, Y., Ogura, N., Tamura, H., Tanaka, S., and Ohno, N. (2002) Relationship between the physical properties of *Candida albicans* cell well beta-glucan and activation of leukocytes *in vitro*. *Int. Immunopharmacol.* **2**, 1109–1122.

Janusz, M.J., Austen, K.F., and Czop, J.K. (1987) Lysosomal enzyme release from human monocytes by particulate activators is mediated by beta-glucan inhibitable receptors. *J. Immunol.* **138**, 3897–3901.

Kreutz, M., Fritsche, J., Ackermann, U., Krause, S.W., and Andreesen, R. (1998a) Retinoic acid inhibits monocyte to macrophage survival and differentiation. *Blood* **91**, 4796–4802.

Kreutz, M., Fritsche, J., Andreesen, R., and Krause, S.W. (1998b) Regulation of cellular retinoic acid binding protein (CRABP II) during human monocyte differentiation *in vitro*. *Biochem. Biophys. Res. Commun.* **248**, 830–834.

Liebert, M., Hubbel, A., Chung, M., Wedemeyer, G., Lomax, M.I., Hegeman, A., Yuan, T.Y., Brozovich, M., Wheelock, M.J., and Grossman, H.B. (1997) Expression of mal is associated with urothelial differentiation *in vitro*: identification by differential display reverse-transcriptase polymerase chain reaction. *Differentiation* **61**, 177–185.

Masuzawa, S., Yoshida, M., Ishibashi, K., Saito, N., Akashi, M., Yoshikawa, N., Suzuki, T., Nameda, S., Miura, N.N., Adachi, Y., and Ohno, N. (2003) Solubilized *Candida* cell wall β-glucan, CSBG, is an epitope of natural human antibody. *Drug Develop. Res.* **58**, 179–189.

Mork, A.C., Helmke, R.J., Martinez, J.R., Michalek, M.T., Patchen, M.L., and Zhang, G.H. (1998) Effects of particulate and soluble (1→3)-beta-glucans on Ca2+ influx in NR8383 alveolar macrophages. *Immunopharmacol.* **40**, 77–89.

Nagi, N., Ohno, N., Adachi, Y., Aketagawa, J., Tamura, H., Shibata, Y. Tanaka, S., and Yadomae, T. (1993) Application of *Limulus* test (G pathway) for the detection of different conformers of (1→3)-β-D-glucans. *Biol. Pharm. Bull.* **16**, 822–828.

Obayashi, T., Yoshida, M., Mori, T., Goto, H., Yasuoka, A., Iwasaki, H., Teshima, H., Kohno, S., Horiuchi, A., Ito, A., *et al.* (1995) Plasma (1→3)-beta-D-glucan measurement in diagnosis of invasive deep mycosis and fungal febrile episodes. *Lancet* **345**, 17–20.

Ohno, N., Adachi, Y., Ohsawa, M., Saito, K., Oikawa, S., and Yadomae, T. (1986) Conformation changes of the two different conformers of grifolan in sodium hydroxide, urea or dimethylsulfoxide solution. *Chem. Pharm. Bull.* **35**, 2108–2113.

Ohno, N., Miura, T., Miura, N.N., Adachi, Y., and Yadomae, T. (1996) Effect of various β-D-glucans on vascular permeability in mice. *Pham. Pharmcol. Lett.* **6**, 115–118.

Ohno, N., Uchiyama, M., Tsuzuki, A., Tokunaka, K., Miura, N.N., Adachi, Y., Aizawa, M.W., Tamura, H., Tanaka, S., and Yadomae, T. (1999) Solubilization of yeast cell-wall beta-(13)-D-glucan by sodium hypochlorite oxidation and dimethyl sulfoxide extraction. *Carbohydr. Res.* **316**, 161–172.

Rubin, R.H. (1993) Fungal and bacterial infections in the immunocompromised host. *Eur. J. Clin. Microbiol. Infect. Dis.* **12**, S42–48.

Schlaepfer, D.D., Hauck, C.R., and Sieg, D.J. (1999) Signaling through focal adhesion kinase. *Prog. Biophys. Mol. Biol.* **71**, 435–478.

Suzuki, T., Ohno, N., Saito, K., and Yadomae, T. (1992) Activation of the complement system by (1→3)-beta-D-glucans having different degrees of branching and different ultrastructures. *J. Pharmacobiodyn.* **15**, 277–285.

Suzuki, T., Ohno, N., Chiba, N., Miura, N.N., Adachi, Y., and Yadomae, T. (1996) Immunopharmacological activity of the purified insoluble glucan, zymocel, in mice. *J. Pharm. Pharmacol.* **8**, 1243–1248.

Tokunaka, K., Ohno, N., Adachi, Y., Tanaka, S., Tamura, H., and Yadomae, T. (2000) Immunopharmacological and immunotoxicological activities of a water-soluble (1→3)-beta-D-glucan, CSBG from *Candida spp. Int. J. Immunopharmacol.* **22**, 383–394.

Tokunaka, K., Ohno, N., Adachi, Y., Miura, N.N., and Yadomae, T. (2002) Application of *Candida* solubilized cell wall beta-glucan in antitumor immunotherapy against P815 mastocytoma in mice. *Int. Immunopharmacol.* **2**, 59–67.

Zhu, W., Downey, J.S., Gu, J., Di Padova, F., Gram, H., and Han, J. (2000) Regulation of TNF expression by multiple mitogen-activated protein kinase pathways. *J. Immunol.* **164**, 6349–6358.

10 Detection and Measurement of (1→3)-β-*D*-Glucan with Limulus Amebocyte Lysate-Based Reagents

Malcolm A. Finkelman and Hiroshi Tamura

CONTENTS

10.1 Introduction 180
10.2 Structure of (1→3)-β-*D*-Glucan 181
10.3 Measurement of (1→3)-β-*D*-Glucan by LAL 182
    10.3.1 The LAL Cascade 182
    10.3.2 Factor G 183
    10.3.3 Pro-clotting Factor 183
    10.3.4 Preparation of (1→3)-β-*D*-Glucan-Specific LAL 184
        10.3.4.1 Fractionation and Combination of Limulus
                  Coagulation Factors 184
        10.3.4.2 Specific Antibody Blockade of Endotoxin-Sensitive
                  Coagulation Factor C 184
        10.3.4.3 Sample Pretreatment with an Endotoxin-Neutralizing
                  Peptide 184
    10.3.5 (1→3)-β-*D*-Glucan Structure and Factor G Activation 185
        10.3.5.1 Glycosidic Linkage Specificity 185
        10.3.5.2 Molecular Weight 186
        10.3.5.3 Single versus Triple Helix 186
        10.3.5.4 Branching 186
10.4 (1→3)-β-*D*-Glucan-Specific Photometric Techniques 187
10.5 Applications of (1→3)-β-*D*-Glucan-Specific LAL 189
10.6 Summary 189
References 192

10.1 INTRODUCTION

(1→3)-β-D-glucan (BG) is generally characterized, in its simplest form, as a homopolymer of glucose molecules, linked through β-(1→3)-D-glucosidic linkages (Saito et al., 1968). A highly heterogeneous material, BG may occur with variable molecular weight, branching, substitution, and quaternary structure. BG has become the object of increasing interest and investigation over the last 20 years, due to a growing understanding of its role as a biological response modifier (Yadomae, 1992; Bohn and BeMiller, 1995; Yan, 1999; Brown and Gordon, 2003). BGs elicit potent biological responses across several kingdoms, including plants, insects, and animals (Jacobs et al., 2003; Jiang et al., 2004; Brown et al., 2003).

This polymer has been shown to be produced by a variety of organisms, including fungi, algae (laminarin), bacteria (curdlan), and plants (callose). In fungi, it is a key component of the cell walls of most families, including medically important fungi such as *Candida* and *Aspergillus*, but not *Zygomycetes*. It is also one of the major components of the cyst wall of *Pneumocystics carini* (Yasuoka et al., 1986). In some fungal species, it can comprise more than 50% of the cell wall dry weight, although this is highly variable (Fleet and Manners, 1976; Nguyen et al., 1998). It is one of a group of beta-linked polysaccharides that play important roles in the mechanical strength and rigidity of fungi cell walls (Cabib et al., 1988; Kollar et al.,1997). In plants, BG is observed to be synthesized in the apical meristem, in specialized vascular connections called plasmodesmata, in wound response tissue synthesis (Jacobs et al., 2003), and in specialized structures in seeds (Yim and Bradford, 1998). Commercial exploitation of BG of alginic origin (Laminarin) for industrial applications, and BG of bacterial origin (curdlan) as an additive in the processed food industry, are well established (Spicer et al., 1999). Numerous medical applications have been proposed, and developed, for BGs. Among these; antitumor applications have been a particularly strong focus, especially in Japan (Ohno et al., 1986a, 1986b, 1986c). Other applications of BGs include the stimulation of general immunity and immuno-protective effects (Hong et al., 2003; Brown and Gordon, 2003; Li et al., 2004) and wound healing (Wei et al., 2002). The relationship between BG structure and immunomodulatory activity has been reviewed (Yadomae and Ohno, 1996; Brown and Gordon, 2003).

As traces of plant and fungal matter are ubiquitous in virtually all environments, BG is present, at some level, almost everywhere. Human exposure occurs through breathing, ingestion, fungal colonization and infection, systemic administration of BG-containing pharmaceuticals, and through invasive use of cellulosic medical devices such as surgical sponges, gauze packings, and dialysis membranes (Kimura et al., 1995; Nakao et al.,1997; Usami et al., 2002; Taniguchi et al., 1990). Researchers are devoting increased resources to the investigation of BG exposure for the purposes of clarifying both the basic elements of cellular BG-sensing and physiological response, as well as the determination of environmental levels that may harm health. In addition, the estimation of blood levels of BG is being used as a diagnostic tool for the detection of fungal infection (Obayashi et al., 1992; Odabasi et al., 2004). Given the expanding work detailing biological responses to BG exposure, and the goal of reliably estimating levels of BG, there has been a need to develop

Detection and Measurement of (1→3)-β-D-Glucan

and utilize sensitive, specific, and reproducible methods for the detection and quantitation of BG.

Over the last 20 years, a number of methods for the detection and quantitation of BG have been developed and reported. These tests have had various levels of specificity, sensitivity, and applicability to samples where BG might comprise only a small fraction of the material under examination. These include such techniques as dye binding (aniline blue [Kauss, 1985], chlorantine fast green BLL [Pearce, 1986; Shedletzky et al., 1997], congo red [Ogawa et al., 1994], calcofluor white [Borg-von Zeppelin and Wagner, 1995], etc.); enzymatic digestion with (1→3)-β-D-glucanase (Fontaine et al., 2000); NMR spectroscopy (Kim et al., 2000; Lowe et al., 2001); antibody-based ELISA (Douwes et al., 1996); glucan binding protein (GBP)-based techniques (Tamura et al., 1997); and surface plasmon resonance (Rice et al., 2002). These have been reported and discussed elsewhere, and the reader is referred to the bibliographic references.

This chapter reviews the use of *Limulus* Amebocyte Lysate (LAL) and *Tachypleus* Amebocyte Lysate (TAL)-based techniques for the detection of BG. For the purposes of abbreviation, glucan-specific LAL (or TAL) will be referred to as LAL.

10.2 STRUCTURE OF (1→3)-β-*D*-GLUCAN

In its simplest form, BG consists of a (1→3)-β-*D*-linked glucose polymer, with no branching (Figure 10.1). In addition to the (1→3)-β-*D*-linked backbone, BG generally contains branching linkages to other glycans. These often appear as (1→2)-β, (1→4)-β, (1→6)-β links to glycan side chains, such as glucan or mannan (Inoue et al., 1983; James et al., 1990; Kim et al., 2000; Schmid et al., 2001).

Typically, BG is characterized as a complex molecular entity (Kollar et al., 1997). It is rarely encountered in a relatively pure form. Partial exceptions to this are laminarin, an algal polysaccharide, and curdlan, a product of *Alcaligenes faecalis*. These have polydisperse molecular weights and consist, essentially, of linear BG. They may be relatively easily prepared, in high purity, from their source material. With these exceptions, BG is usually encountered as polysaccharide that is heterogeneous in molecular weight, ranging from oligosaccharides of a few thousand Daltons, to large polysaccharides of hundreds of thousands or millions of Daltons

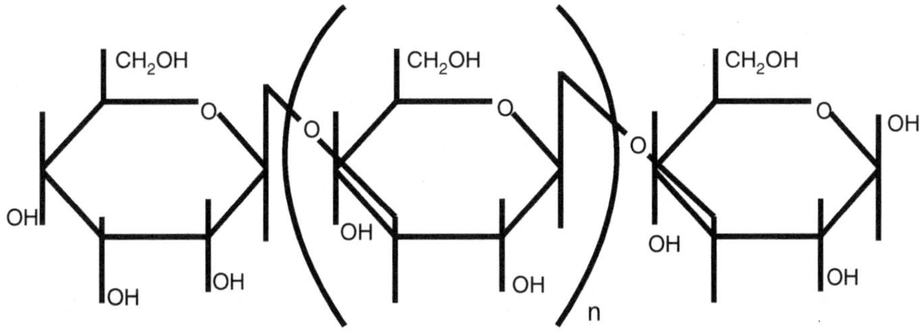

FIGURE 10.1 (1→3)-β-*D*-glucan structure.

(Mork et al., 1998; Muller et al., 1997; Ishibashi et al., 2002). BG may also appear as particulate material, as BG is relatively insoluble in aqueous media. As an example, the yeast *Sacculus* is approximately 50%, by dry weight, BG, which requires aggressive treatment to extract (Sandula et al., 1995). BG may be solubilized, to various extents, by a variety of techniques including hot water, alkaline solutions, acids, and the use of organic solvents (Muller et al., 1997; Miura et al., 2003; Ha et al., 2002). The purity, molecular weight, and structural attributes of the resulting BG preparations are dependent upon their processing history. The relative insolubility of BG, its molecular weight heterogeneity, and its diverse structural presentations create special challenges to all measurement techniques, including the use of LAL-based preparations.

10.3 MEASUREMENT OF (1→3)-β-D-GLUCAN BY LAL

10.3.1 THE LAL CASCADE

The LAL cascade is shown in Figure 10.2. As a result of its sensitivity for both endotoxin and BG, the LAL cascade cannot be said to be a specific assay for either, unless specially prepared as such. Further, where both endotoxin and BG are present in a sample, synergistic activation of LAL is observed, leading to measurement artifacts (Roslansky and Novitsky, 1991). In ligand-specific preparations of LAL, sensitivity to activation by either LPS or BG is eliminated. Some manufacturers of LAL offer such preparations in order to permit the reliable detection of either LPS or BG (Obayashi et al., 1985; Tamura et al., 1994).

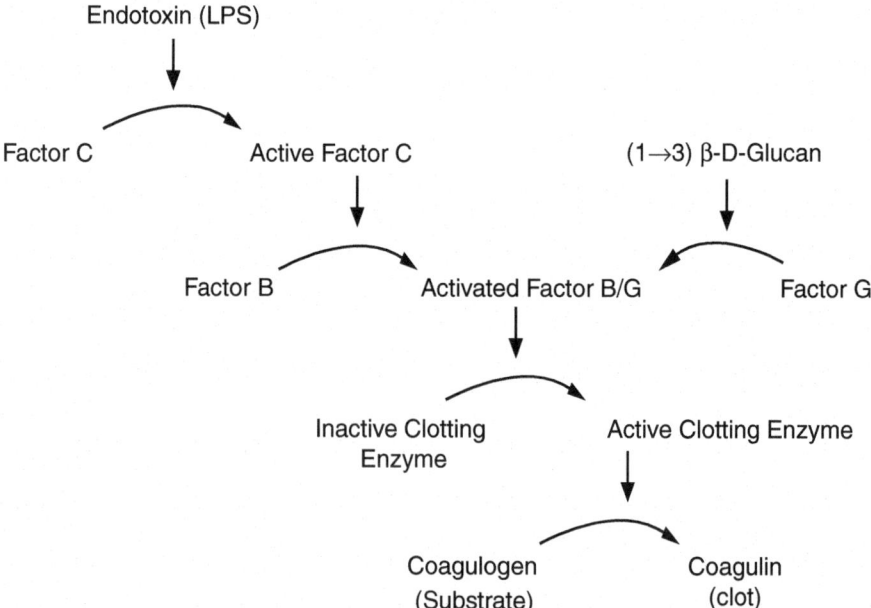

FIGURE 10.2 LAL cascade.

10.3.2 Factor G

The purification and characterization of the BG-sensitive LAL cascade factor was pioneered in Japan by Iwanaga and coworkers (Iwanaga et al., 1992; 1998; Muta et al., 1995). Factor G is a serine protease zymogen, of approximately 110 kD MW. It is heterodimeric, comprising noncovalently associated alpha and beta domains of differing molecular weights and functions (Takaki et al., 2002). The alpha domain, which is the glucan recognition and binding domain, has a reported MW of 72 kDa, while the beta domain, which has the protease activity, has an estimated 37 kDa MW. Both the protein sequence and genes of both subunits have been characterized in detail.

The DNA coding for both subunits of Factor G has been analyzed (Seki et al., 1994). Individual genes were found to code for each of the two subunits. Subunit α is a complex protein comprised of 654 amino acids and a molecular weight of 73.9 kDa (Seki et al., 1994). The subunit β gene codes for a 30.8 kD mature protein, comprised of 278 amino acids, exhibiting homology to a serine protease, with a trypsin-like catalytic site (Seki et al., 1994).

Subunit α was shown to exhibit multiple domains. Two key domains are a pair of C-terminal repeated elements that have a high level of homology to xylanase Z (91.3% homology). The BG binding domain has been characterized by Takaki and coworkers (2002) and has been shown to consist of the two repeated xylanase Z-like elements. Each element was shown to bind to BG, but the affinity increased by two orders of magnitude as the tandem repeat structure (K_a of 8×10^8/M) (Takaki et al., 2002). The interaction of BG structures with the Factor G alpha subunit results in autocatalytic cleavage between the Arg_{150}-Glu_{151} residues in the alpha subunit and the Arg15-Ile16 residues in the beta subunit. This autocatalytic activation is only achieved through the interaction of the alpha subunit and glucan homoploymers with (1→3)-β-linkages (Muta et al., 1995).

The beta subunit of Factor G was observed to have 40.5% homology to Factor B of the LAL cascade. Factor B is a serine protease zymogen that cleaves and activates the ultimate serine protease zymogen of the LAL cascade, pro-clotting enzyme (Muta et al., 1990). A similar reactivity toward pro-clotting enzyme was established for activated Factor G subunit β (Muta et al., 1995). The activation of pro-clotting enzyme permits the pro-clotting enzyme-mediated partial proteolysis of coagulogen, forming coagulin, the clot forming protein of the LAL cascade. This observation established the mechanism for the BG-induced gelation of LAL (Morita et al., 1981; Muta et al., 1995).

10.3.3 Pro-clotting Factor

This serine protease zymogen is comprised of a single polypeptide chain of 375 amino acids. Cleavage of a 29 residue signal sequence yields a 346 amino acid mature protein with a 38.2 kD molecular weight (Muta et al., 1995). Activation is achieved by proteolytic cleavage between residues Arg[98] and Ile[99] of the mature protein. This proteolysis results in two polypeptides, a 98 amino acid L chain, and a 248 amino acid H chain. In the active form, the L and H chains are covalently associated through a disulfide linkage between Cys89 and Cys219, of the L and H chains, respectively (Muta et al., 1990). Once activated, this enzyme can cleave chromogenic or fluorogenic peptide substrates, providing for a simple, spectrophotometric method for the measurement of its activity (Muta et al., 1993). Boc-Leu-Gly-Arg-pNA was identified as a particularly effective substrate for monitoring the activity of pro-clotting enzyme (Iwanaga et al., 1978). Incorporated into suitably modified LAL preparations, this chromogen has been utilized for the detection of BG-based activation of the LAL cascade.

10.3.4 Preparation of (1→3)-β-D-Glucan-Specific LAL

A variety of methods for preparing BG-specific LAL have been reported and established. These include the following.

10.3.4.1 Fractionation and Combination of *Limulus* Coagulation Factors

BG-sensitive coagulation factor (Factor G) and pro-clotting enzyme have been partially purified not only with column chromatography but also with adsorption to dedicated membranes. Then these factors have been appropriately recombined to accomplish the formulation of BG-specific LAL. Column chromatography techniques included affinity and cation exchange, employing dextran-sulphate Sepharose CL6B (Obayashi et al., 1985), and SP-Toyopearl 650C-carriers (Kitagawa et al., 1991), respectively. One of the resulting products, the BG specific chromogenic reagent, Gluspecy, was the first product that was successfully developed and launched (Seikagaku Corporation, 1993) based on a recombined coagulation technology. This led to the development of a novel serological diagnostic product for invasive fungal infection (Fungitec™ G-test, Seikagaku Corporation [Obayashi et al., 1995]). In the U.S., a BG-specific test kit of the same kind, Glucatell™, has been developed by Associates of Cape Cod, Inc. (Ostrosky-Zeichner et al., 2003; Odabasi et al., 2004). The specific elimination of Factor C, coupled with the introduction of the chromogenic peptide Boc-Leu-Gly-Arg-pNA, permitted simple chromogenic assays to be used to measure BG.

10.3.4.2 Specific Antibody Blockade of Endotoxin-Sensitive Coagulation Factor C

Monoclonal antibodies against the endotoxin-sensitive coagulation Factor C, or to activated Factor C, have been added to the conventional LAL, as an effective blocker of generation of endotoxin-mediated LAL activation. (Tanaka et al., 1993; Yoshida et al., 1996; Zhang et al., 1994). This simple process allows for the preparation of BG-specific LAL, lacking reactivity to endotoxin. This type of approach has not, however, been employed to create commercial products.

10.3.4.3 Sample Pretreatment with an Endotoxin-Neutralizing Peptide

This technique involves pretreating a sample to be tested with an antibacterial, endotoxin-neutralizing peptide, prior to the conventional *Limulus* reaction. Through this process, endotoxins may be inactivated in the presence of polymixin B (Zhang et al., 1994; Mori et al., 1997), or tachyplesin/polyphemusin (Nakajima et al., 1994). The resulting preparation becomes BG-specific. A BG-specific test, manufactured by Wako Pure Chemical Industries, Ltd., is used as a generic adjunct method for the diagnosis of invasive fungal infection. This test is based on turbidimetric monitoring of LAL gelation, after pretreatment with polymixin B, using a specialized instrument (Toxinometer, Wako Pure Chemical Industries, Ltd.).

10.3.5 (1→3)-β-*D*-Glucan Structure and Factor G Activation

The key factors in determining the reactivity of BG with Factor G include the glucopyranose moiety, the (1→3)-β-*D*-linkage, the molecular weight of the polysaccharide, and the triple helix/single helix configuration. These are discussed below.

10.3.5.1 Glycosidic Linkage Specificity

A number of factors have been shown to be of importance in the activation of LAL factor G. These include glycosyl moiety, glycosidic linkage, molecular weight, branching, and tertiary and quaternary structure.

The glycosyl moiety and linkage specificity of Factor G activation have been intensively investigated. Biochemical evaluation of a number of glycan structures has identified BG as the sole structure that is able to activate Factor G (Saito et al., 1991; Tanaka et al., 1991; Aketagawa et al., 1993). Tanaka and colleagues (1991) investigated the ability of a number of polysaccharide structures to activate Factor G. Polysaccharides, such as yeast mannan, as well as carboxymethyl-cellulose, galactan, and xylan, were observed to activate Factor G, but with a million-fold lower activation capability. Digestion with purified (1→3)-β-*D*-glucanase extinguished the Factor G activating activity of these polysaccharides, implicating trace contamination with BG as the source of the activity (Obayashi et al., 1992). This observation is critical, as the extreme sensitivity of Factor G to activation by BG can result in BG contamination-related activation being confused with activation by the predominant structure. In such cases, pretreatment of the polysaccharide preparation with purified (1→3)-β-*D*-glucanase is recommended to verify the source of the activity. Given the difficulties associated with the elimination of BG contamination, for this sort of analysis, much care must be given to the preparation of pure materials.

Detailed analysis of the binding domain elements of the alpha subunit of Factor G by Muta and colleagues (Takaki et al., 2002) has shown that short elements of xylan, a (1→4)-β-*D*-xylose polymer, can bind to Factor G and inhibit the binding of BG. Factor G activation by xylan, including (1→3)-β-*D*-xylan, has not, however, been demonstrated (Tanaka et al., 1991). The absence of activation by either (1→3)-β-*D*-xylan or (1→3)-β-*D*-galactan, led Tanaka and colleagues (1991) to suggest that the presence of a hydroxymethyl group at C-5 is an important determinant for factor G activation.

10.3.5.2 Molecular Weight

The reactivity of Factor G with BG of different molecular weights is quite complex. It has been demonstrated that the most reactive glucans are of large molecular weight, on the order of 100 kD and higher (Tanaka et al., 1991). Lower molecular weight BGs have progressively less reactivity. Activity varies over approximately 3 orders of magnitude, between BGs of 6,8 kD and 200 kD. At low molecular weight, optimally 5.8 kD or less, inhibition of reactivity occurs (Tanaka et al., 1993). This latter observation is the basis of commercial glucan activation-blocking preparations that serve to eliminate BG interference in LAL-based bacterial endotoxin detection. These "glucan blockers" utilize soluble hydrolysates of BG to inhibit Factor G activation. Their use permits the differentiation of BG-related activation of standard LAL from endotoxin-related activation (Huszar et al., 2002).

Muta and colleagues (1995) have proposed a mechanism by which high molecular weight BG binds multiple Factor G heterodimers, through their alpha subunits, leading to an aggregation-dependent self-activation process. This model would account for the ability of low molecular weight BG to inhibit activation by blocking the aggregation and proximity-based self-activation.

Studies of the glucan influence upon endotoxin measurement have reported synergy, leading to an overestimation of LPS in the presence of BG (Roslansky and Novitsky, 1991). Such synergies represent the additive effects of Factor G activation and point to the need to investigate the potential presence of either ligand when contamination is observed through non-specific LAL activation.

10.3.5.3 Single versus Triple Helix

The role of the quaternary structure of BG in the activation of Factor G has been evaluated (Ohno et al., 1990; Saito et al., 1990, 1991, and 1992; Aketagawa et al., 1993). The predominate form of BG is a triple helical structure (Kopecka et al., 1986). The triple helical form may be converted to the single helical form by a variety of techniques, including exposure to high pH, solvents, and heat (Saito et al., 1990, 1991). Analysis of BG so treated has shown that the single helical form has a greater ability to activate Factor G, on the order of 10- to 100-fold (Saito et al., 1991 and 1992; Aketagawa et al., 1993; Aketagawa et al., 1994). Accordingly, the processing history of a BG preparation may influence its reactivity in LAL-based assays. Investigators must document, carefully, the methods used, in order to permit comparisons of the reactivity of different preparations.

10.3.5.4 Branching

Although the key quality for LAL reactivity is the (1→3)-β-D-backbone (Ohno et al., 1990; Tanaka et al., 1991), branching is also a factor. Tanaka and colleagues (1991) observed that both linear as well as branched BGs generally activated Factor G with high efficiency. This was not absolute, however, as they observed that a high molecular weight BG, schizophyllan, was a poor activator. Curdlan and pachyman, two highly linear BGs that differ in the amount of (1→6)-β branch points, have been tested with a BG specific preparation of *Tachypleus* lysate. Curdlan, a product of *Alcaligenes faecalis var. myxogenes*, is a linear BG with minimal branching. Pachyman has a similar structure, but contains more branch points. The relative activity of the two polysaccharides has been evaluated with Factor G from *Tachypleus* and the curdlan has been shown to be more reactive (Kitagawa et al., 1991). Several additional studies have developed data showing that increased branching is correlated with a reduction of LAL reactivity (Ohno et al., 1990; Nagi et al., 1993). Working with a series of increasingly branched BGs derived from different fungal sources, and including the unbranched BG, curdlan, Ohno and colleagues (1990) observed a trend of reduced LAL reactivity, with increased branching. Optimum reactivity ranged over 100-fold from curdlan, with no branching and maximum activity at 100 nanograms/mL, to OL-2, an extract of *Omphalia lapidescens* with 66% branching and maximum activity at 10 micrograms/mL (Ohno et al., 1990). Examination of the branching effect with a series of BGs, all derived from *Omphalia lapidescens*, but with different levels of branching, confirmed the effect (Saito et al., 1992). The effect of branching in the activation of Factor G requires more study with highly characterized BGs, with well-understood branching differences.

10.4 (1→3)-β-*D*-GLUCAN-SPECIFIC PHOTOMETRIC TECHNIQUES

The ability to prepare a BG-specific LAL preparation has permitted the development of easy-to-use photometric techniques. These consist of both microplate-based and tube-based techniques. The former approach is the most widespread. Both chromogenic and turbidimetric methods have been developed using the techniques described above. These techniques allow the measurement of BG in a variety of matrices, including aqueous extracts and blood or blood fractions.

One approach has involved a turbidimetric technique, in which the formation of particles of gelling coagulin, the natural substrate for pro-clotting factor, is monitored. Inclusion of polymixin B, an endotoxin neutralizing peptide, permits BG specificity. This technique has been used by the WAKO Pure Chemical Company in the preparation of a commercial product. Another approach, adopted by Seikagaku Corporation and Associates of Cape Cod, Inc., has utilized the addition of a chromogenic peptide, Boc-Leucine-Glycine-Arginine-para-nitroanilide, to Factor C depleted LAL. Hydrolysis of the bond between the terminal amino acid and the nitroanilide leaving group produces a chromophore that absorbs in the 405 nanometer range. This permits the reaction to be monitored spectrophotometrically. Several chromogenic method approaches have been developed for the analysis of BG in solution. These include both endpoint and kinetic methods. An example of a standard curve performed with a BG-specific chromogenic reagent, Glucatell, is illustrated in Figure 10.3. Limits of detection down to nearly a picogram per mL are possible, due to the extreme amplification properties of the LAL cascade.

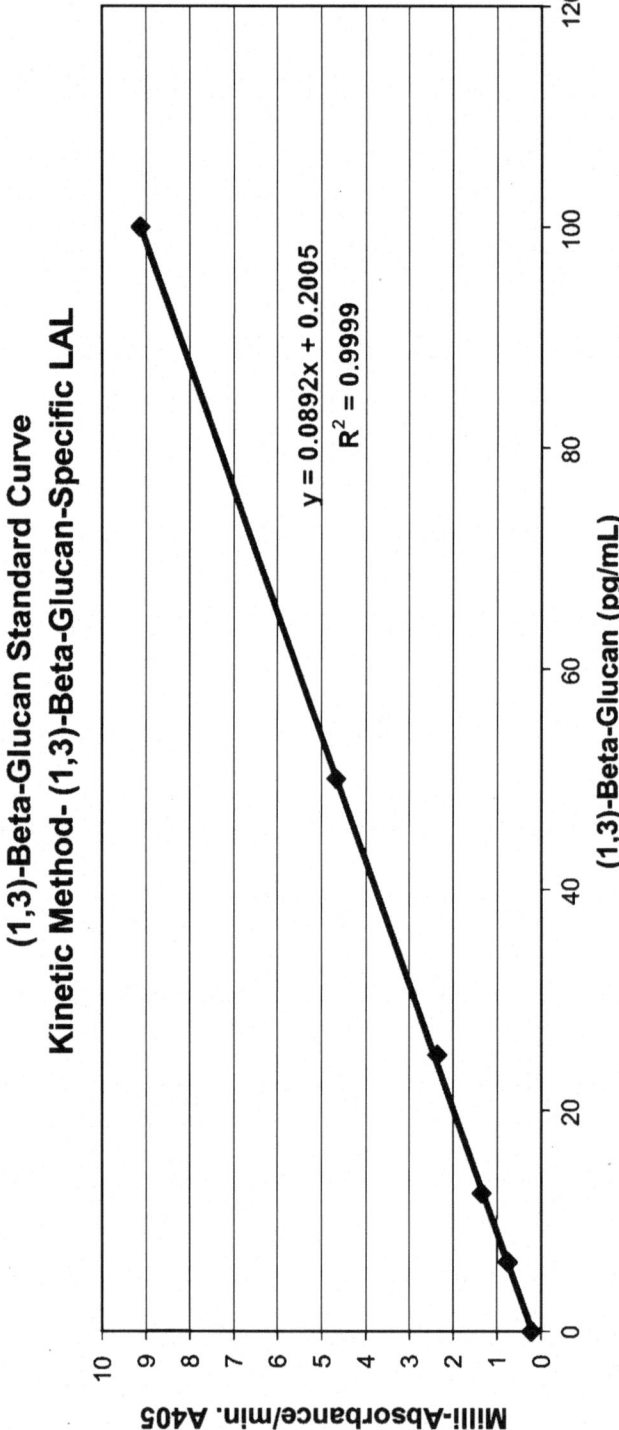

FIGURE 10.3 (1→3)-β-*D*-glucan standard curve generated using (1→3)-β-*D*-glucan specific LAL.

10.5 APPLICATIONS OF (1→3)-β-*D*-GLUCAN-SPECIFIC LAL

The availability of commercial BG-specific reagents has permitted facile analysis of a variety of materials for its presence. Analysis for the presence of BG in blood has assumed a high level of importance in the diagnosis of fungal infection, and this is one of the most well developed applications of BG-specific LAL. The diagnostic topic has been well covered elsewhere, including within this volume. The reagents are also finding increasing utilization in the detection of BG contamination in medical products. The latter area is developing rapidly due to the increasing understanding of the importance and function of BG as an immunomodulatory material. The product types observed to have BG contamination include drugs, blood fractionation products, biologics, and cellulose-containing medical devices (Ikemura et al., 1989; Finkelman and Lempitiski, 2002; Kato et al., 2001; Usami et al., 2002). In many cases, the BG contamination may be introduced due to raw materials, recombinant fungal or yeast fermentation methods, and processing materials. For the latter, cellulosic depth filters are one of the key contributors of BG contamination. These are widely used in the drug and biologic processing industry for clarification. As BG is a commonly observed constituent of plant material, termed callose, it is widely observed to leach from depth filters. Analysis of cellulosic filter extracts have shown that all of the BG-specific LAL signal present in the filtrate is susceptible to digestion by (1→3)-β-*D*-glucanase (Figure 10.4). Similarly contaminated leachate has been observed in filtrates from a common laboratory sterilizing filter (Anderson et al., 2002). Additional studies of leachates from common medical devices such as surgical sponges, dressings, and menstrual tampons (Figure 10.5) have been performed using BG-specific LAL and demonstrate its utility in identifying hitherto unrecognized biologically active contaminants.

10.6 SUMMARY

BG-specific preparations of LAL have been produced using a variety of techniques. These reagents are highly specific for (1→3)-β-*D*-linked glucans, and molecular weight, conformation, branching, and quaternary structure influence their reactivity. The availability of these BG-specific reagents permits applications in diagnosis, drug development, contamination control, and environmental monitoring. The latter areas are developing rapidly due to the increasing understanding of the importance of the biological response to BG, including its effects as an immunomodulatory material. Although the adverse effects of BG have not been extensively examined, a variety of studies have been performed involving intravenous chronic toxicity, pulmonary toxicity, allergy and asthma, and inflammatory cell activation (Fogelmark et al., 1994; Rylander et al., 2000; Takahashi et al., 2001).

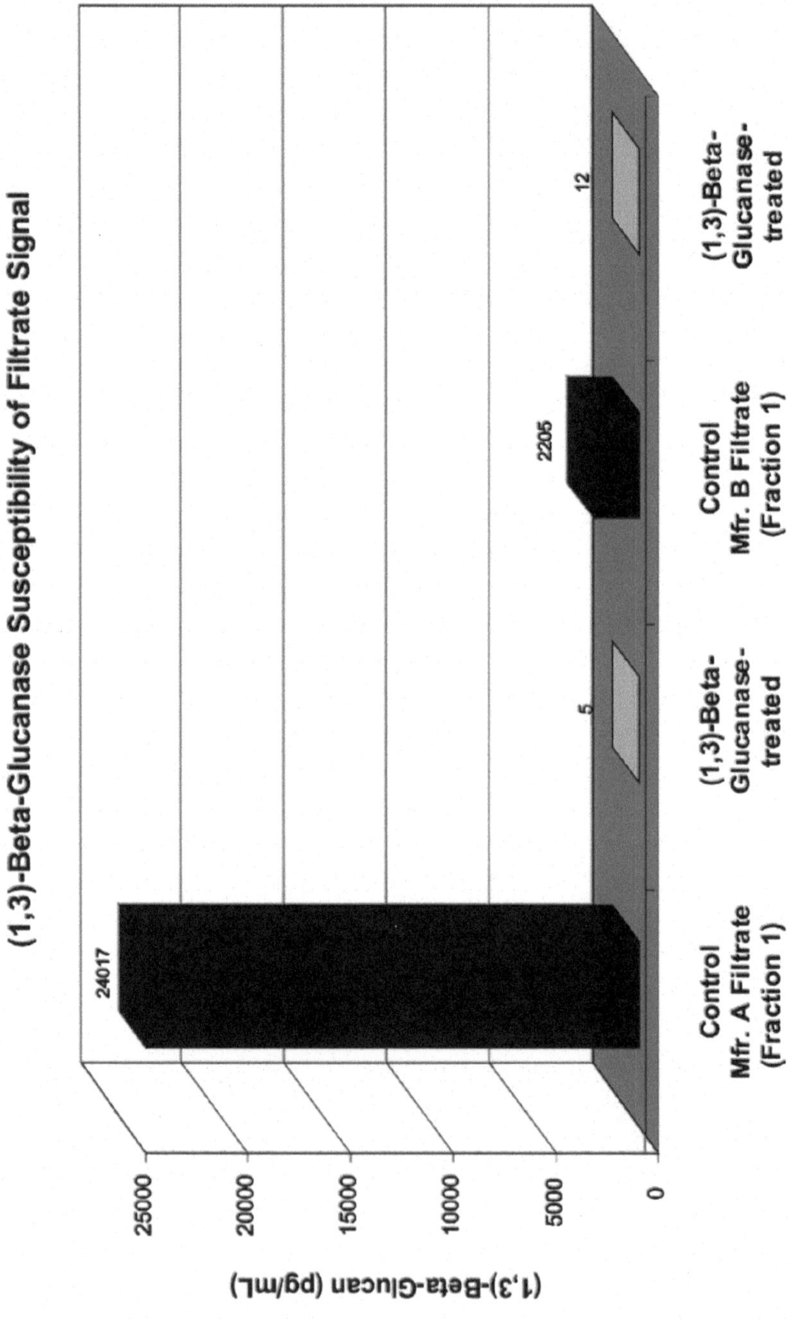

FIGURE 10.4 Glucan-specific LAL analysis of filtrate for (1→3)-β-*D*-glucan leachate using a (1→3)-β-*D*-glucan-specific LAL reagent.

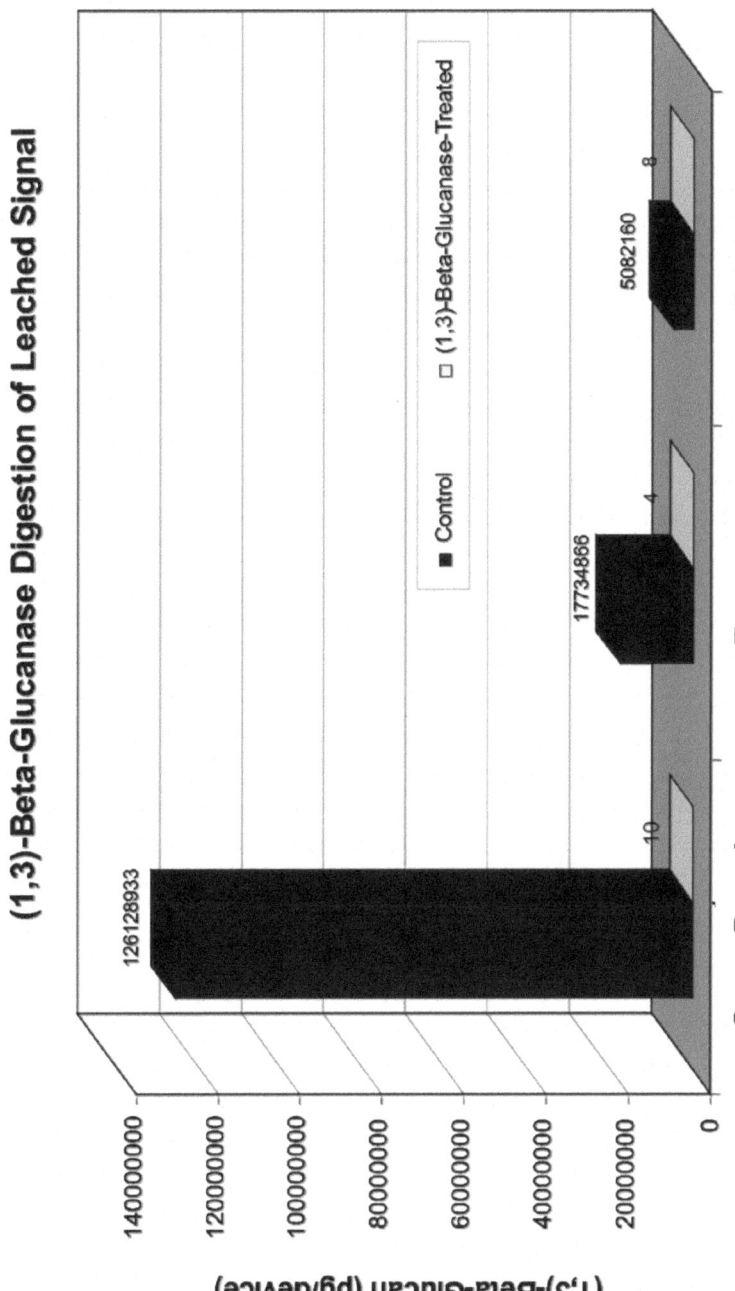

FIGURE 10.5 Analysis of leachate from cotton-containing medical devices using a (1→3)-β-D-glucan-specific LAL reagent.

REFERENCES

Aketagawa, J., Tanaka, S., Tamura, H., Shibata, Y., and Saito, H. (1993). Activation of *Limulus* coagulation Factor G by several (1→3)-β-D-glucans. Comparison of the potency of glucans of identical degree of polymerization but different conformations. *J. Biochem.* 113, 683–686.

Aketagawa, J., Tamura, H., and Tanaka, S. (1994). Measurement of (1-3)-β-D-glucan using Gluspecy (G-Test). In *Third Glucan Inhalation Toxicity Workshop*, R. Rylander and H. Goto, eds., Kompendiet, Goteberg: Committee on Organic Dusts, ICOH, 1/94, pp. 4–17.

Anderson, J., Eller, M., Finkelman, M., Birx, D., Schlesinger-Frankel, S., and Marovich, M. (2002). False positive endotoxin results in a dendritic cell product caused by (1→3)-β-D-glucans acquired from a sterilizing cellulose filter. *Cytotherapy* 6, 557–559.

Bohn, J.A. and BeMiller, J.N. (1995). (1→3)-β-D-glucans as biological response modifiers: a review of structure function activity relationships. *Carbohydr. Polymers* 28, 3–14.

Borg-von Zeppelin, M. and Wagner, T. (1995). Fluorescence assay for the detection of adherent *Candida* yeasts to target cells in microtest plates. *Mycoses* 38, 339–347.

Brown, G.D., Herre, J., Williams, D.L., Willment, J.A., Marshall, A.S., and Gordon, S. (2003). Dectin-1 mediates the biological effects of beta-glucans. *J. Exp. Med.* 197, 1119–1124.

Brown, G.D. and Gordon, S. (2003). Fungal beta-glucans and mammalian immunity. *Immunity* 19, 311–315.

Cabib, E., Bowers, B., Sburlati, A., and Silverman, S.J. (1988). Fungal cell wall synthesis: the construction of a biological structure. *Microbiol. Sci.* 5, 370–375.

Douwes, J., Doekes, G., Montijn, R., Heerdrik, D., and Brunekreef, B. (1996). Measurement of beta-(1→3)-glucans in occupational and home environments with an inhibition enzyme immunoassay. *Appl. Environ. Microbiol.* 62, 3176–3182.

Finkelman, M. and Lempitski, S. (2002). Cotton-containing medical devices leach large quantities of (1→3)-β-D-glucan, a pro-inflammatory biological response modifier. Abstracts of the 7th International Endotoxin Society Meeting, Arlington, VA.

Fleet, G.H. and Manners, D.J. (1976). Isolation and composition of an alkali-insoluble glucan from the cell walls of *Saccharomyces cerevisiae*. *J. Gen. Microbiol.* 94, 180–192.

Fogelmark, B., Goto, H., Yuasa, K., Marchat, B., and Rylander, R. (1992). Acute pulmonary toxicity of inhaled beta-1,3-glucan and endotoxin. *Agents Actions* 35, 50–56.

Fontaine, T., Simenel, C., Dubreucq, G., Adam, O., Delepierre, M., Lemoine, J., Vorgias, C.E., Diaquin, M., and Latge, J.P. (2000). Molecular organization of the alkali-insoluble fraction of *Aspergillus fumigatus* cell wall. *J. Biol. Chem.* 275, 27594–27607.

Ha, C.H., Lim, K.H., Kim, Y.T., Kim, C.W., and Chang, H.I. (2002). Analysis of alkali-soluble glucan produced by *Saccharomyces cerevisiae* wild-type and mutants. *Appl. Microbiol. Biotechnol.* 58, 370–377.

Hong, F., Hansen, R.D., Yen, J., Allendorf, D.J., Baran, J.T., Ostroff, G.R., and Ross, G.D. (2003). Beta-glucan functions as an adjuvant for monoclonal antibody immunotherapy by recruiting tumoricidal granulocytes as killer cells. *Cancer Res.* 63, 9023–9031.

Huszar, G., Jenei, B., Szabo, G., and Medgyesi, G.A. (2002). Detection of pyrogens in intravenous IgG preparations. *Biologicals* 30, 77–83.

Ikemura, K., Ikegami, K., Shimazu, T., Yoshioka, T., and Sugimoto, T. (1989). False-positive result in *Limulus* test caused by *Limulus* Amebocyte Lysate-reactive material in immunoglobulin products. *J. Clin. Microbiol.* 27, 1965–1968.

Inoue, K., Hohno, M., and Kadoya, S. (1983). Structure-antitumor activity relationship of a D-manno-D-glucan from *Microellobosporia grisea*: effect of periodate modification on antitumor activity. *Carbo. Res.* 123, 305–314.
Ishibashi, K., Miura, N., Adachi, Y., Ogura, N., Tamura, H., Tanaka, S., and Ohno, N. (2002). Relationship between the physical properties of *Candida albicans* cell wall β-glucan and activation of leukocytes *in vitro*. *Internat. Immunopharm.* 2, 1109–1122.
Iwanaga, S., Morita, T., Harada, T., Nakamura, S., Niwa, M., Takada, K., Kimura, T., and Sakakibara, S. (1978). Chromogenic substrates for horseshoe crab clotting enzyme. Its application for the assay of bacterial endotoxins, *Haemostasis* 7, 183–188.
Iwanaga, S., Miyata, T., Tokunaga, and Muta, T. (1992). Molecular mechanism of hemolymph clotting system in *Limulus*. *Thrombosis Res.* 68, 1–32.
Iwanaga, S., Kawabata, S., and Muta, T. (1998). New types of clotting factors and defense molecules found in hoeseshoe crab hemolymph: their structures and functions. *J. Biochem.* 123, 1–15.
Jacobs, A.K., Lipka, V., Burton, R.A., Panstruga, R., Strizhov, N., Schultz-Lefert, P., and Fincher, G.B. (2003). An *Arabidopsis* callose synthetase, GSL5, is required for wound and papillary callose formation. *Plant Cell* 15, 2503–2513.
James, P.G., Cherniak, R., Jones, R.G., Stortz, C.A., and Reiss, E. (1990). Cell-wall glucans of *Cryptococcus neoformans* Cap 67. *Carbo. Res.* 198, 23–38.
Jiang, H., Ma, C., Lu, Z.Q., and Kanost, M.R. (2004). Beta-(1→3)-glucan recognition protein-2 (beta-GRP-2) from *Manduca sexta*; an acute-phase protein that binds beta-(1→3)-glucan and lipoteichoic acid to aggregate fungi and bacteria and stimulate phenolperoxidase activation. *Insect Biochem. Mol. Biol.* 34, 89–100.
Kato, A., Takita, T., Furuhashi, M., Takahashi, T., Maruyama, Y., and Hishida, A. (2001). Elevation of blood (1→3)-β-D-glucan concentrations in hemodialysis patients. *Nephron* 89, 15–19.
Kauss, H. (1985). Callose biosynthesis as a Ca2+ regulated process and possible relations to the induction of other metabolic changes. *J. Cell Sci. Suppl.* 2, 89–103.
Kim, Y.T., Kim, E.H., Cheong, C., Williams, D.L., Kim, C.W., and Lim, S.T. (2000). Structural characterization of β-D-(1→3, 1→6)-linked glucans using NMR spectroscopy. *Carbo. Res.* 328, 331–341.
Kimura, Y., Nakao, A., Tamura, H., Tanaka, S., and Takagi, H. (1995). Clinical and experimental studies of the *Limulus* test after digestive surgery. *Surg. Today; Jpn. J. Surg.* 25, 790–794.
Kitagawa, T., Tsuboi, I., Kimura, S., and Sasamoto, Y. (1991). Rapid method for preparing a beta-glucan-specific sensitive fraction from *Limulus* (*Tachypleus tridentatus*) amebocyte lysate. *J. Chromatog.* 567, 267–273.
Kollar, R. Rheinhold, B.B., Petrakova, E., Yeh, H.J., Ashwell, G., Robbins, P., and Cabib, E. (1997). Architecture of the yeast cell wall. Beta-(1→6)-glucan interconnects mannoprotein, Beta-(1→3)-glucan, and chitin. *J. Biol. Chem.* 272, 17762–17775.
Kopecka M. and Kreger, D.R. (1986). Assembly of microfibrils *in vivo* and *in vitro* from (1→3)-β-D-glucan glucan synthesized by protoplasts of *Saccharomyces cerevisiae*. *Arch. Microbiol.* 143, 387–395.
Levin, J. and Bang, F.B. (1964). The role of endotoxin in the extracellular coagulation in the *Limulus*. *Bull. Johns Hopkins Hosp.* 115, 265–274.
Levin, J. and Bang, F.B. (1968). Clottable protein in *Limulus*: its localization and kinetics of its coagulation by endotoxin. *Thromb. Diathes. Haemorrh.* 19, 186–197.
Li, C., Ha, T., Kelley, J., Gao, X., Qiu, Y., Kao, R.I., Browder, W., and Williams, D.L. (2004). Modulating toll-like receptor mediated signaling by (1→3)-β-D-glucan rapidly induces cardioprotection. *Cardiovasc. Res.* 61, 538–547.

Lindsay, G.K., Roslansky, P.F., and Novitsky, T.J. (1989). Single-step, chromogenic *Limulus* amebocyte lysate assay for endotoxin. *J. Clin. Microbiol.* 27, 947–951.

Lowe, E.P., Rice, P.J., Ha, T., Li, C., Kelley, J., Ensley, H., Lopez-Perez, J., Kalbfleisch, J., Lowman, D., Margl, P., Browder, J.W., and Williams, D.L. (2001). A (1→3)-β-D-glucan-linked heptasaccharide is the unit ligand for glucan pattern recognition receptors on human monocytes. *Microbes Infect.* 3, 789–797.

Miura, N.N., Adachi, Y., Yadomae, T., Tamura, H., Tanaka, S., and Ohno, N. (2003). Structure and biological activities of beta-glucans from yeast and mycelial forms of *Candida albicans*. *Microbiol. Immunol.* 47, 173–182.

Mori, T., Ikemoto, H., Matsumura, M., Yoshida, M., Inada, K., Endo, S., Ito, A., Watanabe, S., Yamaguchi, H., Mitsuya, M., Kodama, M., Tani, T., Yokota, T., Kobayashi, T., Kambayashi, J., Nakamura, T., Masaoka, T., Teshima, H., Yoshinaga, T., Kohno, S., Hara, K., and Miyazaki, S. (1997). Evaluation of plasma (1→3)-beta-D-glucan measurement by the kinetic turbidimetric Limulus test, for the clinical diagnosis of mycotic infections. *Eur. J. Clin. Chem. Clin. Biochem.* 35, 553–560.

Morita, T., Tanaka, S., Nakamura, T., and Iwanaga, S. (1981). A new (1→3)-β-D-glucan-mediated coagulation pathway found in Limulus amebocytes. *FEBS Lett.* 129, 318–321.

Mork, A.C., Helmke, R.J., Martinez, J.R., Michaelek, M.T., Patchen, M.I., and Zhang, G.H. (1998). Effects of particulate and soluble (1→3)-β-glucans on Ca2+ influx in NR8383 macrophages. *Immunopharmacol.* 40, 77–89.

Muller, A., Ensley, H., Pretus, H., McNamee, R., Jones, E., McLaughlin, E., Chandley, W., Browder, W., Lowman, D., and Williams, D. (1997). The application of various protic acids in the extraction of (1→3)-β-D-glucan from *Saccharomyces cerevisiae*. *Carbo. Res.* 299, 203–208.

Muta, T., Hashimoto, R., Miyata, T., Nishimura, H., Toh, Y., and Iwanaga, S. (1990). Proclotting enzyme from horseshoe crab hemocytes. cDNA cloning, disulfide locations, and subcellular localization. *J. Biol. Chem.* 265, 22426–22433.

Muta, T., Nakamura, T., Hashimoto, R., Morita, T., and Iwanaga, S. (1993). Limulus proclotting enzyme. *Methods Enzymol.* 223, 352–358.

Muta, T., Seki, N., Takaki, Y., Hashimoto, R., Oda, T., Iwanaga, A., Tokunaga, F., and Iwanaga, S. (1995). Purified horseshoe crab factor G. Reconstitution and characterization of the (1→3)-beta-D-glucan-sensitive serine protease cascade. *J. Biol. Chem.* 270, 892–897.

Nagi, N., Ohno, N., Adachi, Y., Aketagawa, J., Tamura, H., Shibata, Y., Tanaka, S., and Yadomae, T. (1993). Application of Limulus tests (G Pathway) for the detection of different conformers of (1→3)-β-D-glucans. *Biol. Phar. Bull.* 16, 822–828.

Nakajima, H., Yamamoto, H., and Cho, J. (1994). β-glucans detection reagents and methods of detecting 'β-glucans'. U.S. Patent No. 5,571,683.

Nakao, A., Yasui, M., Kawagoe, T., Tamura, H., Tanaka, S., and Takagi, H. (1997). False-positive endotoxemia derives from gauze glucan after hepatectomy for hepatocellular carcinoma with cirrhosis. *Hepato-Gastroenterol.* 44, 1413–1418.

Nguyen, T.H., Fleet, G.H., and Rogers, P.L. (1998). Composition of the cell walls of several yeast species. *Appl. Microbiol. Biotechnol.* 50, 206–212.

Obayashi, T., Tamura, H., Tanaka, S., Ohki, M., Takahashi, S., Arai, M., Masuda, M., and Kawai, T. (1985). A new chromogenic endotoxin-specific assay using recombined *Limulus* coagulation enzymes and its clinical applications. *Clin. Chim. Acta.* 149, 55–65.

Obayashi, T., Yoshida, M., Tamura, H., Aketagawa, J., Tanaka, S., and Kawai, T. (1992). Determination of plasma (1→3)-β-D-glucan: a new diagnostic aid to deep mycosis. *J. Med. Vet. Mycol.* 30, 275–280.

Obayashi, T., Yoshida, M., Mori, T., Goto, H., Yasuoka, A., Iwasaki, H., Teshima, H., Kohno, S., Horiuchi, A., Ito, A., Yamaguchi, H., Shimada, K., and Kawai, T. (1995). Plasma (1→3)-beta-D-glucan measurement in diagnosis of invasive deep mycosis and fungal febrile episodes. *Lancet* 345, 17–20.

Odabasi, Z., Mattiuzzi, G., Estey, E., Kantarjian, H., Saeki, F., Ridge, R., Ketchum, P., Finkelman, M., Rex, J., and Ostrosky-Zeichner, L. (2004). Beta-D-glucan as a diagnostic adjunct for invasive fungal infections: validation, cut-off development, and performance in patients with acute myelogenous leukemia and myelodysplastic syndrome. *Clin. Inf. Dis.* 39, 199–205.

Ogawa, K., Dohmaru, T., and Yui, T. (1994). Dependence of complex formation of (1→3)-β-D-glucan with Congo red on temperature in alkaline solutions. *Biosci. Biotechnol. Biochem.* 58, 1870–1872.

Ohno, N., Suzuki, I., and Yadomae, T. (1986a). Structure and anti-tumor activity of a (1→3)-β-D-glucan isolated from the culture filtrate of *Sclerotina sclerotium* IFO 9395. *Chem. Pharm. Bull.* 34, 1362–1365.

Ohno, N., Adachi, Y., Suzuki, I., Sato, K., Oikawa, S., and Yadomae, T. (1986b). Characterization of the anti-tumor glucan obtained from liquid cultured *Grifola frondosa*. *Chem. Pharm. Bull.* 34, 1709–1715.

Ohno, N., Hayashi, K., Iino, I., Suzuki, S., Oikawa, K., Sato, K., Suzuki, Y., and Yadomae, T. (1986c). *Chem. Pharm. Bull.* 34, 2149–2154.

Ohno, N., Emori, Y., Yadomae, T., Saito, K., Masuda, A., and Oikawa, S. (1990). Reactivity of *Limulus* amebocyte Lysate towards (1→3)-β-D-glucans. *Carbo. Res.* 207, 311–318.

Ostrosky-Zeichner, L., Alexander, B., Kett, D., Vazquez, J., Pappas, P., Saeki, F., Ketchum, P.A., Wingard, J.R., Schiff, R., Tamura, H., Finkelman, M., and Rex, J. (2003). Multicenter clinical evaluation of the (13)-β-D-glucan (BG) assay (Glucatell™) as an aid to diagnosis of invasive fungal infections (IFI) in humans. Abstract M1034A, Interscience Conference on Antimicrobial and Anti-Cancer Therapy (Chicago).

Pearce, R.B. (1986). Chlorantine fast green BBL as a stain for callose in oak phloem. *Stain Technol.* 61, 47–50.

Rice, P.J., Kelley, J.L., Kogan, G., Ensley, H.E., Kalbfleisch, J.H., Browder, I.W., and Williams, D.L. (2002). Human monocyte scavenger receptors are pattern recognition receptors for (1→3)-β-D-glucans. *J. Leukoc. Biol.* 72(1), 140–146.

Roslansky, P.F. and Novitsky, T.J. (1991). Sensitivity of *Limulus* amebocyte lysate (LAL) to LAL-reactive glucans. *J. Clin. Microbiol.* 29, 2477–2483.

Rylander, R. and Lin, R.H. (2000). (1→3)-beta-D-glucan: relationship to indoor air-related symptoms, allergy, and asthma. *Toxicol.* 152, 47–52.

Saito, H., Misaki, A., and Harada, T. (1968). A composition of the structure of curdlan and pachyman. *Agr. Biol. Chem.* 32, 1261–1269.

Saito, H., Yoshioka, Y., Yoloi, M., and Yamada, J. (1990). Distinct gelation mechanism between linear and branched (1→3)-β-D-glucans as revealed by high resolution solid state ^{13}C NMR. *Biopolymers* 29, 1689–1698.

Saito, H., Yoshioka, Y., Uehara, N., Aketagawa, J., Tanaka, S., and Shibata, Y. (1991). Relationship between conformation and biological response for (1→3)-β-D-glucans in the activation of coagulation Factor G from *Limulus* amebocyte lysate and host-mediated anti-tumor activity. Demonstration of single helix conformation as a stimulant. *Carbo. Res.* 217, 181–190.

Saito, K., Nishijima, M., Ohno, N., Nagi, N., Yadomae, T., and Miyazaki, T. (1992). Activation of complement and *Limulus* coagulation systems by an alkali-soluble glucan isolated from *Omphalia lapidescens* and its less-branched derivatives. (Studies on fungal polysaccharide XXXIX). *Chem. Pharm. Bull.* 40, 1227–1230.

Sandula, J., Machova, E., and Hribalova, V. (1995). Mitogenic activity of particulate yeast beta-(1→3)-D-glucan and its water soluble derivatives. *Int. J. Biol. Macromol.* 17, 323–326.

Schmid, F., Stone, R.A., McDougall, B.M., Bacic, A., Martin, K.L., Brownlee, R.T., Chai, E., and Seviour, R.J. (2001). Structure of epiglucan, a highly side-chain/branched (1→3; 1→6)-beta-glucan from the micro fungus *Epicoccum nigrum* Ehernb. Ex Schlecht. *Carbo. Res.* 331, 163–171.

Seki, N., Muta, T., Oda, T., Iwaki, D., Kuma, K., Miyata, T., and Iwanaga, S. (1994). Horseshoe crab (1→3)-β-D-glucan-sensitive coagulation factor G. A serine protease zymogen heterodimer with similarities to beta-glucan binding proteins. *J. Biol. Chem.* 269, 1370–1374.

Shedletzky, E., Unger, C., and Delmer, D.P. (1997). A microtiter-based fluorescence assay for (1→3)-β-glucan synthases. *Anal. Biochem.* 249, 88–93.

Soderhall, K., Levin, J., and Armstrong, P. (1985). The effects of (1→3)-β-D-glucans on blood coagulation and amebocyte release in the Horseshoe Crab, *Limulus polyphemus*. *Biol. Bull.* 169, 661–674.

Spicer, E.J., Goldenthal, E.I., and Ikeda, T. (1999). A toxicological assessment of curdlan. *Food Chem. Toxicol.* 37, 455–479.

Takahashi, H., Ohno, N., Adachi, Y., and Yadomae, T. (2001). Association of immunological disorders in lethal side effect of NSAIDs on beta-glucan-administered mice. *FEMS Immunol. Med. Microbiol.* 31, 1–14.

Takaki, Y., Seki, N., Kawabata, S., Iwanaga, S., and Muta, T. (2002). Duplicated binding sites for (1→3)-β-D-glucan in the horseshoe crab coagulation factor G. *J. Biol. Chem.* 277, 14281–14287.

Tamura, H., Arimoto, Y., Tanaka, S., Yoshida, M., Obayashi, T., and Kawai, T. (1994). Automated kinetic assay for endotoxin and (1→3)-β-D-glucan in human blood. *Clin. Chim. Acta* 226, 109–112.

Tamura, H., Tanaka, S., Ikeda, T., Obayashi, T., and Hashimoto, Y. (1997). Plasma (1→3)-β-D-glucan assay and immunohistochemical staining of (1→3)-β-D-glucan in the fungal cell walls using a novel horseshoe crab protein (T-GBP) that specifically binds to (1→3)-β-D-glucan. *J. Clin. Lab. Anal.* 11, 104–109.

Tanaka, S., Aketagawa, J., Takahashi, S., Shibata, Y., Tsumuraya, Y., and Hashimoto, Y. (1991). Activation of a *Limulus* coagulation factor G by (1→3)-β-D-glucans. *Carbo. Res.* 218, 167–174.

Tanaka, S., Aketagawa, J., Takahashi, S., Shibata, Y., Tsumuraya, T., and Hashimoto, Y. (1993). Inhibition of high molecular weight (1→3)-β-D-glucan-dependent activation of *Limulus* coagulation factor G by laminarin oligosaccharides and curdlan degradation products. *Carbo. Res.* 244, 115–127.

Tanaka, S. (1993). Method for determining (1→3)-β-D-glucan. U.S. Patent No. 5,266,461.

Taniguchi, T., Katsushima, S., Lee, K., Hidaka, A., Konishi, J., Ideguchi, H., and Kawaguchi, Y. (1990). Endotoxemia in patients on hemodialysis. *Nephron.* 56, 44–49.

Usami, M., Ohata, A., Horiuchi, T., Nagasawa, K., Wakabayashi, T., and Tanaka, S. (2002). Positive (1→3)-β-D-glucan in blood components and release of (1→3)-β-D-glucan from depth-type membrane filters for blood processing. *Transfusion* 42, 1189–1195.

Wei, D., Williams, D., and Browder, W. (2002). Activation of AP-1 and SP1 correlates with wound growth factor gene expression in glucan-treated fibroblasts. *Int. Immunopharmacol.* 2, 1163–1172.

Yadomae, T. (1992). Immunopharmacological activities of β-glucans: Structure-activity relationships in relation to various conformations. *Jpn. J. Med. Mycol.* 33, 267–77.

Yadomae, T. and Ohno, N. (1996). Structure-activity relationship of immunomodulating (1→3)-β-D-glucans. *Recent Res. Devel. Chem. Pharm. Sci.* 1, 22–38.

Yan, J., Vetvicka, V., Xia, Y., Coxon, A., Carroll, M.C., Mayadas, T.N., and Ross, G.D. (1999). Beta-glucan, a "specific" biologic response modifier that uses antibodies for cytotoxic recognition by leukocyte complement receptor 3 (CD11b/CD18). *J. Immunol.* 163, 3045–3052.

Yasuoka, A., Tachikawa, N., Shimada, K., Kimura, S., and Oka, S. (1986). (1→3)-β-D-Glucan as a quantitative serological marker for *Pneumocystis carinii* pneumonia. *Clin. Diag. Lab. Immunol.* 3, 197–199.

Yim, K.O. and Bradford, K.J. (1998). Callose deposition is responsible for apoplastic semipermeability of the endosperm envelope of muskmelon seeds. *Plant Physiol.* 118, 83–90.

Yoshida, M., Inada, K., Endo, S., Yamashita, H., Iwanari, H., Sekiguchi, K., Kawamura, Y., Ito, Y., and Chiba, M. (1996). An assay method of (1,3)-beta-D-glucan to diagnose invasive mycoses. A utilization of monoclonal antibody to the activated factor C in blood coagulation system of horseshoe crab. In *The Proceedings of the International Symposium on Fungal Cells in Biodefense Mechanism*, August 4–6, Sendai, Miyagi, Japan.

Zhang, G.H., Baek, L., Buchardt, O., and Koch, C. (1994). Differential blocking of coagulation-activating pathways of *Limulus* amebocyte lysate. *J. Clin. Microbiol.* 32, 1537–1541.

11 Clinical Utilization of the Measurement of (1→3)-β-D-Glucan in Blood

Taminori Obayashi

CONTENTS

11.1 Introduction 199
11.2 Methodology 200
11.3 Clinical Implications 203
References 206

11.1 INTRODUCTION

(1→3)-β-D-glucan (referred to as β-glucan in the following discussion) is a characteristic, as well as the most abundant cell-wall constituent, of fungi, with the exception of *Zygomycetes*. All other microorganisms, including bacteria, rickettsiae, and viruses, lack this polysaccharide. Moreover, animals including humans do not synthesize β-glucan in the body or absorb it through intact intestinal walls. Therefore, β-glucan is a good diagnostic indicator of fungi, in that, the detection of β-glucan in the bloodstream of febrile patients indicates that the invading organism is a fungus, especially when the patient is immunocompromised.

An assay of β-glucan was materialized by the discovery of Factor G, a blood coagulation factor in the amebocyte lysate of the horseshoe crab that is extremely sensitive to β-glucan (Morita et al., 1981). The amebocytes are hemocytes containing a set of coagulation enzymes, of which endotoxin was thought to be the sole potent activator. Thus, the lysate was originally used to detect endotoxin (Levin et al., 1968) and was named the *Limulus* amebocyte lysate test, or *Limulus* test, after the American horseshoe crab, *Limulus polyphemus*. The test was believed to be specific for gramnegative bacteria until about 20 years ago, when it was found to be equally sensitive to trace amounts of β-glucan (Kakinuma et al., 1981). This is thought to be because the horseshoe crab recognizes the occurrence of external bleeding by contact with such microbial substances, which are copious in the surroundings. At the same time they are able to defend themselves from invasion by fungi and gram-negative bacteria by clotting their blood *in situ*. The initial coagulation processes of these two substances, however, differ. Endotoxin sets off a coagulation cascade by activating Factor C, while β-glucan activates Factor G (Figure 11.1). Thus, Factor G was used in coordination with the common downstream coagulation enzymes to develop a test kit for determining β-glucan in blood. Furthermore, patients with deep mycosis were shown, for the first time, to have high levels of β-glucan in their blood (Obayashi et al., 1985 and 1992). The method is extremely sensitive and is now widely accepted by clinicians in Japan for the care of immunocompromised patients, in whom the surveillance of opportunistic fungal infection is essential. Moreover, this method is especially helpful in corroborating the diagnosis of *Pneumocystis Carinii* pneumonia, in which plasma β-glucan is invariably high, to such an extent that its elevation is now deemed a diagnostic criterion by some experts.

11.2 METHODOLOGY

There are two types of β-glucan assay kits on the market: the colorimetric (Fungitec G test MK, Seikagaku Corporation, Tokyo) and the turbidimetric (Glucan Test Wako, Wako Pure Chemical, Tokyo). The colorimetric method employs a Factor C-free lysate prepared by recombination of the fractionated amebocyte lysate of the Japanese horseshoe crab *Tachypleus tridentatus*, whereas the turbudimetric employs natural whole lysate, securing the specificity by removing endotoxin from samples with polymyxin B. The recombined lysate contains higher concentrations of Factor G than the natural lysate and is about 16 times more sensitive than the latter in terms of its detection limit (Table 11.1). This has been proved using the standard β-glucan originating from *Candida albicans* (*Candida* Standard β-Glucan [CSBG]) (Ohno et al., 1999) dissolved in distilled water. Currently, however, the tests are not standardized. They not only employ different standards with different activities but also express the results in weight. Particular attention must be paid to the weight unit of the turbidimetry test, as it is an arbitrary unit, used to equate the apparent sensitivity to that of the colorimetric method. Thus, a mere comparison of the determined data is meaningless. Measurements using the colorimetric method for blood samples are three to six times higher, and there is only a moderate correlation between them. Quite recently another colorimetric test has been commercialized (β-Glucan Test Maruha, Maruha, Tokyo), adding a third method, or fourth if CSBG (i.e., standard glucans) is included. In order to avoid further confusion, a universal standard β-glucan should be established. Among those standard glucans, CSBG is the best candidate because it is derived from a major opportunistic fungal pathogen and because its physicochemical as well as biological characteristics are well documented (Tokunaka et al., 2000). The data reported herein are based on Seikagaku's colorimetry.

The assay kit has been confirmed to respond directly to polysaccharides extracted using hot water from various pathogenic fungi but not to extracts from Gram-positive bacteria or lipopolysaccharides (endotoxins) from Gram-negative bacteria.* The fungi confirmed to respond include *Candida albicans, Aspergillus fumigatus, A.*

---

* After completion of this manuscript, a brand new colorimetric assay kit for β-glucan, Fungitell, was launched by Associates of Cape Cod, Inc. (East Falmouth, MA, USA), using amebocyte lysate of Limulus polyphemus.

Clinical Utilization of the Measurement of (1→3)-β-D-Glucan in Blood

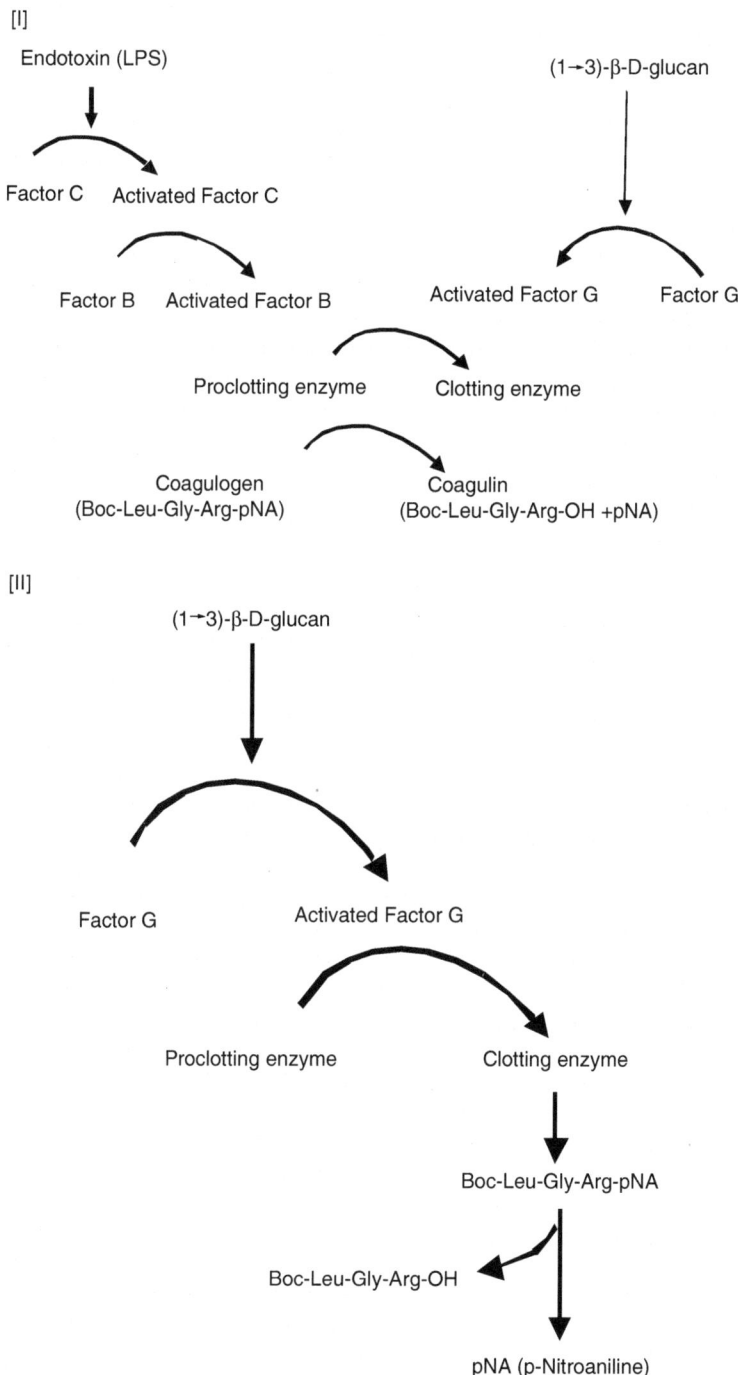

FIGURE 11.1 Principles of the Limulus test on hemocoagulation cascade of the Horseshoe crab [I] and of Fungitec G test [II].

TABLE 11.1
Comparison of β-Glucan Assay Kits

	Conversion to CSBG*	Converted Minimal Detection Limit
Fungitec G Test MK	1.0 pg = 1.8 pg	7.0 pg/mL
Glucan Test Wako	1.0 pg = 18.6 pg	111.6 pg/mL

*CSBG: Candida standard β-glucan.

flavus, Cryptococus neoformans, Trichophyton rubrum, T. mentagrophytes, Microsporum canis, and *Saccharomyces cerevisiae,* although *Cryptococus neoformans* was 10 to 100 times less active than the others (Figure 11.2) (Obayashi, 1997). False positive reactions were reported to occur when samples were heavily hemolysed or contained high concentrations of immunoglobulins, due to the nonspecific increase in absorbance. False positive results were also obtained for several days after hemodialysis, major abdominal surgery, or artificial cardiopulmonary circulation (Figure 11.3), due to the inherent presence of β-glucan in cellulose-based dialysis membranes and gauze sponges (Nakao et al., 1997) or from contaminated medical devices. Likewise, caution must be exercised with the parenteral use of albumin or globulin preparations as well as with amino acid solutions, which are often contaminated with β-glucan (Usami et al., 2002) and may result in a slight,

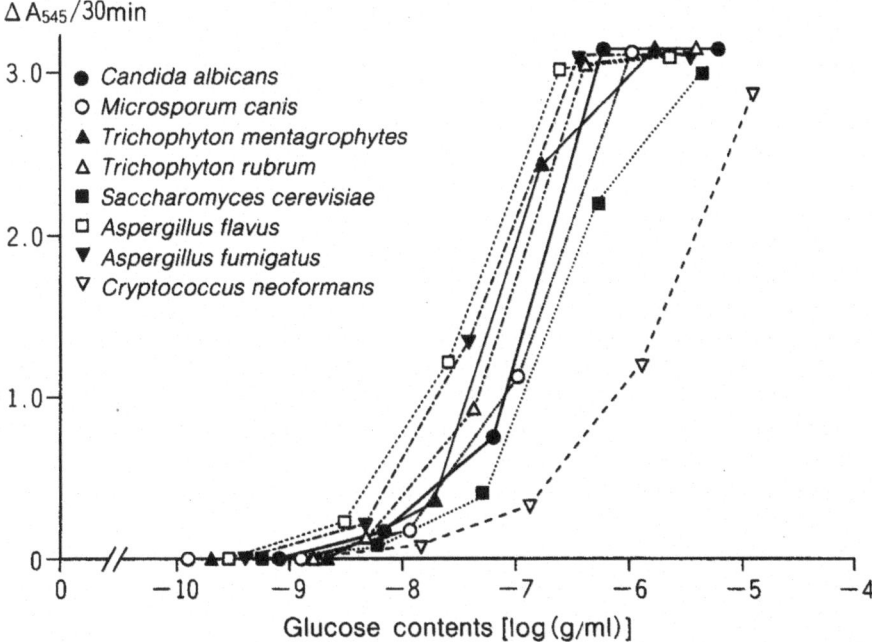

FIGURE 11.2 Reactivity of various fungal polysaccharides to Fungitec G Test.

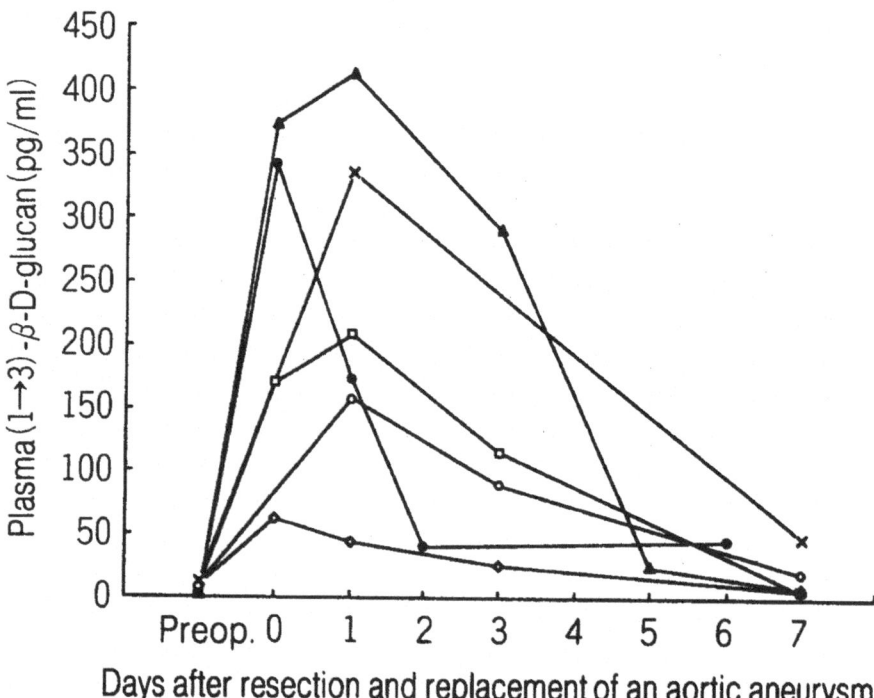

FIGURE 11.3 Postoperative profile of (1→3)-β-D-glucan concentrations in blood in six patients who received artificial cardiopulmonary circulation for resection of a thoracic aortic aneurysm.

transient rise immediately after infusion. The practice of subcutaneous or intramuscular injections of β-glucan preparations as biological response modifiers to treat cancer results in substantially high levels of β-glucan in the blood.

11.3 CLINICAL IMPLICATIONS

Plasma β-glucan concentrations in normal subjects are less than 10 pg/ml (n = 60) (Obayashi et al., 1995). They increase in systemic fungal infections and have been shown to be high in autopsy-proven deep mycoses, fungemias, and antimycotic-responsive febrile episodes (Table 11.2). Thus far, fungi reported to be involved in the elevation of β-glucan in blood include *Aspergillus*, *Candida*, *Trichosporon*, *Cryptococcus*, *Saccharomyces*, *Fusarium*, *Acremonium*, and *Ochroconis gallopavum*. However, pulmonary cryptococcosis rarely seem to be associated with hyper-β-glucanemia unless it is invasive or extensive. This may be due to a possible low content of β-glucan in the cell walls of *Cryptococcus*. This low β-glucan content has been inferred from the lower *in vitro* Factor G-activating titer of crude polysaccharides from this fungi described above (Figure 11.2). Consequently, to test for cryptococcosis, a latex agglucination test for glucronoxylomannan, abundant in the

TABLE 11.2
(1→3)-β-D-Glucan in Blood in Deep Mycoses

Autopsy-Proven	Underlying Disease	β-Glucan (pg/mL)	Pathological Diagnosis
1	ALL	42.9	Systemic Asperigillosis
2	APL	145.1	Systemic Aspergillosis
3	AML	50.9	Pulmonary Aspergillosis
4	ALL	139.0	Pulmonary Aspergillosis
5	ALL	21.4	Syntemic Fungal Infection
6	CML	82.0	Syntemic Fungal Infection
7	APL	205.4	Syntemic Fungal Infection
8	AA	21.1	Fungal Pneumonia
9	AML	43.4	Fungal Pneumonia
10	AML	44.9	Fungal Enterocolitis
Fungemia			**Isolates**
1	ALL	304.6	*Candida albicans*
2	CLL	48.5	*C. albicans*
3	All	205.7	*C. albicans*
4	AML	24.6	*C. tropicalis*
5	ALL	54.6	*C. tropicalis*
6	AA	362.6	*C. glabrata*
7	NHL	49.1	*C. glabrata*
8	CML	112.0	*C. parapsilosis*
9	MM	402.0	*C. guilliermondii*
10	APL	87.6	*C. krusei*
11	AML	39.0	*C. krusei*
12	ALL	27.6	*Candida spp.*
13	AA	32.6	*Trichosporon beigelli*
14	AIHA	525.6	*Cryptococcus neoformans*
Antifungal-Effective			**Antifungals**
1	AML	645.6	fluconazol
2	ALL	76.5	miconazol
3–i	ALL	84.0	AMPH-B
–ii		264.4	AMPH-B
–iii		91.9	Systemic candidiasis by autopsy

capsule, should be employed. This test, however, crossreacts with *Trichosporon* when polyclonal antibodies are used.

For candidiasis, the β-glucan assay excels other *Candida* antigen tests, not only in that it is by far more sensitive, but also in that it covers all species of *Candida*, including *C. glabrata* and *C. krusei*, both of which often evade detection by the latter on account of an antigenic difference. *Pneumocystis carinii*, once considered a protozoa, has recently been deemed a fungi on the basis of DNA homology (Edman et al., 1988; Ypma-Wong et al., 1992). In fact, *Pneumocystis carinii* pneumonia is almost always associated with hyper-β-glucanemia (Yasuoka et al., 1996; Teramoto et al., 2000). When it is coupled with an increased serum KL-6, a marker of interstitial pneumonia, along with the ground-glass appearance on chest X-rays and marked hypoxia, the diagnosis is almost certain, even if it is difficult to obtain sputum for direct confirmation. Although slightly less sensitive for *Aspergillus* than for *Candida*, the glucan assay is also helpful in managing aspergillosis. This is because *Aspergillus* is rarely cultured from blood, in spite of its propensity for vascular invasion. In addition, the sensitivity of other antigen detection methods is still too low to be considered satisfactory. Aspergillomas, however, are usually not associated with hyper-β-glucanemia, unless there is an invasion of the cavitary walls by the fungi. In addition to these common fungal pathogens, many other rare fungal species have been reported to cause hyper-β-glucanemia, further attesting to the utility of the β-glucan assay (Yoshida et al., 1997).

Therefore, the glucan assay is instrumental to the early recognition of deep mycoses by various fungi. It is especially applicable in the surveillance of immunologically compromised hosts. Although it cannot identify the species, it permits a prompt institution of an antimycotic measure. In addition, it serves as a good monitor for antifungal treatment; the elevated β-glucan level gradually decreases to normal as the treatment effectively eliminates the causative fungi (Figure 11.4). There are some cases, however, in which β-glucan levels, after the initial decline, remain high despite the apparent resolution of the underlying fungal infections. Though the basis for this remains unclear, the same phenomenon has been observed for other antigen detection systems, including those for *Cryptococcus* and *Streptococcus pneumoniae*. An animal study showed that most of the β-glucan infused into

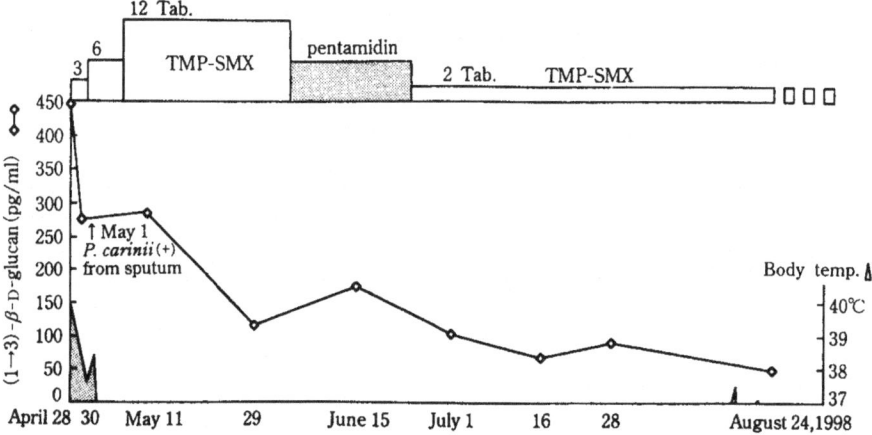

FIGURE 11.4 Temporal profile of blood (1→3)-β-D-glucan concentrations with *Pneumocystis carinii* pneumonia in a patient with AIDS.

the bloodstream is incorporated into the liver and spleen, and that the β-glucan level lingers above normal for a prolonged period of time after the acute fall from the peak formed after injection. Thus, β-glucan appears to be constantly released into the circulation from the reticuloendothelial system, once it has been saturated with the polysaccharide.

On the other hand, the extremely high sensitivity of the β-glucan assay allows us to negate the possibility of a systemic fungal infection when the test is negative. Thus, the use of antifungal agents in patients with fevers of unknown etiology can be safely withheld. This avoids an unnecessary, blind administration of antifungals, which lessens the chance of inducing resistant strains of pathogenic fungi.

Finally and curiously, there have been occasional patients who exhibit extraordinarily high β-glucan concentrations in the blood, on the order of 10 ng/ml, without any indication of systemic infection or history of injections of β-glucan. This rare finding remains to be explained.

REFERENCES

Edman, J.C., Kovacs, J.A., Masur, H., Santi, D.V., Elwood, H.J., and Sogin, M.L. (1988). Ribosomal RNA sequence shows *Pneumocystis carinii* to be a member of the fungi. *Nature* 334, 519–552.

Kakinuma, A., Asano, T., Torii, H., and Sugino, Y. (1981). Gelation of *Limulus* amebocyte lysate by an antitumor (1→3)-β-D-glucan. *Biochem. Biophys. Res. Commun.* 101, 434–439.

Levin, J. and Bang, F.B. (1968). Clottable protein in *Limulus*: its localization and kinetics of its coagulation by endotoxin. *Thromb. Diath. Haemorrh.* 19, 186–197.

Morita, T., Tanaka, S., Nakamura, T., and Iwanaga, S. (1981). A new (1→3)-β-D-glucan-mediated coagulation pathway found in *Limulus* amebocytes. *FEBS Lett.* 129, 318–321.

Nakao, A., Yasui, M., Kawagoe, T., Tamura, H., Tanaka, S., and Takagi, H. (1997). False-positive endotoxemia derives from gauze glucan after hepatectomy for hepatocellular carcinoma with cirrhosis. *Hepato-Gastroenterol.* 44, 1413–1418.

Obayashi, T., Tamura, H., Tanaka, S., Ohki, M., Takahashi, S., Arai, M., Masuda, M., and Kawai, T. (1985). A new chromogenic endotoxin-specific assay using recombined *Limulus* coagulation enzymes and its clinical applications. *Clin. Chim. Acta.* 149, 55–65.

Obayashi, T., Yoshida, M., Tamura, H., Aketagawa, J., Tanaka, S., and Kawai, T. (1992). Determination of plasma (1→3)-β-D-glucan: a new diagnostic aid to deep mycosis. *J. Med. Veterin. Mycol.* 30, 275–280.

Obayashi, T., Yoshida, M., Mori, T., Goto, H., Yasuoka, A., Iwasaki, H., Teshima, H., Kohno, S., Horiuchi, A., Ito, A., Yamaguchi, H., Shimada, K., and Kawai, T. (1995). Plasma (1→3)-β-D-glucan measurement in diagnosis of invasive deep mycosis and fungal febrile episodes. *Lancet* 345, 17–20.

Obayashi, T. (1997). (1→3)-β-D-glucanemia in deep mycosis. *Mediators Inflam.* 6, 271–273.

Ohno, N., Uchiyama, M., Tsuzuki, A., Tokunaka, K., Miura, N.N., Adachi, Y., Aizawa, M.W., Tamura, H., Tanaka, S., and Yadomae, T. (1999). Solubilization of yeast cell-wall β-(1→3)-D-glucan by sodium hypochlorite oxidation and dimethyl sulfoxide extraction. *Carbo. Res.* 316, 161–172.

Teramoto, S., Sawaki, D., Okada, S., and Ouchi, Y. (2000). Markedly increased plasma (1→3)-β-D-glucan is a diagnostic and therapeutic indicator of *Pneumocystis carinii* pneumonia in a non-AIDS patient. *J. Med. Microbiol.* 49, 393–394.

Tokunaka, K., Ohno, N., Adachi, Y., Tanaka, S., Tamura, H., and Yadomae, T. (2000). Immunopharmacological and immunotoxicological activities of a water-soluble (1→3)-β-D-glucan, CSBG from *Candida* spp. *Int. J. Immunopharmac.* 22, 383–394.

Usami, M., Ohta, A., Horiuchi, T., Nagasawa, K., Wakabayashi, T., and Tanaka, S. (2002). Positive (1→3)-β-D-glucan in blood components and release of (1→3)-β-D-glucan from depth-type membrane filters for blood processing. *Transfusion* 42, 1189–1195.

Yasuoka, A., Tachikawa, N., Shimada, K., Kimura, S., and Oka, S. (1996). (1→3)-β-D-glucan as a quantitative serological marker for *Pneumocystis carinii* pneumonia. *Clin. Diagn. Lab. Immunol.* 3, 197–199.

Yoshida, M., Obayashi, T., Iwama, A., Ito, M., Tsunoda, S., Suzuki, T., Muroi, K., Ohta, M., Sakamoto, S., and Miura, Y. (1997), Detection of plasma (1→3)-β-D-glucan in patients with *Fusarium*, *Trichosporon*, *Saccharomyces*, and *Acremonium* fungaemias. *J. Med. Vet. Mycol.* 35, 371–374.

Ypma-Wong, M.F., Fonzi, W.A., and Sypherd, P.S. (1992). Fungus-specific translocation elongation factor 3 gene present in *Pneumocystis carinii. Infect. Immun.* 60, 4140–4145.

Index

(1→3)-β-*D*-glucan-sensitive coagulation factor.
  See Factor G
(1→3)-β-*D*-glucan-specific LAL, applications of, 189
(1→3)-β-*D*-glucan-specific photometric techniques, 187
(1→3)-β-*D*-glucans
  activation of signaling pathways by, 23
  animal inhalation studies, 55
  biological responses due to, 180
  bond linkages of, 67
  breakdown of, 55
  characterizing physiochemical properties of, 67
  conformation, 78 (*See also* conformation)
  control of exposure to, 47
  degree of polymerization and antitumor activity of, 68
  dependency of molecular weight on reactivity of Factor G, 186
  dose effects of, 57
  effects of exposure on inflammatory markers, 39
  effects of exposure on peak flow variability, 38
  effects on respiratory health of, 129
  exposure to airborne, 54
  health effects of exposures to, 36
  human inhalation studies, 56
  inflammatory effects of, 66
  measurement of by LAL, 182
  measuring in indoor air environments, 45
  nature of receptors for, 15
  protective effects of, 60
  solubility of, 69
  structural characterization of, 4
  structures, 129, 181
    Factor G activation and, 185
  toxicological characteristics of, 75
(1→3)-β-glucans. *See also* (1→3)-β-*D*-glucans
  Dectin-1 and biological activities of, 103
  effect on pulmonary responses to endotoxin, 84
  higher order structure of, 167
  in fungal cell walls, 110
  molecular interaction of, 167

(1→6)-linked homoglycans, solubility of, 69
1,3-β-glucan. *See also* (1→3)-β-glucans
  receptors, signaling on plasma membranes, 100
^{13}C NMR, structural characterization of (1→3)-β-*D*-glucans by, 4
^{1}H NMR, 4
8 *M* urea, use of to solubilize (1→3)-β-*D*-glucans, 77

A

Acetylated-LDL, 21, 98
Adhesive molecules, 171
Adjuvant activity of β-glucans, 127
AGRUs, 2
AIDS, 109
Airway irritation, following organic dust exposures, 71
Albumin, evaluation of in bronchoalveolar lavage, 72
Alcaligenes faecalis var. myxogenes, 187
Alveolar epithelial cells, stimulation of by β-glucans, 100
Alveolar macrophage chemiluminescence.
  See AM-CL
AM-CL, 72
  effect of combined zymosan-LDS exposure on, 86
Amphotericin B, 110
Anaphylatoxin, dependency of PBMC activation by CSBG, 165
Anhydroglucose repeat units. *See* AGRUs
Aniline blue, 80
Animals, biological responses in due to (1→3)-β-*D*-glucans, 180
Annealing, 80
Anomeric protons, 9
Anti-CSBG antibody, 167
Antibodies
  production of and (1→3)-β-*D*-glucans, 56
  significance of in blood, 168
Antigen presenting cells, 138
Antimycotic-responsive febrile episodes, plasma β-glucan concentrations in, 203

Antitumor activity
  relationship between β-glucan dosage and duration of, 118
  relationship of particulate and soluble β-glucans to, 163
Antitumor effects of SSG, 111
Arachidonic acid, production and release of by particulate β-glucans, 162
Aspergillus sp., 200
Asthma, 35, 128
  endotoxin exposures and, 59
Atopy
  effects of (1→3)-β-*D*-glucan exposures on, 44, 130
  Th2 up-regulation and, 59
Auricularia auricula judae, 68
Autoimmune diseases, gender and HLA differences in frequency of onset of, 153

B

B-glucan assay, 203, 205
B-glucan binding protein. *See* GBP
B-glucan elicitor receptors, 96
B-glucan receptors, requirement of for signal transduction, 164
B-glucan recognition proteins, 96
  immune response of insects and, 97
B-Glucan Test Maruha, 200
B-glucans, 2, 144. *See also* (1→3)-β-*D*-glucans
  adjuvant activity of, 127, 134
  antibodies against, 153, 155
  biological activities of, 153
  changes in activity due to dissolution methods, 175
  deep mycoses and, 162
  effect on antibody response to ovalbumin, 133
  effect on PLN weight and cell numbers, 133
  effects of nitric oxide in lethal side effects caused by, 152
  immunological mechanism of, 137
  in fungal cell walls, 110
  lethal toxicity due to concomitant administration of NSAIDs, 145
  metabolism of, 123
  sensitivity to endotoxin due to concomitant administration with indometacin, 151
  solubilization of from *Candida* cells, 118
  stimulation of alveolar epithelial cells by, 100
  study of blood concentrations using Limulus assay, 114

Bacterial DNA, toll-like receptors and recognition mediation of, 102
Binding of (1→3)-β-*D*-glucan receptors, 15
  influence of polymer makeup on, 16
Bioaerosol exposure, health effects of, 38
Biodefense reactions, cellular immunity and, 166
Biological activity, role of glucan conformation in, 79
Biological response modifiers, 84, 110
  1,3-β-glucans as, 95
  injection of β-glucan preparations as, 203
Blood concentrations
  (1→3)β-*D*-glucans, 180
  study of β-glucans in using Limulus assay, 114
Blood lymphocytes, (1→3)-β-*D*-glucan exposure and, 41, 45
Blue-green algae, 128
Boc-Leu-Gly-Arg-pNA, 184, 187
Bombyx mori, 97
Bond linkage, (1→3)-β-*D*-glucans, 67
Branched 1,3-β-glucans, 181. *See also* β-glucans
Branched glucans, 8
  adjuvant activity of, 135
  solubility of, 70, 77
Branching
  activation of Factor G due to, 186
  degree of, 68
Breathing frequency, following organic dust exposures, 71
Bronchoalveolar lavage, following organic dust exposures, 71

C

C-type lectins, 99
Cancer patients, administration of β-glucans to, 114
Candia, 95
Candia albicans, Dectin-1 and, 103
Candida albicans, 167, 200
Candida cells
  distribution in the body during fungal mycoses, 116
  distribution of following administration of soluble β-glucans, 123
  measurement of accumulation in organs, 117
  solubilization of β-glucans from, 118
  tritium-labeled, 110
Candida solubilized β-glucan. *See* CSBG
Candidiasis
  β-glucan assay for, 204
  CSBG involvement in onset and progression of, 166

Index

Carbohydrate polymers, 13
Carbohydrate recognition molecules, 96
Carbon-13 nuclear magnetic resonance spectroscopy. *See* ^{13}C NMR
Carboxymethyl cellulose. *See* CM-cellulose
Carrageenan, 99
CD4$^+$ T cells, 138
Cell walls, glucans in, 2
Cellular function modulation, 17
Cellular immunity, biodefense reactions against infection and, 166
Cellulosic depth filters, as (1→3)-β-*D*-glucan contaminators, 189
Chemokines, 24, 171
Chromogenic photometric technique, 187
Cigarette smoke, effects of exposure when combined with (1→3)-β-*D*-glucans, 55
Ciliary neutrophic factor. *See* CNTF
CM-cellulose, 21
CNTF, 25
Coagulation Factor C, specific antibody blockade of endotoxin-sensitive, 184
Collagen synthesis, 23
Colorimetric G test MK, 200
Column chromatography, 184
Complex biomatrices, analysis and quantification of glucans in, 11
Conformation, glucans, 69. *See also* specific conformations
  pulmonary inflammation due to, 78
Correlated spectroscopy. *See* COSY
COSY, 6
COX inhibition, 143
  NSAID drug development and, 154
CR3, 15, 20, 98, 100, 129, 168
CRABP II, 174
Cryptococcosis, 203
Cryptococus, 95
Cryptococus neoformans, 202
CSBG, 161, 163, 200
  cellular interaction and lymphocyte activating effects of, 175
  increase in OX40L due to stimulation by, 174
  involvement in onset and progression of candidiasis, 166
  preparation and biological activity of, 164
  stimulation and NF-κB gene, 173
Curdlan, 45, 55, 76, 180
  antitumor activity of, 69
  hypochlorous acid oxidation of, 118
  inflammatory response to inhalation of, 66
  reactivity of, 187
Cyclooxygenase inhibition. *See* COX inhibition

Cytokines
  effect of glucans on expression of, 24
  inflammatory, induction of by CSBG, 165
  up-regulation of Th2 induced by (1→3)-β-*D*-glucans, 59
  zymosan-mediated inflammatory production, 101

D

D-glucopyranose, 67
DBA/1, 153
  anti-CSBG antibodies in, 168
DBA/2, 153
  anti-CSBG antibodies in, 168
Dectin-1, 20, 55, 99, 102
  biological activities of fungal 1,3-β-glucans and, 103
Dectin-TLR2 cooperation, 23
Deep mycoses, 155, 161, 175
  in immunocompromised hosts, 109
  plasma β-glucan concentrations in, 203
Degradation of (1→3)-β-*D*-glucans, 76
Detection of (1→3)β-*D*-glucans, 181
Detergent insoluble membrane. *See* DIM
Deuterium oxide, 4
Dextran sulfate, 99
Differential viscometry, 11
DIM, 100
Dimethyl sulfoxide. *See* DMSO
DMSO
  obtaining soluble β-glucan by treatment with, 163
  use of to solubilize (1→3)-β-*D*-glucans, 77
DMSO-d$_6$, 4, 13
DNA microarray method, analysis of gene expression in leukocytes activating CSBG, 169
Dose-response relationship
  (1→3)-β-*D*-glucans, 61
  zymosan A-induced pulmonary inflammation, 73
Drosophila CI scavenger receptor, 21
Drosophila melanogaster, 97
Drug side effects, 145
Dust, (1→3)-β-*D*-glucans in, 46, 54
Dust mites, 47

E

ECP, effect of (1→3)-β-*D*-glucan inhalation on, 56

ELISA, 41, 49, 127
  measurement of serum OA-specific IgE by, 132
Endotoxin-neutralizing peptide, 184
Endotoxin-sensitive coagulation Factor C, 184
Endotoxins, 66, 128
  coagulation cascade caused by activation of Factor C, 200
  correlation with glucan levels in occupational environment studies, 39
  increased sensitivity due to concomitant administration of β-glucans and NSAIDs, 151
  inhaled, 27
  interaction with (1→3)-β-glucans, 84
  neutrophilia due to, 55
  shock due to, 155
Environmental factors, increase in asthma and allergic diseases due to, 128
Eosinophilic cationic protein. *See* ECP
ERK, 173
Exposure assessment, strategy for (1→3)-β-D-glucans, 45
Exposure grouping, 46
Exposure specificity, 54

F

Factor B of LAL cascade, 183
Factor G, 183, 199
  glycosidic linkage specificity and activation of, 185
Farming, (1→3)-β-D-glucan exposure in, 42
FcR, signaling mediation by, 168
Fgr proteins, 100
Fibroblasts
  glucan treatment of, 24
  growth factor, 25
Field studies, glucan exposure and respiratory effects, 36
Flagellin, toll-like receptors and recognition mediation of, 102
Fluconazole, 110
Freezing, inhibition of annealing process by, 81
Fungal β-glucans, 2, 110. *See also* β-glucans
  recognition of, 96
Fungal glucans
  activation of signaling pathways by, 23
  physicochemical characterization of, 4
Fungal infections
  defensive role of Dectin-1 in, 103
  plasma β-glucan concentrations in, 203
Fungemias, plasma β-glucan concentrations in, 203

Fungi, β-glucan as a diagnostic indicator of, 199
Fungitec G-test, 184

G

Galactan, Factor G activation by, 185
Galactosylceramide, 98
Gangliosides, 100
Ganoderm lucidum, 68
Garbage workers
  effects of (1→3)-β-D-glucan exposures in, 40, 43
  exposure levels of (1→3)-β-D-glucans, 54
GBP, 97
Gel permeation chromatographic analysis. *See* GPC analysis
Gene expression, analysis of in leukocytes activating CSBG, 169
Glucan activation-blocking preparations, 186
Glucan assay, 203, 205
Glucan phosphate, 18, 21
Glucan receptors, recognition of glucan polymers by, 17
Glucans. *See also* specific glucans
  activation of proinflammatory and immunoregulatory signaling pathways by, 22
  analysis and quantification of, 11
  anti-inflammatory activity of, 25
  binding sites of, 20
  conformation of, 69
  cytokine and growth factor expression, 24
  effect of degree of branching on biological activity of, 68
  exposure to in environmental settings, 14, 36
  higher order structure of, 166
  immunobiology of, 14
  immunomodulating activity of, 68
  inhaled, 27
  ligands, anti-inflammatory role of, 25
  molecular weight analysis of, 11
  pleiotrophic nature of, 23
  polymer molecular weight, structure and conformation, 16
  recognition and binding of by membrane receptors, 15
  sampling of in air, 36, 54
  solubility of as determining factor for inflammatory potency, 78
  structural characterization of by NMR, 4
  system effects of, 16
Glucatell, 184
Glucronoxylomannan, 203
Gluspecy, 184

Index

Glycosidic linkage specificity, 185
Glycosphingolipids. *See* GSLs
Glycosylsphingolipids, 98
GPC analysis, 11
GPC/MALLS, 11
Grifolan, 45, 56, 114
   helical transformation of, 167
Growth factor, effect of glucans on expression of, 24
GSLs, signaling and, 100

H

H_2O_2 production by particulate β-glucans, 162
Health effects, (1→3)-β-D-glucan exposure, 41
Hepatic system, SSG use for tumors in, 111
Heptaglucoside, (1→3)-β-linked, 16
HETCOR, 6
Heteroglycans, solubility of, 69
Heteronuclear correlation spectroscopy. *See* HETCOR
Heteronuclear multiple bond correlation. *See* HMBC
HMBC, 7
Homoglycans, solubility of, 69
Host cells, interaction with microbial products, 15
House dust mites, 47
Human challenge studies, (1→3)-β-D-glucan exposure, 42
Human monocyte β-glucan receptors, 16
Hydroxyl signals, assignment of, 6
Hypchlorous acid oxides, 118
Hyper-β-glucanemia, 205

I

IgE antibodies, 127
   (1→3)-β-D-glucans and production of, 56, 58
   assay for ovalbumin-specific, 132
IgG antibodies, 127
   (1→3)-β-D-glucans and production of, 56, 58
   assay for ovalbumin-specific, 132
IκBα degradation, 22
Immune response
   β-glucan recognition proteins in insects, 97
   changes during concomitant administration of β-glucans and NSAIDs, 149
   glucan molecular weight and, 68
Immune system, receptors in, 15
Immunobiology, 14
Immunocompromised individuals, 109
Immunological reaction, physiological state of glucans determining, 136
Immunopharmacological activity, SSG, 112
Immunoregulatory mediators, 16, 23
Immunosuppressants, 155
Immunotyrosine-based activation motif. *See* ITAM
Indometacin, 145
   lethal side effects when administered concomitant to β-glucans, 149
   sensitivity to endotoxin due to concomitant administration with β-glucans, 151
Indoor air quality, 36
   (1→3)-β-D-glucans and, 130
Indoor dampness, 35
Inflammation, NF-κB activation and, 171
Inflammatory cells, effect of (1→3)-β-D-glucan inhalation on, 56
Inflammatory cytokines. *See also* interleukins; TNFα
   production of, 165
   stimulation with OX-CA or CSBG and levels of, 171
Inflammatory markers, 45
   effects of (1→3)-β-D-glucan exposures on, 39, 43
Inflammatory parameters, changes during concomitant administration of β-glucans and NSAIDs, 149
Inhalation risk assessment, (1→3)-β-D-glucans, 55
INOS, 152
Insects, biological responses in due to (1→3)-β-D-glucans, 180
Integrins, 171
Interleukins
   effect of (1→3)-β-D-glucan inhalation on, 56
   effects of water-soluble glucans on, 25
   inhibition of IL-8 production by OX-CA, 164
Intracellular signaling pathways, activation of proinflammatory and immunoregulatory pathways, 22
IRAK signaling protein, 22
ITAM, 20
Itraconazole, 110

K

Krestin, 110

L

LacCer, 98. *See also* lactosylceramide
   neutrophils and, 101
Lactate dehydrogenase. *See* LDH

Lactosylceramide, 15, 98
  -mediated leukocyte activation, 100
Laminarin, 17, 180
  branching frequency of, 9
  use of to inhibit phagocytosis of zymosan, 101
LDH, lung cell damage determination through assay of, 71
Lentinan, 110, 155
  structure, 129
Leukocytes
  1,3-β-glucan receptors on plasma membrane of, 97
  analysis of gene expression in, 169
  infiltration of during concomitant administration of β-glucans and indometacin, 149
  polymorphonuclear, 121
  response to 1,3-β-glucans, 104
Lgr proteins, 100
Lichenan, D-glucosyl-modified, antitumor activity of, 69
Limulus amebocyte lysate (LAL) assay, 36, 49, 66, 181, 199
  (1→3)β-D-glucan-specific, preparation of, 184
  applications of (1→3)-β-D-glucan-specific, 189
  LAL cascade, 182
  measurement of (1→3)-β-D-glucans by, 182
  reactivity of solubilized β-glucans, 118
  study of blood β-glucan concentrations using, 114
  use to diagnose deep mycoses, 110
Limulus coagulation factors, fractionation and combination of, 184
Limulus factor G reactivity, 163
Linear glucans, NMR spectroscopy of, 8
Linkage specificity, Factor G activation, 185
Lipoarabinomannans, 15
Lipopolysaccharides. *See* LPS
Lipoproteins, 15
  toll-like receptors and recognition mediation of, 102
Lipoteichoic acid, 15
Liver. *See* hepatic system
Lower airway effects of (1→3)-β-D-glucan exposures, 44. *See also* asthma
LPS, 15, 66, 99, 128
  combined exposure with zymosan, 85
  increase in sensitivity to, 151
  indometacin and, 146
  LAL cascade and, 182
  toll-like receptors and recognition mediation of, 102

Lung function
  effect of NaOH-treated zymosan on, 78
  effects of (1→3)-β-D-glucan exposures on, 36, 44
Lymphocytes. *See also* blood lymphocytes
  effect of (1→3)-β-D-glucans on, 55, 57, 60

M

Mac^{-1} receptors, 55
MacroGard, 129
Macrophage scavenger receptors, 21
Macrophages, 55
  phagocytic defense mechanisms in, 129
  production of ROS by, 103
Mannoprotein, in fungal cell walls, 110
MAPK cascade, 173
MEKK3, 173
Membrane receptors
  binding of (1→3)-β-D-glucans to, 16
  Dectin-1 and scavenger receptors, 20
  recognition and binding of glucans by, 15
Micafungin, 110
Microbial products
  cell wall agents, 53
  interaction with host cells, 15
  lethal toxicity due to concomitant administration with NSAIDs, 146
Microphages, TNFα production by particulate β-glucans, 162
Microsporium canis, 202
Mite exposures, 47
MLK3, 173
Mold exposure, 35, 54
  increase in infections due to, 60
Molecular conformation. *See* conformation
Molecular water, removal of from glucan preparations for NMR, 6
Molecular weight
  effect on antitumor activity of (1→3)-β-D-glucans, 68
  effect on reactivity of Factor G, 186
Monocyte 20-$_kD_A$ receptor, 97
MPO, effect of (1→3)-β-D-glucan inhalation on, 56
Multi-angle laser light scattering. *See* GPC/MALLS
Mycobacterium tuberculosis, protective anti-infection effects of SSG, 136
MyD88 signaling protein, 22
Myeloperoxidase. *See* MPO

Index

N

Nabumetone, 146
NADPH oxidase, 100
NaOH, use of to solubilize (1→3)-β-D-glucans, 77
NaOH-treated zymosan, inflammatory response to, 80
Natural killer cells, 98, 129
   Dectin-1 and, 99
Neutrophilic airway inflammation, effects of (1→3)-β-D-glucan exposures on, 45, 55
Neutrophils, LacCer on plasma membranes of, 101
NF-IL6, preventing activation of, 25
NF-κB activation, 171
NFκB signaling pathway, 22
   preventing activation of, 25
Nitric oxide
   effects of in lethal side effects caused by β-glucans, 152
   isozymes (*See* iNOS)
   release from AMs following organic dust exposure, 72
NMR, structural characterization of (1→3)-β-D-glucans by, 4
Nonsteroidal anti-inflammatory agents. *See* NSAIDs
NSAIDs, 143
   development of COX2 specific, 154
   digestive tract disorders caused by, 144
   lethal toxicity due to concomitant administration of microbial components, 145
Nuclear magnetic resonance spectroscopy. *See* NMR
Nuclear translocation, 22

O

Omentum, involvement of in processing foreign objects in peritoneal cavity, 116, 123
Omphalia lapidescens, branching of (1→3)β-D-glucans derived from, 187
Opportunistic infections, role of anti-CSBG antibody in, 168
Organ distribution of SSG, 111
Organic dust toxic syndrome, 71
Organic dusts, exposure, pulmonary reactions to, 70
Organic dusts, exposure, role of (1→3)-β-D-glucans in pulmonary inflammatory symptoms, 67
Organic solvents, use of to solubilize (1→3)-β-D-glucans, 77
Organic waste, levels of (1→3)-β-D-glucans in, 48
Ovalbumin, 56, 127
   assay for serum IgE and IgG, 132
   effect of β-glucans on antibody response to, 133
OX-CA, 161, 163
   inhibition of IL-8 production by, 164
   proinflammatory activity of, 175
   stimulation and NF-κB gene, 173
   stimulation of, 169
OX40L, increase in expression of due to CSBG stimulation, 174
Oxidant production, following organic dust exposures, 72

P

P38RK, 173
Pachyman, 76
   reactivity of, 187
PAMPs, 15
Paper industry, glucan exposures in, 40
Particulate β-glucan. *See* OX-CA
Pasmodesmata, 180
Pathogen-associated molecular patterns. *See* PAMPs
Pathogenic fungi, glucans in cell walls of, 2
Pattern recognition receptors. *See* PRRs
PBMC, 56
   blunting effect of (1→3)-β-D-glucan on, 58
   dependency of anaphylatoxin on activation by CSBG, 165
PDGFs, 25
Pentifylline, 24
Peptidoglycan, 15
Peripheral blood monocytic cells. *See* PBMC
Peritoneal cavity. *See also* hepatic system
   Candida cells in, 117
Personal protection equipment, 48
Pesstalotia sp. 815, 68
Phagocytic defense mechanisms, 129
Phagocytosis, CR3-mediated, 100
Photometric techniques, (1→3)-β-D-glucan-specific, 187
PKC-mediated phosphorylation, 100
Plants, biological responses in due to (1→3)-β-D-glucans, 180
Plasma membranes, signaling via 1,3-β-glucan receptors on, 100
Platelet-derived growth factors. *See* PDGFs
PLN assay, 127, 131

Pneumocystis carinii, 205
    corroboration of diagnosis by plasma
        β-glucan, 200
    Dectin-1 and, 103
Polymerization, degree of, 68
Polymicrobial sepsis, glucan role in decreasing morbidity due to, 25
Polymixin B, 185
    photometric techniques and, 187
Polymorphonuclear leukocyte, infiltration following organic dust exposures, 71
Polymorphonuclear leukocytes, 121
    stimulation of tumoricidal activity by β-glucans in, 129
Polysaccharides
    activation of Factor G with, 185
    solubility of, 69
Popliteal lymph node assay. *See* PLN assay
Pro-clotting factor, 183
Proinflammatory mediators, 16, 23
Properdin, 70
Prostaglandins, 143
Proton nuclear magnetic resonance spectroscopy. *See* ^{1}H NMR
PRRs, 15, 20, 99
    toll-like receptors, 22
Pulmonary inflammation
    annealing of zymosan and changes in, 80
    dose-response of zymosan A-induced, 73
    role of glucan conformation in, 79

Q

Quantitation of (1→3)β-*D*-glucans, 181

R

Receptor ligand interactions, 18
Receptor signals, 104
Receptors, 1,3-β-glucan, signaling on plasma membranes, 100
Recognition proteins, 97
RES, effect of intravenously administered zymosan on, 70
Respiratory diseases, 35
Reticulo-endothelial system. *See* RES
Retinoic acid binding protein II. *See* CRABP II
Risk assessment, 53
    (1→3)-β-*D*-glucan, 61

S

Saccharomyces cerevisiae, 4, 202

Saccharomyces cerevisiae, zymosan from, 70
Sacculus, 182
Sampling methods, (1→3)-β-*D*-glucans in, 46
SAPK, 173
Saprophytic fungi, glucans in cell walls of, 2
Sawmills, glucan exposures in, 39
Scavenger receptors, 21, 98
Schizophyllan, 18, 110
    effect of molecular weight on antitumor activity of, 68
Scleroglucans, 18. *See also* SSG
    branching frequency of, 8
Sclerotinia sclerotiorum IFO 9395, 110
Sepsis, glucan role in decreasing morbidity due to, 25
Serine/threonine-phosphorylation, 100
Serum IFN-γ production, 163
Side-chain branching, effect of on biological activity of glucans, 68
Signal assignment, 6
Signaling molecules, 22, 100
    toll-like receptors and, 102
Single helical glucans, 24
    antitumor activity of, 69
    nitric oxide production induced by, 145
    vs. triple helical glucans, 186
Soluble β-glucan recognition proteins, 96
Solublization of (1→3)-β-*D*-glucans, 76
Sonifilan, 114, 155
    helical transformation of, 167
Source control, 48
Soybeans, β-glucan elicitor receptors in, 96
SP-D, 96
SPG, β-glucan dosage and duration of antitutmor activity, 118
Spleen. *See* hepatic system
SRS, 151
SSG
    adjuvant activities of, 130
    hypochlorous acid oxidation of, 118
    lethal toxicity when administered concomitantly with NSAIDs, 146
    protective anti-infection effects of, 136
    structure, 129
    tritium-labeled, 110
Surfactant protein D. *See* SP-D
Suspended particulate matter, modification of immune response due to, 128
Systemic inflammatory response syndrome. *See* SRS

T

T-cells

Index

activation of by CSBG stimulation, 174
CD4+, 138
Dectin-1 and stimulation of, 99
Tachyplesin/polyphemusin, 185
Tachypleus amebocyte lysate assay, 181
Terbinafine, 110
TGFs, 25
Th1 cells, effects of CSBG production on, 166
Th2 up-regulation, induction of by (1→3)-β-
    D-glucans, 59
TLRs, 22
    contribution of to zymosan-mediated
        inflammatory cytokine production,
        101
TNFα, 24
    blunting effect of (1→3)-β-D-glucan on, 58
    effect of (1→3)-β-D-glucan inhalation on, 56
    effect of combined zymosan-LDS exposure on
        production of, 86
    increase in due to Dectin-1, 103
    production by microphages due to particulate
        β-glucans, 162
    production of by OX-CA, 174
    production of under complement inactivated
        conditions, 165
Toll-like receptors. *See* TLRs
Transcription factor activation, 23, 27
Transforming growth factors. *See* TGFs
Translocation, elimination of microorganisms due
    to, 152
Trichophyton sp., 202
Triple helical glucans, 24, 167
    antitumor activity of, 69
    pulmonary inflammation and, 79
    vs. single helical glucans, 186
Tritium-labeled cells, use for in studies of β-
    glucans, 110
Tumor necrosis factor α. *See* TNFα
Turbidimetric photometric technique, 187, 200
Type 3 complement receptor. *See* CR3
Tyrosine-phosphorylation, 100

U

U937 promonocytic cells, 17
Upper airway effects of (1→3)-β-D-glucan
    exposures, 44. *See also* health effects

V

Vascular endothelial cell growth factor. *See* VEGF
Vascular permeability
    cytokines and, 171
    of particulate β-glucans, 162
VEGF, 25

W

Waste management, glucan exposures in workers,
    40
Water soluble glucans, conformations of, 3
Wood industry, glucan exposures in, 39, 54
Wound growth factor gene, 25
Wound response tissue synthesis, 180

X

Xylan, Factor G activation by, 185

Y

Yeast, 1,6- and 1,3-β-glucans in, 95
Yeast heptaglucoside, (1→3)-β-linked, 16

Z

Zymogen, 183
Zymosan, 2, 20, 70, 97, 100, 163
    -mediated inflammatory cytokine production,
        101
    combined exposure with LPS, 85
    conformation status of reannealing, 80
    hypochlorous acid oxidation of, 118
    inflammation caused by, 75
Zymosan A
    inflammatory response to inhalation of, 67
    pulmonary inflammation due to, 83
        dose-response relationship, 73
        recovery from exposure to, 74